T0142328

Lecture Notes in Networks and Systems 734

The series "Lecture Notes in Networks and Systems" publishes the latest developments in Networks and Systems—quickly, informally and with high quality. Original research reported in proceedings and post-proceedings represents the core of LNNS.

Volumes published in LNNS embrace all aspects and subfields of, as well as new challenges in, Networks and Systems.

The series contains proceedings and edited volumes in systems and networks, spanning the areas of Cyber-Physical Systems, Autonomous Systems, Sensor Networks, Control Systems, Energy Systems, Automotive Systems, Biological Systems, Vehicular Networking and Connected Vehicles, Aerospace Systems, Automation, Manufacturing, Smart Grids, Nonlinear Systems, Power Systems, Robotics, Social Systems, Economic Systems and other. Of particular value to both the contributors and the readership are the short publication timeframe and the world-wide distribution and exposure which enable both a wide and rapid dissemination of research output.

The series covers the theory, applications, and perspectives on the state of the art and future developments relevant to systems and networks, decision making, control, complex processes and related areas, as embedded in the fields of interdisciplinary and applied sciences, engineering, computer science, physics, economics, social, and life sciences, as well as the paradigms and methodologies behind them.

Indexed by SCOPUS, INSPEC, WTI Frankfurt eG, zbMATH, SCImago.

All books published in the series are submitted for consideration in Web of Science.

For proposals from Asia please contact Aninda Bose (aninda.bose@springer.com).

Ngoc Thanh Nguyen · Hoa Le-Minh ·
Cong-Phap Huynh · Quang-Vu Nguyen
Editors

The 12th Conference on Information Technology and Its Applications

Proceedings of the International Conference CITA 2023

Editors
Ngoc Thanh Nguyen
Wroclaw University of Science and Technology
Wroclaw, Poland

Cong-Phap Huynh
The University of Danang – Vietnam-Korea University of Information and Communication Technology
Danang, Vietnam

Hoa Le-Minh
Northumbria University
Newcastle, UK

Quang-Vu Nguyen
The University of Danang – Vietnam-Korea University of Information and Communication Technology
Danang, Vietnam

ISSN 2367-3370 ISSN 2367-3389 (electronic)
Lecture Notes in Networks and Systems
ISBN 978-3-031-36885-1 ISBN 978-3-031-36886-8 (eBook)
https://doi.org/10.1007/978-3-031-36886-8

This Springer imprint is published by the registered company Springer Nature Switzerland AG
The registered company address is: Gewerbestrasse 11, 6330 Cham, Switzerland

Preface

Conference on Information Technology and its Applications (CITA) is an annual scientific conference on information technology and its applications in all fields. CITA has initiated and organized since 2012. The main objective of the conference is to create a forum to gather and connect Vietnamese and international researchers, scientists and specialists to participate in the fields of information technology and its applications.

All papers submitted to CITA are reviewed seriously, closely and thoroughly by 02–04 reviewers with appropriate expertise, with professional advice from reputable scientists in the fields of information and communication technology, digital economy such as Prof. Dr. Nguyen Ngoc Thanh (Poland), Assoc. Prof. Dr. Le Minh Hoa (UK), Prof, Nguyen Thanh Thuy (Vietnam), ... The average rate of papers accepted by CITA is below 50%.

Over the past 12 years of establishment and development, CITA has been receiving the companionship and support from Vietnam Association of Faculties-Institutes-Schools-Universities of ICT (FISU Vietnam), the community of scientists, a network of training and research units in Vietnam and international. CITA has been playing an important and significant role in promoting the development of science, technology, training in ICT and digital economy of the Central and Central Highlands region in particular and the whole country in general.

The 12th CITA, 2023, is hosted by the Vietnam - Korea University of Information and Communication Technology (VKU), taking place on July 28–29, 2023, in Da Nang City, the most beautiful and livable city in Vietnam. For this edition of the conference, we have received in total 154 papers whose authors come from over 20 countries around the world. Each paper was peer reviewed by at least two members of the Program Committee. Only 33 papers of the highest quality were selected for oral presentation and publication in this LNNS volume of the CITA 2023 International Track proceedings. The acceptance rate of CITA 2023 is about 21%.

Papers included in these proceedings cover the following topics : Data Science and Artificial Intelligence, Image and Natural Language Processing, Software Engineering and Information System, Network and Communications and Digital Economy. The accepted and presented papers focus on new trends and challenges facing information technology and its applications. The presenters show how research works can stimulate novel and innovative applications. We hope that you find these results useful and inspiring for your future research work.

We cordially thank our main sponsors: Vietnam - Korea University of Information and Communication Technology (Vietnam), Wrocław University of Science and Technology (Poland) and IEEE Systems, Man and Cybernetics Society as well as FISU Vietnam. Our special thanks are also due to Springer for publishing the proceedings and to all the other sponsors for their kind support.

We express our gratitude to the Steering Committee, Program Chairs, Organizing Chairs and Publicity Chairs for their work toward the conference. We sincerely thank all the members of the Program Committee for their valuable efforts in the review process, which helped us to guarantee the highest quality of the selected papers for the conference. We cordially thank all the authors, for their valuable contributions, and the other participants of the conference. The conference would not have been possible without their support. Thanks are also due to the numerous experts who contributed to making the event a success.

July 2023

Ngoc Thanh Nguyen
Hoa Le-Minh
Cong-Phap Huynh
Quang-Vu Nguyen

Conference Organization

Steering Committee

Ngoc-Thanh Nguyen (Chair)	Wrocław University of Science and Technology, Poland
Nguyen Thanh Thuy (Co-Chair)	Vietnam National University, Hanoi, Vietnam
Hoa Le-Minh	Northumbria University, UK
Dosam Hwang	Yeungnam University, Republic of Korea
Cong-Phap Huynh	The University of Danang – Vietnam-Korea University of Information and Communication Technology, Danang, Vietnam

Organizing Chairman

Cong-Phap Huynh	The University of Danang - Vietnam-Korea University of Information and Communication Technology, Vietnam

Program Committee Chairs

Ngoc-Thanh Nguyen (Chair)	Wrocław University of Science and Technology, Poland
Thanh-Thuy Nguyen	Vietnam National University, Hanoi, Vietnam
Hoa Le-Minh	Northumbria University, UK
Ba-Thanh Truong	The University of Danang – University of Economics
Manh-Thanh Le	Hue University
Thanh-Binh Nguyen	The University of Danang - Vietnam-Korea University of Information and Communication Technology, Vietnam
Phu-Khanh Nguyen	Phenikaa University
Hong-Son Ngo	Phenikaa University

Organizing Committee Chairs

The-Son Tran — The University of Danang - Vietnam-Korea University of Information and Communication Technology, Vietnam
Quang-Vu Nguyen — The University of Danang - Vietnam-Korea University of Information and Communication Technology, Vietnam

Publicity Chairs

Quang-Vu Nguyen — The University of Danang - Vietnam-Korea University of Information and Communication Technology, Vietnam
Quang-Hung Nguyen — Journal of Information and Communication, Vietnam Information and Communications Publisher

Keynote Speakers

Nhien-An Le-Khac — School of Computer Science, University College Dublin, Ireland

Local Organizing Committee

Ngoc-Tho Huynh — The University of Danang - Vietnam-Korea University of Information and Communication Technology, Vietnam
Duc-Hien Nguyen — The University of Danang - Vietnam-Korea University of Information and Communication Technology, Vietnam
Minh-Duc Le-Thi — The University of Danang - Vietnam-Korea University of Information and Communication Technology, Vietnam
Cuu-Long Le-Phuoc — The University of Danang - Vietnam-Korea University of Information and Communication Technology, Vietnam
Anh-Quang Nguyen-Vu — The University of Danang - Vietnam-Korea University of Information and Communication Technology, Vietnam

Minh-Nhut Pham-Nguyen	The University of Danang - Vietnam-Korea University of Information and Communication Technology, Vietnam
Nhu-Thao Le-Ha	The University of Danang - Vietnam-Korea University of Information and Communication Technology, Vietnam
Kim-Ngoc Nguyen-Thi	The University of Danang - Vietnam-Korea University of Information and Communication Technology, Vietnam
Linh-Giang Nguyen	The University of Danang - Vietnam-Korea University of Information and Communication Technology, Vietnam
Ha-Phuong Nguyen	The University of Danang - Vietnam-Korea University of Information and Communication Technology, Vietnam
Ngoc-Thao Huynh Nguyen	The University of Danang - Vietnam-Korea University of Information and Communication Technology, Vietnam
Hong-Hanh Pham-Thi	The University of Danang - Vietnam-Korea University of Information and Communication Technology, Vietnam
Bich-Thao Le-Thi	The University of Danang - Vietnam-Korea University of Information and Communication Technology, Vietnam
Hong-Viet Phan-Thi	The University of Danang - Vietnam-Korea University of Information and Communication Technology, Vietnam

Program Committee

Cong-Phap Huynh	The University of Danang - Vietnam-Korea University of Information and Communication Technology
Dai-Tho Dang	The University of Danang - Vietnam-Korea University of Information and Communication Technology
Dang Vinh	The University of Danang - Vietnam-Korea University of Information and Communication Technology
Doan Ngoc Phi Anh	The University of Danang - University of Economics
Doan Trung Son	Phenikaa University

Hoang Thi Thanh Ha	The University of Danang - University of Economics
Hoang Van Thanh	Quang Binh University
Huu-Ai Duong	The University of Danang - Vietnam-Korea University of Information and Communication Technology
Huu-Duc Hoang	The University of Danang - Vietnam-Korea University of Information and Communication Technology
Huynh Xuan Hiep	Can Tho University
Le Dien Tuan	The University of Danang - University of Economics
Le Phuoc Cuu Long	The University of Danang - Vietnam-Korea University of Information and Communication Technology
Le Thanh Hieu	Hue University
Le Thi Minh Duc	The University of Danang - Vietnam-Korea University of Information and Communication Technology
Le Thi Thu Nga	The University of Danang - Vietnam-Korea University of Information and Communication Technology
Le Xuan Viet	Quy Nhon University
Loc Phuoc Hoang	Khon Kaen University
Marcin Jodłowiec	Wrocław University of Science and Technology
Ngo Hai Quynh	The University of Danang - Vietnam-Korea University of Information and Communication Technology
Ngoc-Tho Huynh	The University of Danang - Vietnam-Korea University of Information and Communication Technology
Nguyen Duc Hien	The University of Danang - Vietnam-Korea University of Information and Communication Technology
Nguyen Duc Man	Duy Tan University
Nguyen Gia Nhu	Duy Tan University
Nguyen Ha Huy Cuong	The Da Nang of University
Nguyen Hoang Vu	The University of Danang - Vietnam-Korea University of Information and Communication Technology
Nguyen Hong Quoc	Hue University of Education
Nguyen Huu Nhat Minh	The University of Danang - Vietnam-Korea University of Information and Communication Technology

Nguyen Le Hung	The University of Danang
Nguyen Ngoc-Thanh	Wroclaw University of Science and Technology
Nguyen Quang-Vu	The University of Danang - Vietnam-Korea University of Information and Communication Technology
Nguyen Thanh Binh	IIASA
Nguyen Thanh Hoai	The University of Danang - Vietnam-Korea University of Information and Communication Technology
Nguyen Thanh-Binh	The University of Danang - Vietnam-Korea University of Information and Communication Technology
Nguyen The Dung	Hue University of Education
Nguyen Thi Kieu Trang	The University of Danang - Vietnam-Korea University of Information and Communication Technology
Nguyen Thi Lan Anh	Hue University of Education
Nguyen Thi Thu Den	The University of Danang - Vietnam-Korea University of Information and Communication Technology
Nguyen Thi Uyen Nhi	The University of Danang - University of Economics
Nguyen Van Binh	The University of Danang - Vietnam-Korea University of Information and Communication Technology
Nguyen Van Khang	Telecom SudParis
Nguyen Van Loi	The University of Danang - Vietnam-Korea University of Information and Communication Technology
Nhien-An Le-Khac	University College Dublin
Ninh Khanh Duy	The University of Danang - University of Science and Technology
Pham Dinh-Lam	Kyonggi University
Pham Minh Tuan	The University of Danang, University of Science and Technology
Pham Nguyen Minh Nhut	The University of Danang - Vietnam-Korea University of Information and Communication Technology
Pham Thi Thu Thuy	Nha Trang University
Pham Van Viet	Quy Nhon University
Pham Xuan Hau	Quang Binh University
Phan Chi Thanh	Quang Tri Teacher Training College
Phan Huyen Trang	Yeungnam University

Phan Thi Lan Anh	The University of Danang - Vietnam-Korea University of Information and Communication Technology
Phan Van Thanh	Quang Binh University
Quang-Hien Dang	The University of Danang - Vietnam-Korea University of Information and Communication Technology
Tang Tan Chien	The University of Danang - University of Science and Technology
Thai Minh Tuan	Can Tho University
Tran Hoang Vu	The University of Danang - University of Technology and Education
Tran The Vu	The University of Danang - VN-UK Institute for Research and Executive Education
Tran The-Son	The University of Danang - Vietnam-Korea University of Information and Communication Technology
Tran Van Cuong	Yeungnam University
Tran Van Dai	The University of Danang - Vietnam-Korea University of Information and Communication Technology
Tran Xuan Tu	VNU University of Engineering and Technology
Trinh Cong Duy	The University of Danang
Truong Hai Bang	Sai Gon International University
Van Hung Trong	The University of Danang - Vietnam-Korea University of Information and Communication Technology
Van-Phi Ho	The University of Danang - Vietnam-Korea University of Information and Communication Technology
Vo Duc-Hoang	The University of Danang - University of Science and Technology
Vo Thi Thanh Thao	Soongsil University
Vo Trung Hung	The University of Danang - University of Technology and Education
Vo Viet Minh Nhat	Hue University
Vuong Cong Dat	The University of Danang - Vietnam-Korea University of Information and Communication Technology

Contents

Network and Communications

Software Engineering and Information System

Data Science and Artificial Intelligence

A New ConvMixer-Based Approach for Diagnosis of Fault Bearing Using Signal Spectrum

Manh-Hung Vu, Van-Quang Nguyen, Thi-Thao Tran, and Van-Truong Pham(✉)

Department of Automation Engineering, School of Electrical and Electronic Engineering, Hanoi University of Science and Technology, Hanoi, Vietnam
truong.phamvan@hust.edu.vn

Abstract. It has been reported that nearly 40% of electrical machine failures are caused by bearing problems. That is why identifying bearing failure is crucial. Deep learning for diagnosing bearing faults has been widely used, like WDCNN, Conv-mixer, and Siamese models. However, good diagnosis takes a significant quantity of training data. In order to overcome this, we propose a new approach that can dramatically improve training performance with a small data set. In particular, we propose to integrate the ConvMixer models to the backbone of Siamese network, and use the few-short learning for more accurate classification even with limited training data. Various experimental results with raw signal inputs and signal spectrum inputs are conducted, and compared with those from traditional models using the same data set provided by Case Western Reserve University (CWRU).

Keywords: Bearing faults diagnosis · Siamese-based Conv-mixer · CWRU data base · Limited data

1 Introduction

The process of locating and detecting faults or weak points in bearings is referred to as bearing problems diagnostics. If these defects are not found and fixed in a timely way, they may result in decreased performance, increased vibration, and even machine failure. Research on fault diagnosis standards in bearings [1] indicated that various signal analysis techniques, including vibration analysis, acoustic emission analysis, and motor current signature analysis, are frequently used to diagnose bearing defects. These methods can reveal the bearing's operational state and assist in locating particular problem types such as inner race fault, outer race fault, ball fault, and roller fault [2].

In the past, various traditional machine learning (ML) techniques, such as Support Vector Machine (SVM) [3], K-Nearest Neighbors (KNN) algorithm [4], and Artificial Neural Network (ANN) [5], were used to identify rolling bearing issues. The requirement to thoroughly understand the signal and the characteristics of errors is a common theme throughout the techniques [6]. We must be

N. T. Nguyen et al. (Eds.): CITA 2023, LNNS 734, pp. 3–14, 2023.
https://doi.org/10.1007/978-3-031-36886-8_1

familiar with electric machinery, such as which errors often make use of which characteristic [7]. Traditional machine learning for bearing diagnostics has certain limitations, such as difficulty in recognizing faults when there is interference and difficulty in examining the characteristics of different problems [8].

To overcome the difficulties when applying traditional ML techniques and achieve even higher accuracy, deep learning neural networks are increasingly being in use. There are several deep-learning based methods that apply to fault diagnosis and achieve significantly high performance such as recurrent neural networks (RNNs), auto-encoders (AE), convolutional neural networks (CNNs), deep learning based transfer learning, deep belief network (DBN). In [9], F. Jia applied a 5-layer auto-encoder to extract features of fault signals in the frequency domain. In a study report in [10], a multi-sensor vibration data fusion approach is employed that integrates frequency-domain and time-domain features using two-layer sparse autoencoders (SAE) and then fed into a three-layer DBN for classification purposes. In 2015, a study [11] reported one of the first uses of RNNs in bearing fault diagnosis. In this study, the fault features that were extracted using the discrete wavelet transform and orthogonal fuzzy neighborhood discriminative analysis were put into an RNN to perform fault detection. In the work of Zhang et al. [12], a CNN with training interference (TICNN) was introduced for bearing health classification, using raw time-varying signals as its input. In [13], a LeNet-5 CNN is employed, which consists of two convolution-pooling layers and two fully-connected layers. In light of [14], the unprocessed vibration data are converted into spectrograms and then fed into a deep fully convolutional neural network (DFCNN). A multi-scale CNN (MS-DCNN) is proposed in [15] to solve the training time problems in most CNN-based methods. To better suppress the impact of high frequency noise, a deep CNN with wide first-layer kernels (WDCNN) mixed with adaptive batch normalization (AdaBN) was adopted in [16]. Overall, modern deep learning technique required a large amount of training data to achieve a high-accuracy diagnosis. But, in [17] A. Zhang et al. proposed a novel deep learning based siamese neural network to classify bearing fault with limited training data.

Motivated from opportunities and challenges of fault bearing analysis, and inspired by advances of deep learning techniques, in this research, we propose a new approach for the diagnosis of fault bearing even in the case of limited training data. In particular, based on the Siamese backbone, we propose to integrate the Convmixer architecture to create a new model, namely Siamese-based Conv-Mixer, combining Siamese and Conv-Mixer, in which, twins network Conv-Mixer includes Deptwise-Pointwise combined with Batch Normalization. Besides, we also propose using the few-short learning prediction method. The proposed model has been validated on the public Case Western Reserve University (CWRU) database. Quantitative assessments in comparison with other approaches and the ablation study have shown that the performance on a limited data set improves greatly when the proposed model is used in conjunction with the few-shot learning prediction approach. For instance, the proposed model achieves an accuracy of 95.48% with just 90 training samples, while all models from earlier research achieve an accuracy of less than 92%. The proposed model has also been tested with having a spectrogram transform on the input and no spectrogram transform on the input on many instances of varying training samples.

2 Related Work

2.1 Conv-Mixer Neural Networks

Based on the network model of CNNs [18] and the Vision Transformer model (ViT) [19], the Conv-Mixer network model is built for better efficiency. The Deptwise and Pointwise convolution layers are part of the Conv-Mixer model's construction. Batch normalization layers are applied after each convolution layer to reduce internal covariate shift (ICS) and speed up the training of deep learning models study [20] indicates. Compared to MLP-Mixer [21] and traditional CNNs [18], Conv-Mixer gives better results for several tasks such as finger knuckle recognition [22], and image segmentation works [23], on the same test due to Normalize with each batch size.

2.2 Siamese Neural Networks

In [24], siamese neural networks contain a twin network that shares the same weights. Siamese Neural Networks (SNN) tasks are to find the similarity between two inputs by comparing their features. A simple structure of the Siamese network is shown in Fig. 1. Using deep learning for diagnosis with a limited amount of data, pair training with a Siamese model is superior to normal training methods, according to [25].

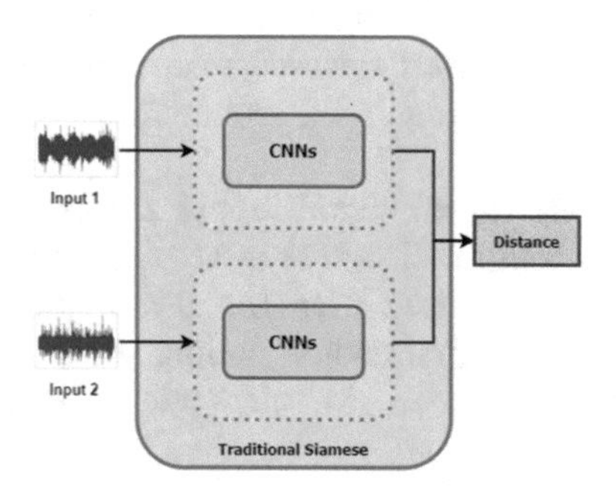

Fig. 1. Traditional Siamese Architecture

In [17] rolling bearing failure study, the authors extended Siamese in combination with WDCNN and gave an accuracy of approximately 92% with 90 training samples and 750 test samples.

3 Methodology

3.1 General Architecture for Failure Diagnosis

We provide a new training architecture for limited data based on the Conv-mixer model and few-shot learning shown in Fig. 2. There are two primary components to the model. The initial step is to use input spectrogram pairs to train the Siamese-based Convmixer model network to identify similarities and differences. The signal is compared to classes in the second step, using few-shot learning to determine which class it belongs to.

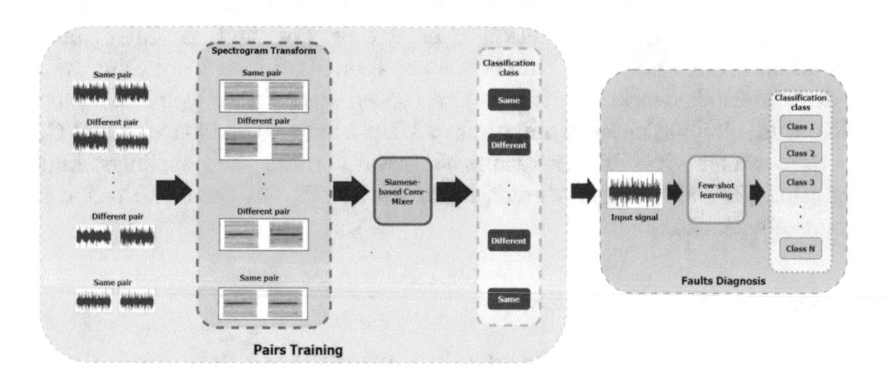

Fig. 2. General architecture for predicting bearing failure

3.2 The Proposed Siamese-Based Conv-Mixer Model

Figure 3 depicts the proposed Siamese-based Conv-Mixer architectural model. Spectrogram-processed signal pairs serve as the input. For effective training, we built a dataset using the CWMR dataset with pairs labeled as 1 if they are different and 0 if they are the same, with around 50% of the pairs being different and similar.

A pair of spectrum transform signal inputs are sent to the network. Their Euclidean distances are computed by (1) after passing through the final Fully-connected layer and are compared with the threshold. This network training aims to push distinct signals apart while pulling pairs of similar signals closer. The distance of two-point (x_1, x_2) in space can be defined as the Euclidean distance:

$$d_f^2(x_1, x_2) = \|x_1 - x_2\| \tag{1}$$

3.3 Diagnosis Network

Following strategy using the model provided in Fig. 3, the Diagnosis model is constructed in accordance with Fig. 4. We create classes called few-shot samples from the train dataset. There will be one or more signals with the same label

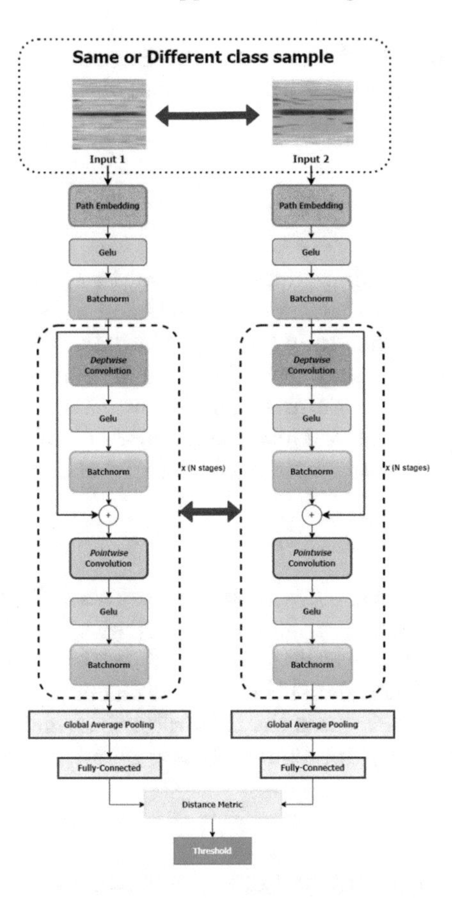

Fig. 3. The proposed Siamese-based Conv-Mixer model

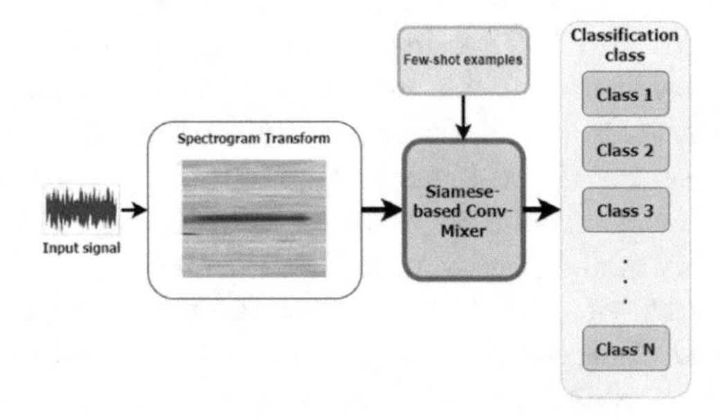

Fig. 4. The proposed Few-shot learning model

in each class. The signal is then transformed into the spectrum. Through the Siamese-based Conv-Mixer described in Sect. 3.2, the signal to be fault-diagnosed is first compared with each signal in each class. To determine which class the signal to diagnose belongs to, average the Euclidean distance displayed (1) against the class and compare it with the threshold.

4 Experiment

4.1 Datasets

The widely used bearing failure dataset created by Case Western Reserve University (CWRU) is used in this research. We tested with data from 90, 300, 600, and 198000 samples for the study with limited data. We build a data set with signal pairings labeled 0 if they have the same fault and label 1 if they are different, in order to train the Siamese-based Conv-Mixer network. Note that the number of pairs with labels of 0 and the number of pairs with labels of 1 must be approximately 50% for the most efficient training of the network.

4.2 Training

To train the Siamese-based conv-mixer network model proposed in Sect. 3.2, we use the Adam optimization method in [26] with learning rate 0.0001. The loss function used is Binary Cross Entropy [27], with the goal of optimizing the Euclidean distance between signal classes. The training process is done with 200 epochs because the goal is to train the network with a small data set, the running time for each epoch is only approximately two minutes.

4.3 Results

In this part, we discuss the results from experiments using various training sample counts and the bearing failure prediction model that was presented in Sect. 3. First, we analyze the Siamese-based Conv-Mixer model's performance using only 90 training examples and 750 testing data. To create pairs of signals for the Siamese-based Conv-Mixer model, 1 random signal is chosen and compared with 5 other random signals. In Fig. 5, the result is given with the left being the input signal, here randomly the signal labeled "Ball 0.014", and on the right the other 5 signals along with their Euclidean distance from the signal "Ball 0.014". With the threshold selected as 0.5 (less than 0.5 is the same class and larger is a different class), we see that with the first 4 signals, the Euclidean distance compared to the signal labeled "Ball 0.014" is larger. 0.5, and the final signal is 0.37 satisfying less than the threshold and in the same class. It is obvious that the signals would be further apart the more diverse they are.

Then, to check the accuracy on the entire test set with 750 samples, the confusion matrix is calculated and shown in Fig. 6(a). The results show that the accuracy is approximately 95.8%. Figure 6(b) for a visual display of the classification of faults according to their labels. Since there were only 90 samples used for training, it is obvious that the classification is accurate for the majority of the areas. However, region 3 with 8 samples, and region 1 with 5 samples still show little divergence.

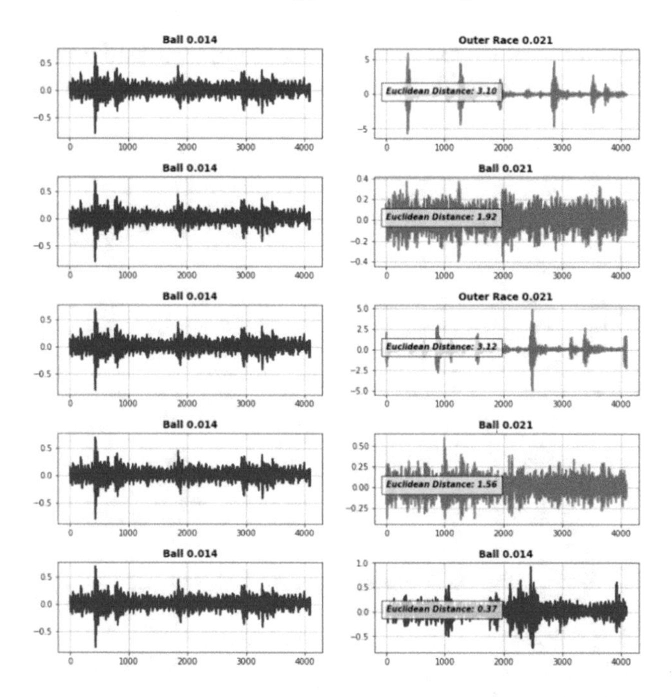

Fig. 5. Representative results when choosing 1 signal compared with 5 other random signals

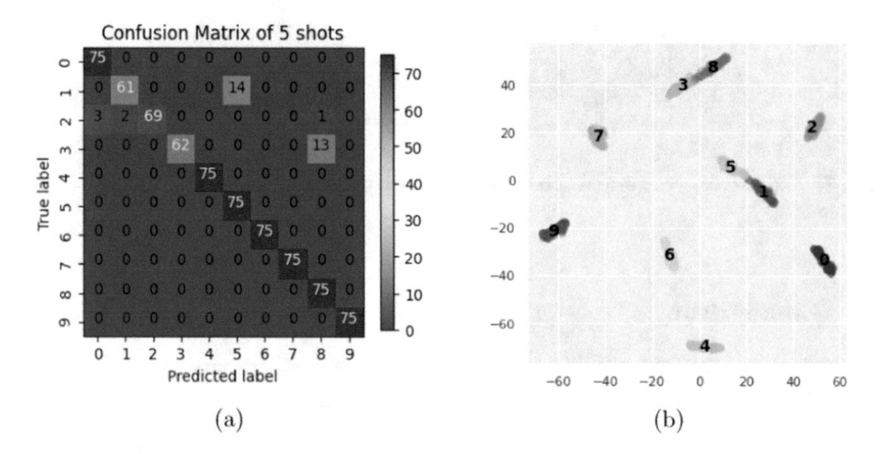

Fig. 6. (a) Confusion matrix when training with 90 samples and (b) Visualization of data while training with 90 samples

Next, we tested with a larger amount of data (19800 samples). By selecting 1 random signal and comparing it with 6 other signals, we derive the Euclidean distance between these signals via Fig. 7. The same amount of test data is 750 samples, the accuracy of the proposed model is approximately 99.84% shown in confusion matrix in Fig. 8(a). Based on Fig. 8(b), we also see a clear classification.

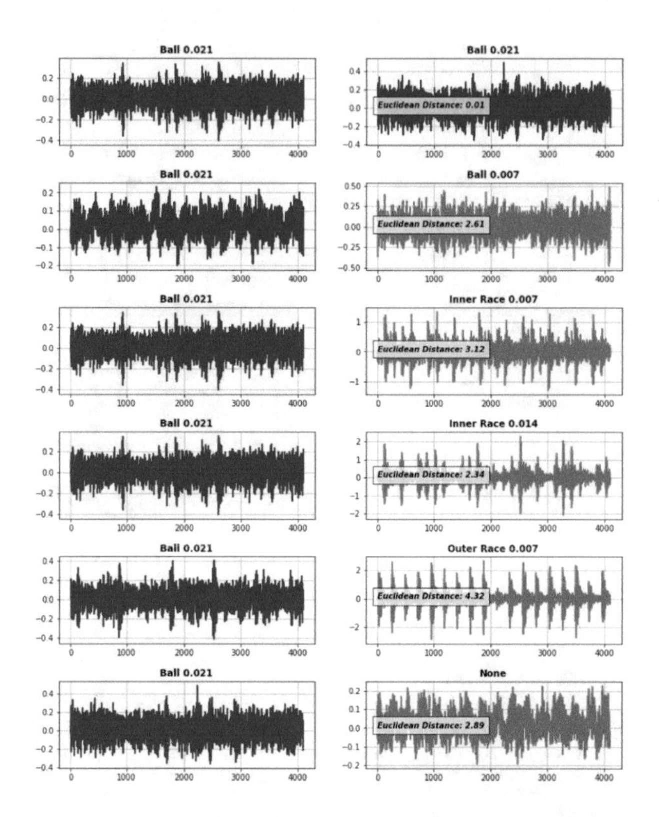

Fig. 7. Representative results when choosing 1 signal compared with 6 other random signals

4.4 Comparison

According to [17], the result of predicting bearing failure with test data set of only 90 samples on the same CWRU dataset, we have confusion matrix and classification in Fig. 9. The proposed model performs much better with 90 training samples as compared to the findings in Fig. 6(a) and Fig. 9(a), Siamese-based Conv-Mixer model shows 95.58% accuracy while the traditional Siamese model only shows approximately 90% accuracy. Based on the comparison in Fig. 6(b) and Fig. 9(b), the data are more clearly categorized. More specifically, we have surveyed some other models and summarized them in Table 1. With just 77.39% accuracy with WDCNN and 80% accuracy using the Conv-Mixer model, traditional deep learning types for a small dataset produce unsatisfactory results.

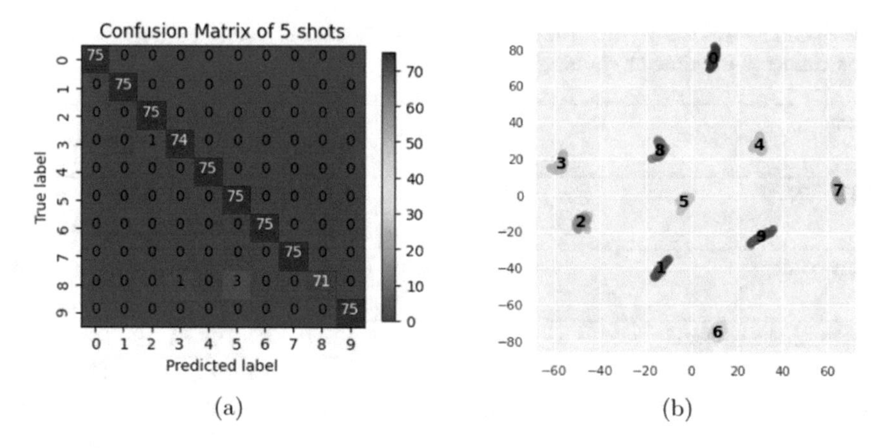

Fig. 8. (a) Confusion matrix when training with 19800 samples and (b) Visualization of data while training with 19800 samples

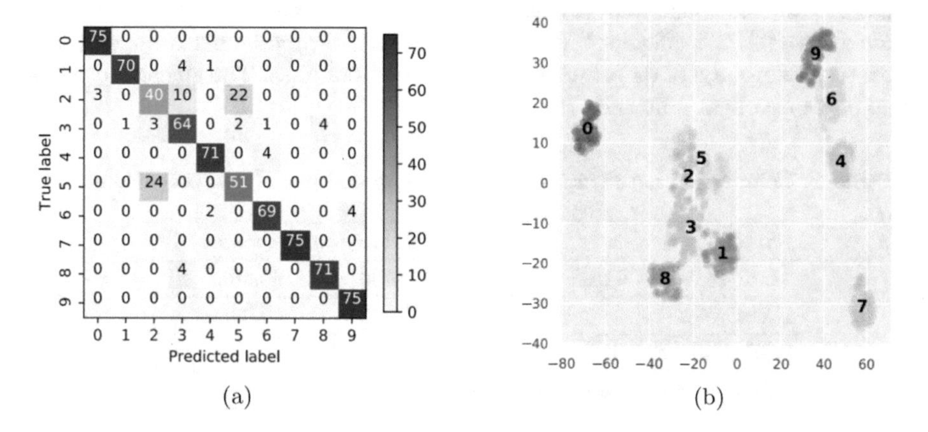

Fig. 9. (a) Confusion matrix when training with 90 samples of using Traditional Siamese and (b) Visualization of data while training with 90 samples of using Traditional Siamese

While the proposed scheme can achieve an accuracy of up to 95.48% with just 90 training samples. The presented method likewise performs well, responding with 99.84% accuracy when there are 19800 training samples. The suggested approach provides efficient results in both small and big data sets, in addition to providing superior outcomes for small data sets.

To further evaluate the performance of the proposed approach, we compare the results of our model with those by other methods reported in recent works in Table 2. The proposed model has a 99.84% efficiency compared to several other rolling bearing defect diagnostic methods. The Multi-scale CNN and Long Short-term Memory (LSTM) in [28] model has a 98.46% efficiency, LAFD-Net in [29] give 99.03% and the Symmetrized Dot Pattern (SDP) + Squeeze-and-Excitation

Table 1. The accuracy (in %) on the same test dataset of comparative models when using different number of training samples

Model	Type of input	Training Samples			
		90	300	600	19800
WDCNN	No spectrum input	77.39	96.59	97.03	99.65
	Spectrum input	82.32	95.78	98.94	99.68
Conv-Mixer	No spectrum input	80.12	96.80	99.73	99.86
	Spectrum input	85.32	96.12	99.20	99.78
Traditional Siamese	No spectrum input	91.37	96.65	96.14	99.97
	Spectrum input	91.84	96.48	97.35	99.53
Siamese-based Conv-Mixer (Ours)	No spectrum input	93.64	95.70	97.63	99.07
	Spectrum input	**95.48**	**97.12**	**99.63**	**99.84**

enabled Convolutional Neural Network (SE-CNN) model mentioned in the [30] article has a 99.75% efficiency. As can be observed, our model performs well both with small amounts of data and with big amounts of data. The suggested model also offers excellent efficiency.

Table 2. Comparison between results by the proposed model and previous models

Model	Parameters	Accuracy (%)
Multi-scale CNN and LSTM [28]	73480	98.46
LAFD-Net [29]	77090	99.03
SDP +SE CNN [30]	10868	99.75
Siamese-based Conv-Mixer (Ours)	51978	**99.84**

5 Conclusion

In this study, we have proposed a novel rolling bearing fault diagnostic model that performs very well with limited training data. The suggested technique addresses a problem in deep learning that demands a significant quantity of data in order to deliver reliable results. The proposed model is built on a combination of Siamese and Conv-Mixer models, assessing pairs of similarities and differences by comparing their characteristics via Euclidean distance using the input spectrum transform of the signal. Additionally, a comparison of several techniques reveals that the Siamese-based Conv-Mixer model outperforms conventional diagnostic techniques in terms of effectiveness.

Acknowledgement. This research is funded by Vietnam National Foundation for Science and Technology Development (NAFOSTED) under grant number 102.05-2021.34.

References

1. Furse, C.M., Kafal, M., Razzaghi, R., Shin, Y.-J.: Fault diagnosis for electrical systems and power networks: a review. IEEE Sens. J. **21**(2), 888–906 (2020)
2. Benbouzid, M.E.H., Kliman, G.B.: What stator current processing-based technique to use for induction motor rotor faults diagnosis? IEEE Trans. Energy Convers. **18**(2), 238–244 (2003)
3. Noble, W.S.: What is a support vector machine? Nat. Biotechnol. **24**(12), 1565–1567 (2006)
4. Soucy, P., Mineau, G.W.: A simple KNN algorithm for text categorization. In: Proceedings 2001 IEEE International Conference on Data Mining, pp. 647–648. IEEE (2001)
5. Saritas, M.M., Yasar, A.: Performance analysis of ANN and Naive Bayes classification algorithm for data classification. Int. J. Intell. Syst. Appl. Eng. **7**(2), 88–91 (2019)
6. Ciliberto, C., et al.: Quantum machine learning: a classical perspective. Proc. Roy. Soc. A Math. Phys. Eng. Sci. **474**(2209), 20170551 (2018)
7. Yang, Y., Yu, D., Cheng, J.: A fault diagnosis approach for roller bearing based on IMF envelope spectrum and SVM. Measurement **40**(9–10), 943–950 (2007)
8. He, M., He, D.: Deep learning based approach for bearing fault diagnosis. IEEE Trans. Ind. Appl. **53**(3), 3057–3065 (2017)
9. Jia, F., Lei, Y., Lin, J., Zhou, X., Lu, N.: Deep neural networks: a promising tool for fault characteristic mining and intelligent diagnosis of rotating machinery with massive data. Mech. Syst. Sig. Process. **72**, 303–315 (2016)
10. Chen, Z., Li, W.: Multisensor feature fusion for bearing fault diagnosis using sparse autoencoder and deep belief network. IEEE Trans. Instrum. Meas. **66**(7), 1693–1702 (2017)
11. Abed, W., Sharma, S., Sutton, R., Motwani, A.: A robust bearing fault detection and diagnosis technique for brushless DC motors under non-stationary operating conditions. J. Control Autom. Electr. Syst. **26**, 241–254 (2015). https://doi.org/10.1007/s40313-015-0173-7
12. Zhang, W., Li, C., Peng, G., Chen, Y., Zhang, Z.: A deep convolutional neural network with new training methods for bearing fault diagnosis under noisy environment and different working load. Mech. Syst. Sig. Process. **100**, 439–453 (2018)
13. Wen, L., Li, X., Gao, L., Zhang, Y.: A new convolutional neural network-based data-driven fault diagnosis method. IEEE Trans. Ind. Electron. **65**(7), 5990–5998 (2017)
14. Zhang, W., Zhang, F., Chen, W., Jiang, Y., Song, D.: Fault state recognition of rolling bearing based fully convolutional network. Comput. Sci. Eng. **21**(5), 55–63 (2018)
15. Zilong, Z., Wei, Q.: Intelligent fault diagnosis of rolling bearing using one-dimensional multi-scale deep convolutional neural network based health state classification. In: 2018 IEEE 15th International Conference on Networking, Sensing and Control (ICNSC), pp. 1–6. IEEE (2018)
16. Zhang, W., Peng, G., Li, C., Chen, Y., Zhang, Z.: A new deep learning model for fault diagnosis with good anti-noise and domain adaptation ability on raw vibration signals. Sensors **17**(2), 425 (2017)
17. Zhang, A., Li, S., Cui, Y., Yang, W., Dong, R., Hu, J.: Limited data rolling bearing fault diagnosis with few-shot learning. IEEE Access **7**, 110895–110904 (2019)
18. O'Shea, K., Nash, R.: An introduction to convolutional neural networks (2015)

19. Yuan, L., et al.: Tokens-to-Token ViT: training vision transformers from scratch on ImageNet. In: Proceedings of the IEEE/CVF International Conference on Computer Vision, pp. 558–567 (2021)
20. Santurkar, S., Tsipras, D., Ilyas, A., Madry, A.: How does batch normalization help optimization? In: Advances in Neural Information Processing Systems, vol. 31 (2018)
21. Tolstikhin, I.O., et al.: MLP-Mixer: an all-MLP architecture for vision. In: Advances in Neural Information Processing Systems, vol. 34, pp. 24261–24272 (2021)
22. Tran, N.-D., Le, H.-H., Pham, V.-T., Tran, T.-T.: KPmixer-a ConvMixer-based network for finger knuckle print recognition. In: 2022 International Conference on Control, Automation and Information Sciences (ICCAIS) (2022)
23. Trinh, M.-N., Nham, D.-H.-N., Pham, V.-T., Tran, T.-T.: An attention-PiDi-UNet and focal active contour loss for biomedical image segmentation. In: 2022 International Conference on Control, Automation and Information Sciences (ICCAIS) (2022)
24. Zhao, X., Ma, M., Shao, F.: Bearing fault diagnosis method based on improved Siamese neural network with small sample. J. Cloud Comput. **11**(1), 1–17 (2022)
25. Chicco, D.: Siamese neural networks: an overview. In: Cartwright, H. (ed.) Artificial Neural Networks. MMB, vol. 2190, pp. 73–94. Springer, New York (2021). https://doi.org/10.1007/978-1-0716-0826-5_3
26. Jais, I.K.M., Ismail, A.R., Nisa, S.Q.: Adam optimization algorithm for wide and deep neural network. Knowl. Eng. Data Sci. **2**(1), 41–46 (2019)
27. Ho, Y., Wookey, S.: The real-world-weight cross-entropy loss function: modeling the costs of mislabeling. IEEE Access **8**, 4806–4813 (2019)
28. Chen, X., Zhang, B., Gao, D.: Bearing fault diagnosis base on multi-scale CNN and LSTM model. J. Intell. Manuf. **32**, 971–987 (2021). https://doi.org/10.1007/s10845-020-01600-2
29. Jian, Y., et al.: LAFD-Net: learning with noisy pseudo labels for semi-supervised bearing fault diagnosis. IEEE Sens. J. **23**(4), 3911–3923 (2023)
30. Wang, H., Xu, J., Yan, R., Gao, R.X.: A new intelligent bearing fault diagnosis method using SDP representation and SE-CNN. IEEE Trans. Instrum. Meas. **69**(5), 2377–2389 (2019)

Differentially-Private Distributed Machine Learning with Partial Worker Attendance: A Flexible and Efficient Approach

Le Trieu Phong[1(✉)] and Tran Thi Phuong[2]

[1] National Institute of Information and Communications Technology (NICT), Tokyo 184-8795, Japan
phong@nict.go.jp

[2] KDDI Research, Inc., Saitama 356-0003, Japan

Abstract. In distributed machine learning, multiple machines or workers collaborate to train a model. However, prior research in cross-silo distributed learning with differential privacy has the drawback of requiring all workers to participate in each training iteration, hindering flexibility and efficiency. To overcome these limitations, we introduce a new algorithm that allows partial worker attendance in the training process, reducing communication costs by over 50% while preserving accuracy on benchmark data. The privacy of the workers is also improved because less data are exchanged between workers.

Keywords: Differential privacy · Partial attendance · Communication efficiency · Distributed machine learning

1 Introduction

1.1 Background

Distributed learning is a method of training machine learning models on multiple machines. It typically involves a parameter server that maintains the model parameters and worker machines that compute gradients on their local datasets. Improved communication efficiency in distributed learning refers to techniques that reduce the amount of data that needs to be communicated between the parameter server and the worker machines in order to train the model.

Communication efficiency plays a crucial role in the success of distributed learning, as it directly impacts the convergence speed of the model. Poor communication can result in slower convergence and prolonged training time. Additionally, it consumes valuable resources such as bandwidth, leading to increased costs and reduced scalability. To address these challenges, various techniques

The work of LTP is partially supported by JST CREST Grant JPMJCR21M1, and JST AIP Accelerated Program Grant JPMJCR22U5, Japan. Parts of this work were done while TTP was at NICT.

N. T. Nguyen et al. (Eds.): CITA 2023, LNNS 734, pp. 15–24, 2023.
https://doi.org/10.1007/978-3-031-36886-8_2

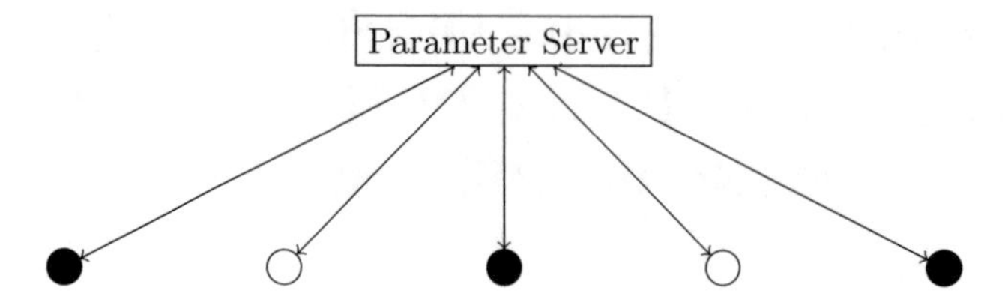

Fig. 1. Distributed learning with partial attendance. In this figure, the parameter server is represented by the rectangle node. The other nodes represent the machines (workers) participating in the distributed learning process. The filled circles indicate the nodes that attend and contribute to the current training iteration. The arrows show the flow of information between the learning nodes and the parameter server.

have been proposed to enhance communication efficiency in distributed learning. One such approach is gradient sparsification, where only a portion of the gradients calculated by the workers is transmitted to the parameter server, rather than all of them. This significantly reduces the amount of data communicated, thereby accelerating the training process. Furthermore, gradient sparsification is often combined with error-feedback methods, which use the unused information from the current training iteration in future iterations, thus preserving the accuracy of the model, as demonstrated in various studies [10–12,15,16,19] (Fig. 1).

Differential privacy [4] is a notion in the field of privacy-preserving data analysis that aims to protect the privacy of individuals in a dataset. In the context of distributed machine learning, it ensures that the models trained on distributed data do not reveal any sensitive information about individual data. This is achieved by adding noise to the communicated vectors before they are sent from the workers to the parameter server. The goal is to ensure that the final model is not biased towards any individual data and that the privacy of the data is preserved. Previous works have examined differential privacy and communication efficiency in the cross-silo context [14] and the cross-device context [1,7,9,17].

In previous work in the cross-silo context of distributed learning [14], it was assumed that each machine must cooperate in each training iteration for the system to work. This is referred to as full attendance of machines. This work builds on [14] by considering the scenario of partial attendance, where not all machines participate in every iteration.

The advantage of a system that allows partial attendance is that it increases the overall efficiency and flexibility of the training process. The system can still function even if some machines are unavailable, reducing the reliance on full attendance. Additionally, it can lower the cost of computation, communication, and energy consumption, making the training more efficient and cost-effective. Moreover, partial attendance can handle scenarios with limited resources or changing network conditions, leading to improved resilience and scalability. Partial attendance has been considered in works such as [13] in the context of decentralized learning which is orthogonal to our scenario. To the best of our knowledge, this is the first examination of partial attendance in cross-silo distributed learning incorporating differential privacy.

Table 1. Comparison with the counterpart in [14] with 14 workers on non-IID data. The smaller the privacy budget, the better; the higher the testing accuracy, the better. The testing accuracies achieved by our algorithm are on par with those obtained through distributed SGD and [14], indicating that our method represents a state-of-the-art solution.

Algorithm	Privacy budget (ϵ, δ)	Testing accuracy	Partial attendance
Distributed SGD	N/A (no privacy)	99.4%	no
Ref. [14] (100% worker attendance)	$(2.99, 4.58 \times 10^{-7})$	98.74%	no
Ours (50% worker attendance)	$(0.64, 4.58 \times 10^{-7})$	98.92%	yes
Ours (25% worker attendance)	$(0.45, 4.58 \times 10^{-7})$	98.59%	yes

1.2 Our Contributions

We propose and assess a differentially-private distributed machine learning algorithm, building upon a recent approach introduced in [14]. However, unlike prior research, our algorithm accommodates partial attendance of machines (workers). Our proposed algorithm is described in Algorithm 1 and possesses the following properties:

- **Partial attendance:** Any worker can choose to participate in the current training iteration or not, making our algorithm more flexible than its counterpart in [14].
- **Improved communication efficiency:** The communication costs in Algorithm 1 have been significantly reduced by over 50%, compared to [14], while still achieving comparable testing accuracy on the same benchmark data.
- **Improved differential privacy:** Each worker has enhanced privacy due to the reduced amount of information sent from each worker, with a lower privacy budget compared to [14].

Concretely, Table 1 presents the comparison between the algorithm in [14] and two variations of the proposed Algorithm 1, with 50% and 25% worker attendance respectively. The privacy budget is measured in terms of (ϵ, δ) values, where lower values indicate higher privacy protection. The testing accuracy is measured as a percentage. The proposed scheme shows improved testing accuracy compared to [14] with 50% worker attendance while maintaining a better privacy budget of $(0.64, 4.58 \times 10^{-7})$. With a stricter setting of 25% worker attendance, the proposed scheme has a lower testing accuracy of 98.59%, and a slightly higher privacy budget of $(0.45, 4.58 \times 10^{-7})$.

1.3 Paper Organization

The structure of this paper is as follows. Section 2 introduces preliminary information on differential privacy. Section 3 outlines our proposed algorithm and calculates its privacy budget. Section 4 presents the results of our experiments. Finally, Sect. 5 concludes the paper.

2 Preliminaries

Two sets d and d' are considered adjacent if they differ by only one entry. Differential privacy states that a randomized mechanism $\mathcal{M}$ is (ϵ, δ)-private if, for any adjacent sets d and d' and any output subset S, the probability of obtaining $\mathcal{M}(d)$ in S is at most $\exp(\epsilon) \cdot \Pr[M(d') \in S] + \delta$:

$$\Pr[M(d) \in S] \leq \exp(\epsilon) \cdot \Pr[M(d') \in S] + \delta.$$

The Laplace mechanism adds Laplace noise with mean 0 and scale b to a function f with sensitivity $\Delta = \max_{x \simeq x'} |f(x) - f(x')|_1$, where the Laplace noise has a PDF of $\frac{1}{2b} \exp\left(-\frac{|x|}{b}\right)$. It has been shown in [4,5] that the mechanism

$$\mathcal{M}(x) = f(x) + \mathsf{Laplace}(0, \frac{\Delta}{\epsilon})$$

provides $(\epsilon, 0)$-differential privacy.

$\mathcal{M} \circ \mathsf{subsample}$ is a mechanism that first subsamples data and then applies $\mathcal{M}$. The following lemma, called the Subsampling Amplification Lemma, is important.

Lemma 1 (Subsampling Amplification [8]). *If $\mathcal{M}$ provides (ϵ, δ)-differential privacy, then $\mathcal{M} \circ \mathsf{subsample}$ provides $(\epsilon_{sub}, \delta_{sub})$-differential privacy with*

$$\begin{cases} \epsilon_{sub} & = \log\left((\exp(\epsilon) - 1)q + 1\right) \\ \delta_{sub} & = \delta q \end{cases}$$

where q is the subsampling probability of $\mathsf{subsample}$.

The following lemma, called the Advanced Composition Lemma, gives an upper bound for privacy budgets. Although it may not be as precise as other methods like [2,3,18], it is sufficient for our needs due to its simplicity and ease of computation.

Lemma 2 (Advanced composition [6]). *If $\mathcal{M}_1, \ldots, \mathcal{M}_K$ are algorithms that can each be chosen based on the outputs of previous algorithms, and each $\mathcal{M}_i$ satisfies (ϵ, δ)-differential privacy, then their composition $(\mathcal{M}_1(d), \ldots, \mathcal{M}_K(d))$ is $(\epsilon', K\delta + \delta')$-differentially private, where ϵ' satisfies the following bound:*

$$\epsilon' \leq \epsilon\sqrt{2K \log(1/\delta')} + K\epsilon(\exp(\epsilon) - 1).$$

3 Our Proposed Algorithm

Our proposed algorithm is described in Algorithm 1. The algorithm is designed to be executed on a cluster of M worker machines, each with a local dataset,

Algorithm 1. Distributed differentially-private SGD with partial attendance

1: **Parameters:** stepsize η; number of workers M; constants C, σ
2: **Initialize:** initial weight $x_0 \in \mathbb{R}^d$; initial vector $e_{0,i} = \mathbf{0} \in \mathbb{R}^d$ on each worker i
3: **for** $t \in \{0, \ldots, T-1\}$ **do**
4: • **on each worker** i **having dataset** Ω_i**:**
5: set Attend_i to True or False
6: sample data $\zeta_{t,i}$ uniformly at random from Ω_i with probability q_i
7: $g_{t,i} = \nabla \mathcal{L}(x_t, \zeta_{t,i}) \in \mathbb{R}^d$ ▷ stochastic gradient
8: $h_{t,i} = \eta g_{t,i} + e_{t,i}$, seen as $h_{t,i} = (h_{t,i,1}, \ldots, h_{t,i,d}) \in \mathbb{R}^d$
9: let I be the indices of top-k components in vector $h_{t,i}$
10: **for** $r \in \{1, \ldots, d\}$ **do**
11: **if** $r \in I$: $\Psi_{t,i,r} = \begin{cases} \frac{h_{t,i,r}}{\max\{1, |h_{t,i,r}|/C\}} + \mathsf{Laplace}(0, 2C\sigma) & \text{if } \mathsf{Attend}_i = \mathsf{True} \\ 0 & \text{if } \mathsf{Attend}_i = \mathsf{False} \end{cases}$
12:
13: **if** $r \notin I$: $\Psi_{t,i,r} = 0$
14: **end for**
15: let $\Psi_{t,i} = (\Psi_{t,i,1}, \ldots, \Psi_{t,i,d}) \in \mathbb{R}^d$
16: push $\Psi_{t,i}$ to the parameter server
17: pull the gradient average $\tilde{\Psi}_t$ from server
18: $x_{t+1} = x_t - \tilde{\Psi}_t$ ▷ weight update in stochastic gradient descent
19: $e_{t+1,i} = h_{t,i} - \Psi_{t,i}$
20: • **on parameter server:**
21: pull $\Psi_{t,i}$ from all workers $1 \le i \le M$
22: compute $\tilde{\Psi}_t = \frac{1}{M}\sum_{i=1}^{M} \Psi_{t,i}$ ▷ averaging all vectors from workers
23: push the average $\tilde{\Psi}_t$ to each worker
24: **end for**

and a parameter server that averages the vectors sent from the workers. The algorithm differs from [14] in that it includes lines 5 and 11 which allow for a boolean flag Attend_i to be set to indicate the attendance of worker i in the training iteration. It is important to note that if $\mathsf{Attend}_i = \mathsf{False}$, meaning the worker does not participate in the training iteration, then $\Psi_{t,i}$ becomes a zero vector.

The algorithm begins by initializing the model parameters and a memory term for each worker. The initial vector $e_{0,i} = \mathbf{0} \in \mathbb{R}^d$ on each worker i means there is no memory initially. In each iteration, each worker selects a sample from its local dataset and computes the gradient of the loss function $\mathcal{L}$ with respect to the model parameters. The worker then adds the memory term to the gradient, where the memory term is computed in line 19 as the difference between the locally computed vector $h_{t,i}$ and the communicated vector $\Psi_{t,i}$. The top-k components from $h_{t,i}$ in magnitude are selected and added with noise sampled from a Laplace distribution. These perturbed gradients are then sent to the parameter server, where they are averaged and the model parameters are updated.

The algorithm protects data privacy by adding noise in line 11 when worker i chooses to attend the training iteration. Dimension reduction is achieved by the sparsifying technique in line 13. The constants C and σ determine the amount of noise added to the gradients and thus the privacy level. Line 11 with $\mathsf{Attend}_i = \mathsf{True}$ of Algorithm 1 satisfies $(1/\sigma, 0)$-differential privacy, as

$$\begin{aligned} &\left| \frac{h_{t,i,r}}{\max\{1, |h_{t,i,r}|/C\}} - \frac{h'_{t,i,r}}{\max\{1, |h'_{t,i,r}|/C\}} \right| \\ \le &\left| \frac{h_{t,i,r}}{\max\{1, |h_{t,i,r}|/C\}} \right| + \left| \frac{h'_{t,i,r}}{\max\{1, |h'_{t,i,r}|/C\}} \right| \\ \le &\, C + C \\ = &\, 2C \end{aligned}$$

implies the corresponding Laplace mechanism has $(1/\sigma, 0)$-differential privacy. The mechanism amplifies through subsampling, as each data point is sampled with probability q_i at line 6, and obeys $(\epsilon_{sub}, 0)$-differential privacy where

$$\epsilon_{sub} = \log\left[(\exp(1/\sigma) - 1)q_i + 1\right]$$

which is computed according to Lemma 1.

Line 11 (with $\mathsf{Attend}_i = \mathsf{False}$) and line 13 always set $\Psi_{t,i,r}$ to zero, thereby perfect privacy is achieved at these lines. Let K be the total number of applying the Laplace mechanism adaptively by a worker at line 11 (with $\mathsf{Attend}_i = \mathsf{True}$). Owing to Lemma 2, the worker i in Algorithm 1 satisfies (ϵ, δ)-differential privacy with

$$\begin{cases} \epsilon \le \epsilon_{sub}\sqrt{2K\log(1/\delta')} + K\epsilon_{sub}(\exp(\epsilon_{sub}) - 1) \\ \delta = \delta' \end{cases} \tag{1}$$

which is later used to compute the privacy budget of our algorithm in Table 1.

4 Experiments

The MNIST8M dataset is a variation of the well-known MNIST (Modified National Institute of Standards and Technology) dataset, which is a benchmark for machine learning and computer vision tasks. The MNIST dataset consists of grayscale images of handwritten digits, with each image having a resolution of 28×28 pixels. The MNIST8M dataset, as its name suggests, contains a significantly larger number of images, with 8,100,000 images in total. One of the reasons why MNIST8M is used is due to its large size, making it an ideal candidate for distributed machine learning. Some data samples are given in Fig. 2.

In our experiments, the MNIST8M dataset is divided among the workers in a non-IID manner, not having the same probability distribution, with the images sorted in ascending order based on their labels. This setting captures the scenario where the data distribution among workers is unbalanced, with some workers having more samples of certain classes than others. This is reflective of real-world scenarios where data distribution is often uneven.

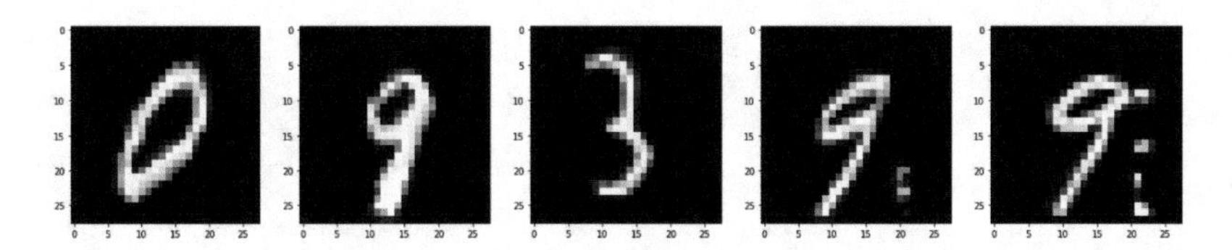

Fig. 2. Random samples from the MNIST8M dataset.

Layer (type)	Output Shape	Param #
Conv2d-1	[-1, 10, 24, 24]	260
Conv2d-2	[-1, 20, 8, 8]	5,020
Dropout2d-3	[-1, 20, 8, 8]	0
Linear-4	[-1, 50]	16,050
Linear-5	[-1, 10]	510

Total params: 21,840
Trainable params: 21,840
Non-trainable params: 0

Fig. 3. Neural network model for training.

For the experiment, each worker employs a simple but powerful neural network model having two convolutional layers (Conv2d-1 and Conv2d-2), one dropout layer (Dropout2d-3), and two fully connected layers (Linear-4 and Linear-5). The first convolutional layer Conv2d-1 has 260 parameters and the second convolutional layer Conv2d-2 has 5,020 parameters. The first fully connected layer Linear-4 has 16,050 parameters and the second fully connected layer Linear-5 has 510 parameters. The total number of parameters in the model is 21,840, all of which are trainable. The model is depicted in Fig. 3.

We implement Algorithm 1 using PyTorch[1], and consider the number of workers $M \in \{14, 21\}$ in the experiments. The training parameters are the same as [14]. Concretely, the learning rate $\eta = 0.05$, $C = 20$, $\sigma = 0.33$, and $k = 1$. Let $\delta = 1/n_{data}^{1.1}$ be the delta part in the privacy budget of each worker, where $n_{data} = 8,100,000/M$ is the number of data items of each worker. The batch size at each worker is $B = 8$, and the sampling probability $q_i = B/n_{data}$. The ϵ part in the privacy budget for worker i is computed by Eq. (1) in which its components are as follows with $M = 14$:

- $\delta = \delta' = 1/n_{data}^{1.1} = 1/(8,100,000/M)^{1.1} \approx 4.58 \times 10^{-7}$.
- $\epsilon_{sub} = \log\left[(\exp(1/\sigma) - 1)q_i + 1\right] \approx 2.724 \times 10^{-4}$.
- $K = r \times (n_{epoch} \times n_{data}/B)$ in which $n_{epoch} = 5$ is the number of epochs (an epoch is a complete iteration over the entire training dataset), and the factor $r = 0.5$ or 0.25 respectively means 50% or 25% worker attendance.

[1] https://pytorch.org/.

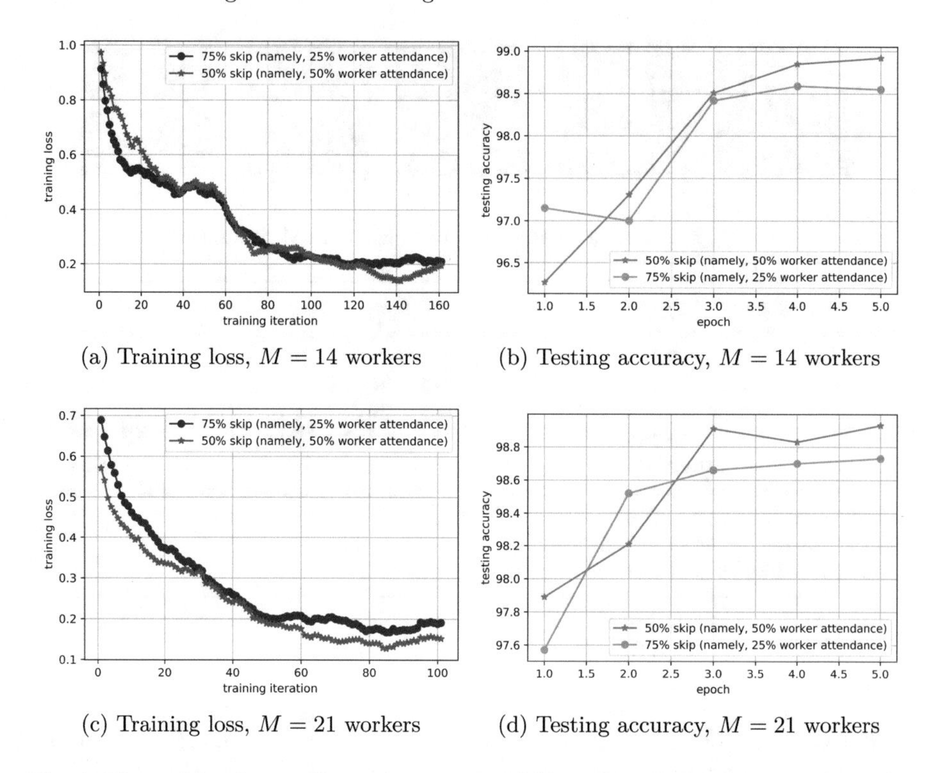

(a) Training loss, $M = 14$ workers

(b) Testing accuracy, $M = 14$ workers

(c) Training loss, $M = 21$ workers

(d) Testing accuracy, $M = 21$ workers

Fig. 4. The training loss and testing accuracy of Algorithm 1, with various attendance ratios and numbers of workers.

The line graphs in Fig. 4(a) and Fig. 4(c) depict the training loss and the line graphs in Fig. 4(b) and Fig. 4(d) display the testing accuracy of the model produced by Algorithm 1 over 5 epochs for the number of workers $M = 14$ and $M = 21$, respectively. The two lines in the plot show how the testing accuracy changes over 5 epochs for two different conditions: one where 50% of the workers are skipped (namely, 50% worker attendance) and another where 75% of the workers are skipped (namely, 25% worker attendance). The higher the testing accuracy, the better the performance of the model. Despite having less than 50% worker attendance, Algorithm 1 still converges and produces a model comparable to that in [14] as stated in Table 1. It is shown in [14] that their algorithm is comparable to distributed SGD, which is state-of-the-art. Therefore, we can conclude that our algorithm is also state-of-the-art in terms of performance.

5 Conclusion

In conclusion, we have designed and evaluated an algorithm for differentially-private distributed machine learning that offers improved privacy protection, communication efficiency, and flexibility compared to the existing algorithm in [14]. Our algorithm allows for partial attendance of workers, reducing the amount of information sent from each worker and hence, providing better privacy protection. The improved communication efficiency has resulted in a reduction of communication costs by over 50% compared to [14] while still achieving comparable testing accuracy on benchmark data.

References

1. Agarwal, N., Suresh, A.T., Yu, F.X., Kumar, S., McMahan, B.: cpSGD: communication-efficient and differentially-private distributed SGD. In: Advances in Neural Information Processing Systems 31: Annual Conference on Neural Information Processing Systems 2018, NeurIPS 2018, pp. 7575–7586 (2018)
2. Balle, B., Barthe, G., Gaboardi, M.: Privacy amplification by subsampling: tight analyses via couplings and divergences. In: Bengio, S., Wallach, H.M., Larochelle, H., Grauman, K., Cesa-Bianchi, N., Garnett, R. (eds.) Advances in Neural Information Processing Systems 31: Annual Conference on Neural Information Processing Systems 2018, NeurIPS 2018, Montréal, Canada, 3–8 December 2018, pp. 6280–6290 (2018)
3. Balle, B., Wang, Y.: Improving the Gaussian mechanism for differential privacy: analytical calibration and optimal denoising. In: Proceedings of the 35th International Conference on Machine Learning, ICML 2018, pp. 403–412 (2018)
4. Dwork, C., McSherry, F., Nissim, K., Smith, A.: Calibrating noise to sensitivity in private data analysis. In: Halevi, S., Rabin, T. (eds.) TCC 2006. LNCS, vol. 3876, pp. 265–284. Springer, Heidelberg (2006). https://doi.org/10.1007/11681878_14
5. Dwork, C., Roth, A.: The algorithmic foundations of differential privacy. Found. Trends Theor. Comput. Sci. **9**(3–4), 211–407 (2014). https://doi.org/10.1561/0400000042
6. Dwork, C., Rothblum, G.N., Vadhan, S.: Boosting and differential privacy. In: Proceedings of the 2010 IEEE 51st Annual Symposium on Foundations of Computer Science, FOCS 2010, pp. 51–60. IEEE Computer Society (2010). https://doi.org/10.1109/FOCS.2010.12
7. Girgis, A.M., Data, D., Diggavi, S., Kairouz, P., Suresh, A.T.: Shuffled model of federated learning: privacy, accuracy and communication trade-offs. IEEE J. Sel. Areas Inf. Theory **2**(1), 464–478 (2021). https://doi.org/10.1109/JSAIT.2021.3056102
8. Li, N., Qardaji, W., Su, D.: On sampling, anonymization, and differential privacy or, k-anonymization meets differential privacy. In: Proceedings of the 7th ACM Symposium on Information, Computer and Communications Security, ASIACCS 2012, pp. 32–33. ACM, New York (2012). https://doi.org/10.1145/2414456.2414474
9. Liu, R., Cao, Y., Chen, H., Guo, R., Yoshikawa, M.: FLAME: differentially private federated learning in the shuffle model. In: Thirty-Fifth AAAI Conference on Artificial Intelligence, AAAI 2021, Thirty-Third Conference on Innovative Applications of Artificial Intelligence, IAAI 2021, The Eleventh Symposium on Educational Advances in Artificial Intelligence, EAAI 2021, Virtual Event, 2–9 February 2021, pp. 8688–8696. AAAI Press (2021)
10. Phong, L.T., Phuong, T.T.: Distributed SignSGD with improved accuracy and network-fault tolerance. IEEE Access **8**, 191839–191849 (2020). https://doi.org/10.1109/ACCESS.2020.3032637

11. Phuong, T.T., Phong, L.T.: Distributed SGD with flexible gradient compression. IEEE Access **8**, 64707–64717 (2020). https://doi.org/10.1109/ACCESS.2020.2984633
12. Phuong, T.T., Phong, L.T.: Communication-efficient distributed SGD with error-feedback, revisited. Int. J. Comput. Intell. Syst. **14**(1), 1373–1387 (2021). https://doi.org/10.2991/ijcis.d.210412.001
13. Phuong, T.T., Phong, L.T.: Decentralized stochastic optimization with random attendance. IEEE Sig. Process. Lett. **29**, 1322–1326 (2022). https://doi.org/10.1109/LSP.2022.3179331
14. Phuong, T.T., Phong, L.T.: Distributed differentially-private learning with communication efficiency. J. Syst. Archit. **128**, 102555 (2022). https://doi.org/10.1016/j.sysarc.2022.102555
15. Sattler, F., Wiedemann, S., Müller, K., Samek, W.: Robust and communication-efficient federated learning from non-i.i.d. data. IEEE Trans. Neural Netw. Learn. Syst. **31**(9), 3400–3413 (2020). https://doi.org/10.1109/TNNLS.2019.2944481
16. Tang, H., Yu, C., Lian, X., Zhang, T., Liu, J.: DoubleSqueeze: parallel stochastic gradient descent with double-pass error-compensated compression. In: Proceedings of the 36th International Conference on Machine Learning, ICML 2019, pp. 6155–6165 (2019)
17. Wang, L., Jia, R., Song, D.: D2P-Fed: differentially private federated learning with efficient communication (2021). https://arxiv.org/abs/2006.13039
18. Wang, Y.X., Balle, B., Kasiviswanathan, S.P.: Subsampled Renyi differential privacy and analytical moments accountant. In: Chaudhuri, K., Sugiyama, M. (eds.) Proceedings of the Twenty-Second International Conference on Artificial Intelligence and Statistics. Proceedings of Machine Learning Research, vol. 89, pp. 1226–1235. PMLR, 16–18 April 2019
19. Zheng, S., Huang, Z., Kwok, J.T.: Communication-efficient distributed blockwise momentum SGD with error-feedback. In: Wallach, H.M., Larochelle, H., Beygelzimer, A., d'Alché-Buc, F., Fox, E.B., Garnett, R. (eds.) Advances in Neural Information Processing Systems 32: Annual Conference on Neural Information Processing Systems 2019, NeurIPS 2019, Vancouver, BC, Canada, 8–14 December 2019, pp. 11446–11456 (2019)

Building Legal Knowledge Map Repository with NLP Toolkits

Hung Q. Ngo[1], Hien D. Nguyen[2,3(✉)], and Nhien-An Le-Khac[4]

[1] Technological University Dublin, Dublin, Ireland
hung.ngo@tudublin.ie
[2] University of Information Techonology, Ho Chi Minh City, Vietnam
hiennd@uit.edu.vn
[3] Vietnam National University, Ho Chi Minh City, Vietnam
[4] School of Computer Science, University College Dublin, Dublin, Ireland
an.lekhac@ucd.ie

Abstract. Today, the legal document system is increasingly strict with different levels of influence and affects activities in many different fields. The increasing number of legal documents interwoven with each other also leads to difficulties in searching and applying in practice. The construction of knowledge maps that involve one or a group of legal documents is an effective approach to represent actual knowledge domains. A legal knowledge graph constructed from laws and legal documents can enable a number of applications, such as question answering, document similarity, and search. In this paper, we describe the process of building a system of knowledge maps for the Vietnamese legal system from the source of about 325,000 legal documents that span all fields of social life. This study also proposes an integrated ontology to represent the legal knowledge from legal documents. This model integrates the ontology of relational knowledge and the graph of key phrases and entities in the form of a concept graph. It can express the semantics of the content of a given legal document. In addition, this study also describes the process of building and exploiting natural language processing tools to build a VLegalKMaps system, which is a repository of Vietnamese legal knowledge maps. We also highlight open challenges in the realization of knowledge graphs in a technical legal system that enables this approach at scale.

Keywords: Knowledge Map · Legal AI · Legal Linked Data · Legal Documents · Knowledge Engineering

1 Introduction

Nowadays, scalable computing platforms, big data, and artificial intelligence (AI) are offering much more than the direct access to huge amounts of information. In law and AI domain, the development of integrated tools for legal research has also increased significantly [11]. In Europe, the complexity of legal systems

N. T. Nguyen et al. (Eds.): CITA 2023, LNNS 734, pp. 25–36, 2023.
https://doi.org/10.1007/978-3-031-36886-8_3

among countries requires legal knowledge sharing between them. Filtz et al. [8] built interlinking legal data to handle legal documents with European Law Identifier (ELI) and the European Case Law Identifier (ECLI). They built web identifiers and vocabularies to describe metadata related to legal documents for a transnational European legal KG. Although no AI applications were presented in this study, possible use case applications of such a knowledge graph in the legal domain are widespread [4], including (1) supporting comparative analyzes of court decisions and different legal interpretations of legislations; (2) enabling the analyzes of the evolution of legislation and jurisdiction; (3) interlinking legal knowledge with other data.

In another study, the CELLAR repository was introduced to provide semantic indexing, advanced search, and data retrieval for multilingual resources of the Publications Office of the European Union [22]. This system is based on a core ontology-based approach (namely Common Metadata Model, CDM) to organize multilingual resources and documents. In general, this system is a multilingual document management system based on the ontology-based approach and RDF(S)/OWL technologies to retrieve an RDF/XML representation of work-level metadata. The CDM model, resources (including general publications, legal resources, legislation, case law, etc.) are identified by Uniform Resource Identifiers (URIs) classified according to the Functional Requirements for Bibliographic Records (FRBR) hierarchy. This proposed CDM modeling approach provides document accessibility and open data services. Another European project is Lynx, which built a legal knowledge graph (LKG) for smart compliance services to assist companies in Europe with compliance needs [10,12]. Lynx approach contains three main phases: data acquisition, data integration, and data exploitation to integrate and link heterogeneous compliance data sources including legislation, case law, regulations, standards, and private contracts [10].

In general, knowledge graph and ontology approaches are used in several studies in the legal and AI domain [12,19,24]. Knowledge graphs (KG) represent real-world facts as structured data. A knowledge graph includes a set of triples (subject, relation, object), where the subject and object entities represent nodes, and their relations are labeled edges in the graph. Several studies proposed ontology-based approaches for the construction of knowledge graphs from legal documents [20,24]. Then, legal knowledge graphs have been used in other legal applications, such as the search engine for German court cases or legal documents [5,23], recommendation system for finding similar cases [7], or question-answering systems [21]. Thus, a support system for intellectual retrieval on law knowledge is very necessary for people [6]. However, generally, in literature there is a lack of a legal knowledge repository large enough to support legal AI applications efficiently.

Therefore, in this paper, we present a new process to build the legal knowledge maps as a linked data repository from legal documents with NLP (Natural Language Processing) Toolkits. It is based on a knowledge map model to represent knowledge in legal documents. This legal knowledge map repository will be a knowledge base to support legal AI applications, such as semantic search engines or auto-answering questions about legal concerns. The contribution of this study is to construct a structure of the VLegalO ontology to represent the content of legal documents. From that, VLegalKMaps resources are organized as linked data to build the knowledge repository of legal documents in a specific sub-domain.

The next section describes the architecture of the knowledge map model, including necessary definitions, ontology construction, and legal knowledge map construction. Section 2 presents the implementation of the legal knowledge map construction from the legal document repository with the assistance of NLP toolkits. We describe the implementation process for validating the proposed model based on the knowledge repository in law and AI domain in Sect. 4. Finally, we conclude the article and give some future research directions in Sect. 5.

2 Legal Knowledge Maps

2.1 Modeling

We propose a new knowledge map model to represent knowledge in legal documents in this paper. This proposed model is built upon the components of previous models, including Rela-model [16,19] and the OAK model [14,15]. Basically, the OAK model is proposed to handle mined knowledge as knowledge maps from data mining tasks in a specific domain like digital agriculture [14].

The proposed model for representing the legal document consists of an ontology as Rela-model combining knowledge maps from legal documents. This model has the structure as follows:

$$\mathbb{K} = (\mathbb{C}, \mathbb{R}, \mathbb{RULES}) + (Key, Rel, weight) \tag{1}$$

In which:

- $(\mathbb{C}, \mathbb{R}, \mathbb{RULES})$ is a structure of the Rela-model [16,19].
- *(Key, Rel, weight)* is a conceptual graph representing the relations between entities or key phrases of legal documents. In which *Key* is a set of entities or key phrases in the law document, *Rel* is a set of arcs representing relationships between entities or key phrases, and *weight* is the number representing the importance of relations in the knowledge maps.

In this proposed knowledge map model, $\mathbb{C}$ and $\mathbb{R}$ contains concepts and relations to present legal documents. *Key* and *Rel* are entities or key phrases, and relationships, which are extracted from legal documents and represented in knowledge maps. In this study, $\mathbb{RULES}$ and *Key* are not used to build the knowledge map repository.

2.2 Hierarchy of Legislation

In Vietnam, under the law on the promulgation of legal documents in 2015 [3] (amending and supplementing in 2020), the system of legal documents has three levels [17] as follows:

- Major legislations passed by the National Assembly is classified in the following order:
  - The Constitution, the highest law of the state;
  - Codes, Laws, and Decree-Laws (Ordinances)
  - Ordinances and Resolutions
- Legal instruments issued by the President of the State and the Minister of the Ministry to implement the main laws, including:
  - Orders and Decisions signed by the State President;
  - Government's documents: Directives and Executive Orders signed by the Prime Minister;
  - Decisions, Directives, and Circulars;
- Local ordinances:
  - Resolutions of People's Councils;
  - Decisions and directives of the People's Committee.
  - International Treaties

In this system, major legislations are the most important because they are national rules, while local ordinances might affect locally at each province/city where it is issued.

2.3 Legal Ontology Design

According to the description of categories and the structure of legislative documents in Sect. 2.2, the deepest and also most complex structure of one legal document has six levels, including document, section, part, chapter, article, clause, and point (Fig. 1). Based on the structure and content of legal documents, we propose an ontology to represent knowledge in legal documents. The diagram in Fig. 2 illuminates the relationships between *LegalItem* and related objects (such as organizations, people, and locations).

This proposed diagram supports connecting legal documents, issued organizations, signed person, affected locations or organizations, and related entities (as shown in Fig. 2).

The VLegalO ontology is implemented by Protege[1], which is a well-known tool for developing and maintaining ontologies. In Protege, object properties are used to create and modify relationships where the source of a relationship is called the domain class, and the destination of the relationship is called the range class.

Furthermore, several classes have been referred to concepts in the Document Components Ontology (DoCO[2]) [4]. DoCO provides a general framework to represent document components, such as chapter, abstract, section, paragraph, sentence, figure, table, glossary, and bibliography.

---

[1] https://protege.stanford.edu/.

[2] http://purl.org/spar/doco.

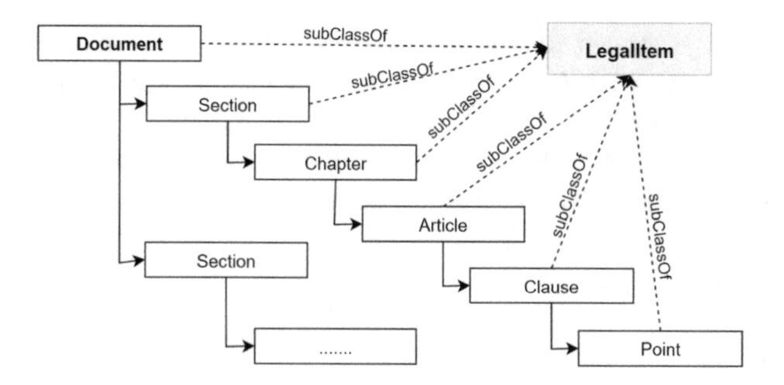

Fig. 1. Structure of a General Legal Document/Law

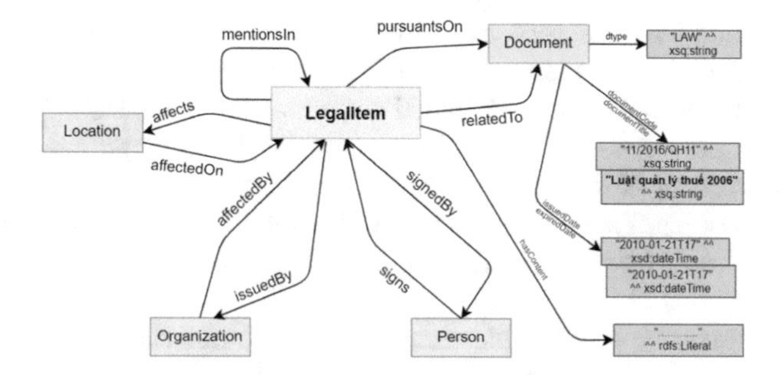

Fig. 2. Legal Ontology for VLegalKMaps

2.4 Legal Knowledge Map Construction

The knowledge maps, called VLegalKMaps, represent each legal document containing two parts, (1) one part for structured items and relations of the documents; and (2) the other part for conceptual graphs of legal knowledge in the documents. According to the theory of conceptual graph and design of the ontology, each legal document will be extracted and transformed into a triple set *(subject, predict, object)*. In which,

- *subject* and *object* are key phrases or entities in the document, and they are *Key* instances, which are linked to the $\mathbb{C}$ set
- *predict* is a pre-defined relation between entities or key phrases in the document, and it is a *Rel* instance, which is linked to the $\mathbb{R}$ set.

Both {Key} and {Rel} are VLegalKMaps resources, which are extracted from legal documents and transferred to knowledge maps. All of them are linked to concepts and relations in the pre-defined VLegalO ontology (Sect. 2.3). For example, *subjects* or *objects* ({Key}) are Chapter, Article elements, or Organization/Location entities in the documents, while *predicts* ({Rel}) are *hasChapter*, *hasArticle*, or *issuedBy* relations.

Based on the proposed model, each legal document is processed to build one legal knowledge map with NLP toolkits. All knowledge map elements (including {Key} and {Rel}) are identified by URIs. Then, all maps extracted from raw data are transformed into RDF triples[3], which is a well-known format for linked data and the foundation of graph databases.

3 Implementation

3.1 Materials

As mentioned in Sect. 2.2, there are many different types of legal documents in three groups; however, the most important documents are major legislative documents, including laws, decrees, and circulars (as shown in Table 1). All raw documents are crawled from many different existing resources, and each document has a code, title, issued date, and issued organization to identify. Nearly 325,000 documents are pre-processed to create structured documents, then transformed into the first version of knowledge maps. Each document will be extracted conceptual graphs of legal knowledge using NLP toolkits and then transformed into the second version of the knowledge map, which represents the legal knowledge in the document.

Legal documents cover a wide range of real-life activities, including administrative apparatus, trade, business, banking & finance, taxes - fees - charges, natural resources - environment, export & import, government finance, sports - health, transport, culture - society, construction - urban planning, investment, employment - wages, information technology, education, administrative violations, real estate, civil rights, insurance, securities, accounting - audit, proceedings, intellectual property, legal services, criminal liability, etc.

Table 1. Raw legal documents for VLegalKMaps

Code	Document Type	*Count*	*Entity*
LAW	Code, Law	494	9,698
DCR	Decree	5,081	74,721
CCL	(Joint) Circulars	16,006	145,075
ORD	Ordinates/Orders	1,200	2,656
DEC	Decisions	173,343	1,166,355
DOC	Documentary	118,227	275,563
NTR	National Technical Regulation	8,049	36,640
INS	Introduction	1,552	12,764
OTH	Others	374	4,255
Total	-	**324,326**	**1,727,727**

[3] https://www.w3.org/TR/n-triples/.

3.2 Vietnamese NLP Toolkits

There are several NLP toolkits that support Vietnamese text processing, such as Underthesea or SpaCy. **Underthesea**[4] is Vietnamese NLP toolkit, which includes the most common tasks in NLP, including sentence segmentation, word segmentation, POS tagging, chunking, dependency parsing, named entity recognition, text classification, and sentiment analysis. **SpaCy**[5] is used as an end-to-end practical tool to implement NLP applications using the Python ecosystem [1]. SpaCy includes the most common NLP tasks, such as tokenizer, morphology, part-of-speech tagger, dependency parser, noun chunking, named entity recognition, coreferences resolution, and entity linker. It has also pretrained NLP models for more than 20 languages. In addition, this toolkit supports fine-tuning and training new models from owned training data.

NLP Tasks for Building VLegalKMaps: Based on the requirements of VLegalKMaps, NLP toolkits need to support following tasks:

- *NP Chunking*, for key phrase extraction.
- *Named Entity Recognition*, for identifying location, organization, person, date, and document entities.
- *Dependency Parsing*, for extract relations in documents.

Basically, SpaCy is the most suitable toolkit for all the above tasks because it includes all well-known foundation NLP tasks, however, it only well supports English. For Vietnamese, it only has the Vietnamese tokenizer[6] and needs to retrain the model for dependency parsing, noun phrase chunking, and named entity recognition. For dependency parsing, this study uses Vietnamese Tree Bank (VTB[7]) from VLSP project [18] for training. This dataset is available as a universal dependency format, therefore, it is used to train the dependency parsing model for Vietnamese SpaCy. For the recognition of Vietnamese named entities, we used the Vietnamese part of EVNECorpus [13] as the dataset, fine-tuned the SpaCy model under SpaCy's guidelines[8], then used the trained model for tagging.

3.3 Building VLegalKMaps as Linked Data

VLegalKMaps is built from raw legal documents based on the VLegalO ontology framework and NLP toolkits. There are four modules in the system. First, the data collection module is used to crawl data and collect data from the Internet. The document is parsed into a structured document, and then key phrases and entities is extracted from each element of the document to create legal knowledge

[4] https://underthesea.readthedocs.io/.
[5] https://spacy.io/usage/linguistic-features.
[6] https://github.com/trungtv/vi_spacy.
[7] https://github.com/UniversalDependencies/UD_Vietnamese-VTB.
[8] https://spacy.io/usage/training.

maps. Finally, all elements of the structured document are converted into RDF triples as linked data. The process to build VLegalKMaps is as follows:

- Crawl raw data from existing resources on the Internet. Raw legal documents are also cleaned to remove noise, such as tables, headers, footers, etc.
- Extract structured documents from raw documents and organize the documents as the structure described in Sect. 2.2.
- Extract entities and keyphrases from each text-based element of structured documents.
- Generate RDF triples and import them into RDF storage.

4 Validations

4.1 Experiment Setup

Basically, each domain creates a group of legal documents and they are linked to real entities in actual life. Listing 1.1 presents the raw RDF triples of legal documents for VLegalKMaps and Fig. 3 shows a part of the legal knowledge map model. The VLegalKMaps resources can be imported into a graph database, like Virtuoso[9] and Apache Jena Fuseki[10] for retrieval.

Listing 1.1: Example Triples of Law on Road Traffic

```
VLegalKMaps:LawDoc349
    rdf:type owl:NamedIndividual ,
             VLegalO:Law ;
    rdfs:label "Law on Road Traffic,
                Law No. 23/2008/QH12." ;
    VLegalO:relatedTo VLegalKMaps:StandardDoc350;
    VLegalO:hasKeyphrase VLegalKMaps:Keyphrase_004 .
            VLegalKMaps:Keyphrase_005 ;
    VLegalO:signedIn VLegalO:HaNoi;
VLegalKMaps:StandardDoc350
    rdf:type   owl:NamedIndividual ,
               VLegalO:VNStandard ;
    rdfs:label "NTR on Traffic Sign and Signals,
               QCVN.41:2019/BGTVT." ;
VLegalKMaps:Keyphrase_004
    rdf:type owl:NamedIndividual ,
            VLegalO:Keyphrase ,
    rdfs:label "Driver license";
    VLegalO:definedIn VLegalKMaps:Article_59;
VLegalKMaps:Keyphrase_004
    rdf:type owl:NamedIndividual ,
            VLegalO:Keyphrase ,
    rdfs:label "Motocycle";
    VLegalO:rawText "Motocycle means a motor vehicle...";
    VLegalO:definedIn VLegalKMaps:StandardDoc350;
[...]
```

[9] https://virtuoso.openlinksw.com/.
[10] https://jena.apache.org/documentation/fuseki2/.

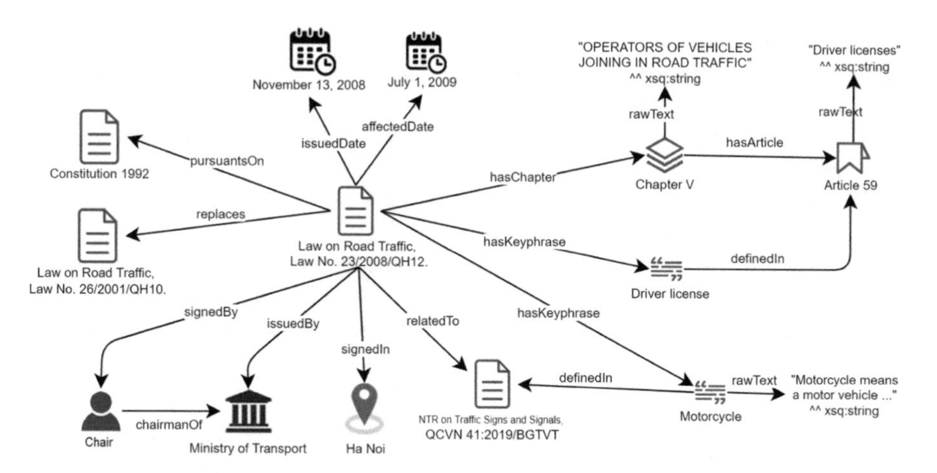

Fig. 3. Example of knowledge representations in VLegalKMaps.

4.2 Case Study

Based on the knowledge base on the road traffic law[11] collected from [2,9] and the architecture proposed in Sect. 3.3, an intelligent search system is designed on Vietnam's road traffic law. This section presents some test results of the system through some kinds of inputted queries.

Example 4.1: Input query q_1 = *"What is expressway?"*

The system will extract keywords from the query q_1: *"What is"*, *"expressway"*. From this, it returns the following results:

"Expressway mean a road reserved only for motor vehicles, with median strips separating carriageways for the two opposite directions of traffic, without crossing at level with any road, furnished with adequate support equipment and devices to ensure uninterrupted and safe traffic, shorten travel time, and with certain points for vehicle exits and entries."

In this query, term *"what is"* is used to classify the query as a kind of declaration of the meaning of a concept. The keyword *"expressway"* helps to find the concept. The result is retrieved from Paragraph 12, Article 3, Law of Road Traffic 2008[12].

Example 4.2: Input query q_2 = *"The fines for using a cell phone while operating a motorbike"*.

The keywords of the query q_2 are *"fines"*, *"using"*, *"cell phone"*, *"driving motorbike"*. Term *"fines"* used to classify the query as penalties and fines for offenses. Term *"operating motorbike"* consists of *"motorbike"* which is similar to term *"motorcycle"*. Therefore, the concepts of query q_2 are *"operating motorcycle"* and *"cell phone"*. The relational keyword is *"use"*. With the concepts and the relation, the rules were used to match them and the result was found. The result is returned:

[11] https://vanbanphapluat.co/law-no-23-2008-qh12-of-november-13-2008-on-road-traffic.

[12] Law Code 23/2008/QH12: Vietnam Law of Road Traffic 2008.

"Through article 6, Decree 100/2019/ND-CP [9]*: Penalties imposed on operators of mopeds and motorcycles (including electric motorcycles) and the like that violate the rules of road traffic.*

A fine ranging from VND 600,000 to VND 1,000,000 shall be imposed upon a vehicle operator who commits any of the following violations:

h) Using an umbrella, cell phone, or audio device other than hearing aid devices while operating the vehicle."

The designed system can perform some common searches on the traffic law. It is effective in finding usual penalties and fines in the road. This system was tested on 95 queries about road traffic codes. The results are shown in Table 2.

Table 2. Results of testing queries on Vietnamese Traffic Law

Kinds	Quantity	Correct	Rate
Queries about concepts/definitions	41	34	83%
Queries about penalties and fines	55	42	76%
Total	**96**	**76**	**79%**

4.3 Statistics and Discussions

The legal document dataset has a huge number of legal documents from the Vietnam legal system, applications of whole all domains from this knowledge repository are still limited. The intelligent searching system is one of the preliminary AI applications in this legal and AI domain. This system aims to support the search for legal information and answer common questions about legal concerns in several specific sub-domains.

Although the knowledge map model based on Rela-model [19] has potential uses in this domain, building a complete knowledge base for all data requires a lot of effort. In addition, it also depends on the specific field of the group of legal documents. For example, the *Weight* parameter in the model above (Sect. 2) requires specific calculations based on each specific area. In this research context, we have modeled the concepts and their relationships to build legal knowledge maps from the original raw legal documents.

5 Conclusions and Future Works

In conclusion, we proposed a legal knowledge map model to achieve legal knowledge from legal documents and support building a knowledge management system for this domain. Based on the proposed model and designed ontology for presenting legal knowledge, legal documents are transformed into the linked data format and imported into the knowledge management system for retrieval. As a result, this study has built a legal knowledge map repository (VLegalKMaps) from a huge number of Vietnamese legal documents. Moreover, the VLegalKMaps resources are used to build an intelligent search system, which supports legal concerns from legal documents. Finally, the VLegalKMaps resources can be downloaded at https://aseados.ucd.ie/vlegal/download/.

Our future work will focus on two areas: (a) Expanding the proposed model to compromise several algorithms to predict tasks based on explainable approaches. In addition, the model of the NLP toolkit will be improved to better address NLP problems on legal documents. (b) Building a question-answering AI application based on the legal knowledge repository with user interfaces, such as the service to interact with several groups of users, taking user questions as input and retrieving existing knowledge in this system.

Moreover, VLegalKMaps resources can be used to build different legal AI applications and services, such as a recommendation system to support the consulting of legal situations, and then developing a reasoning module to provide suitable explanations.

Acknowledgement. This research is funded by Vietnam National University Ho Chi Minh City (VNU-HCM) under grant number DS2023-26-04.

References

1. Altinok, D.: Mastering SpaCy: An End-to-End Practical Guide to Implementing NLP Applications Using the Python Ecosystem. Packt Publishing Ltd. (2021)
2. Assembly, V.N.: Law on Road Traffic, Law No. 23/2008/QH12 (2008)
3. Assembly, V.N.: Law on Promulgation of Legislative Documents, No. 80/2015/QH13 (2015)
4. Constantin, A., Peroni, S., Pettifer, S., Shotton, D., Vitali, F.: The document components ontology (DoCo). Semant. Web **7**(2), 167–181 (2016)
5. Crotti Junior, A., et al.: Knowledge graph-based legal search over German court cases. In: Harth, A., et al. (eds.) ESWC 2020. LNCS, vol. 12124, pp. 293–297. Springer, Cham (2020). https://doi.org/10.1007/978-3-030-62327-2_44
6. Dang, D., Nguyen, H., Ngo, H.E.A.: Information retrieval from legal documents with ontology and graph embeddings approach. In: 36th International Conference on Industrial, Engineering & Other Applications of Applied Intelligent Systems (IEA/AIE 2023), Shanghai, China (2023, in press)
7. Dhani, J.S., Bhatt, R., Ganesan, B., Sirohi, P., Bhatnagar, V.: Similar cases recommendation using legal knowledge graphs. In: ACM SIGKDD International Conference on Knowledge Discovery and Data Mining (2021)
8. Filtz, E., Kirrane, S., Polleres, A.: Interlinking legal data. In: SEMANTiCS Posters & Demos (2018)
9. Vietnam Government: Decree on Administrative penalties for road traffic and rail transport offences, No. 100/2019/ND-CP (2019)
10. Kaltenboeck, M., Boil, P., Verhoeven, P., Sageder, C., Montiel-Ponsoda, E., Calleja-Ibáñez, P.: Using a legal knowledge graph for multilingual compliance services in labor law, contract management, and geothermal energy. In: Curry, E., Auer, S., Berre, A.J., Metzger, A., Perez, M.S., Zillner, S. (eds.) Technologies and Applications for Big Data Value, pp. 253–271. Springer, Cham (2022). https://doi.org/10.1007/978-3-030-78307-5_12
11. Lettieri, N.: Knowledge machineries. Introducing the instrument-enabled future of legal research and practice. In: Knowledge of the Law in the Big Data Age, pp. 10–23. IOS Press (2019)

12. Montiel-Ponsoda, E., Rodríguez-Doncel, V.: Lynx: building the legal knowledge graph for smart compliance services in multilingual Europe. In: Proceedings of the 1st Workshop on LREC (Language Resources and Technologies for the Legal Knowledge Graph) Workshop, pp. 19–22 (2018)
13. Ngo, Q.H., Dien, D., Winiwarter, W.: Building English-Vietnamese named entity corpus with aligned bilingual news articles. In: Proceedings of the Fifth Workshop on South and Southeast Asian Natural Language Processing, pp. 85–93 (2014)
14. Ngo, Q.H., Kechadi, T., Le-Khac, N.-A.: OAK: ontology-based knowledge map model for digital agriculture. In: Dang, T.K., Küng, J., Takizawa, M., Chung, T.M. (eds.) FDSE 2020. LNCS, vol. 12466, pp. 245–259. Springer, Cham (2020). https://doi.org/10.1007/978-3-030-63924-2_14
15. Ngo, Q.H., Kechadi, T., Le-Khac, N.A.: Knowledge representation in digital agriculture: a step towards standardised model. Comput. Electron. Agric. **199**, 107127 (2022)
16. Nguyen, H.D., Pham, V.T., Le, T.T., Tran, D.H.: A mathematical approach for representing knowledge about relations and its application. In: Proceedings of 7th International Conference on Knowledge and Systems Engineering (KSE 2015), pp. 324–327. IEEE (2015)
17. Nguyen, P.K.: How to conduct research in Vietnamese law: overview of the legal system of the socialist republic of Vietnam. Int. J. Leg. Inf. **27**(3), 307–331 (1999)
18. Nguyen, P.T., Vu, X.L., Nguyen, T.M.H., Le, H.P., et al.: Building a large syntactically-annotated corpus of Vietnamese. In: The Third Linguistic Annotation Workshop-The LAW III, p. 6p (2009)
19. Nguyen, T.H., Nguyen, H.D., Pham, V.T., Tran, D.A., Selamat, A.: Legal-Onto: an ontology-based model for representing the knowledge of a legal document. In: Proceedings of 17th Evaluation of Novel Approaches to Software Engineering (ENASE 2022), Online Streaming, pp. 426–434 (2022)
20. Pham, V.T., Nguyen, H.D., Le, T., et al.: Ontology-based solution for building an intelligent searching system on traffic law documents. In: Proceedings of 15th International Conference on Agents and Artificial Intelligence (ICAART 2023), Lisbon, Portugal, pp. 217–224 (2023)
21. Sovrano, F., Palmirani, M., Vitali, F.: Legal knowledge extraction for knowledge graph based question-answering. In: Legal Knowledge and Information Systems, pp. 143–153. IOS Press (2020)
22. Thelen, S.: The ontology-based approach of the publications office of the EU for document accessibility and open data services. In: Kő, A., Francesconi, E. (eds.) Electronic Government and the Information Systems Perspective: 4th International Conference, EGOVIS 2015, Valencia, Spain, 1–3 September 2015, Proceedings, vol. 9265, p. 29. Springer, Cham (2015). https://doi.org/10.1007/978-3-319-22389-6_3
23. Wang, Z., et al.: IFlyLegal: a Chinese legal system for consultation, law searching, and document analysis. In: Proceedings of EMNLP-IJCNLP, pp. 97–102 (2019)
24. Yu, H., Li, H., et al.: A knowledge graph construction approach for legal domain. Tehnički vjesnik **28**(2), 357–362 (2021)

Classification of Ransomware Families Based on Hashing Techniques

Tran Duc Le[1,2], Ba Luong Le[1], Truong Duy Dinh[3](✉), and Van Dai Pham[4]

[1] The University of Danang - University of Science and Technology, 54 Nguyen Luong Bang, Da Nang, Vietnam
letranduc@dut.udn.vn, tran.duc.le@uqtr.ca
[2] Université du Québec à Trois-Rivières, Trois-Rivières, Canada
[3] Posts and Telecommunications Institute of Technology, 122 Hoang Quoc Viet, Hanoi, Vietnam
duydt@ptit.edu.vn
[4] Department of Information Technology, Swinburne Vietnam, FPT University, Hanoi, Vietnam
daipv11@fe.edu.vn

Abstract. The primary objective of this research is to propose a novel method for analyzing malware through the utilization of hashing techniques. The proposed approach integrates the use of Import Hash, Fuzzy Hash, and Section Level Fuzzy Hash (SLFH) to create a highly optimized, efficient, and accurate technique to classify ransomware families. To test the proposed methodology, we collected a comprehensive dataset from reputable sources and manually labelled each sample to augment the reliability and precision of our analysis. During the development of the proposed methodology, we introduced new steps and conditions to identify ransomware families, resulting in the highest performance level. The major contributions of this research include the combination of various hashing techniques and the proposal of a hash comparison strategy that facilitates the comparison of section hashes between ransomware and the pre-build database.

Keywords: Ransomware · Import Hash · Fuzzy Hash · File Level Section Hashing

1 Introduction

The contemporary landscape of cybersecurity is characterized by a proliferation of novel malware and attack vectors, which present numerous challenges to security experts. According to recent findings published by Checkpoint's cybersecurity service [1], global cyber attacks against enterprise networks rose by 50% when compared to 2020, with the education-research sector experiencing an

N. T. Nguyen et al. (Eds.): CITA 2023, LNNS 734, pp. 37–49, 2023.
https://doi.org/10.1007/978-3-031-36886-8_4

increase of up to 75% in targeted attacks. Against this backdrop, it is imperative to investigate effective malware analysis methods that are capable of handling the ever-growing volume of malware.

In light of the swift development and diffusion of malware variants, there is an urgent need to devise solutions that can effectively analyze and classify them with high precision and speed. While numerous methods have been proposed in the literature, it is crucial to continue exploring novel techniques and refining existing ones to keep pace with the evolving threat landscape and to address the ever-growing demand for robust and efficient malware analysis methods.

Malware analysis can be broadly classified into two categories: static and dynamic analysis. Static analysis involves extracting information and code from the malicious sample for analysis and offers the key advantages of being both time-efficient and easy to implement. In contrast, dynamic analysis aims to execute and analyze the actual behavior of the malware, which requires the use of appropriately configured virtual machines or sandboxes and necessitates the expertise of experienced analysts. While dynamic analysis can yield more accurate results, static analysis remains an essential technique for analysts dealing with large volumes of malware samples, where comprehensive and detailed analysis is not always feasible.

The use of hash algorithms in static analysis has garnered significant attention from analysts, who employ a diverse range of hash types to identify and analyze malware. While cryptographic hashes have long been utilized for this purpose, they can only detect identical malicious patterns. Newer hashes, such as Imphash and Fuzzy Hash (File Level Fuzzy Hash - FLFH), offer several advantages and can effectively facilitate the classification of malware by providing a measure of similarity between samples. In particular, Fuzzy Hash can be extracted from Portable Executable (PE)-malware files (refer to Fig. 1) to generate a hash of the internal components of the file, known as Section Level Fuzzy Hash (SLFH). While prior research has explored the use of Imphash, Fuzzy hash, and Fuzzy-Imphash [2], these methods are not without limitations. To address these limitations, integrating Fuzzy hash at a deeper level of the file has been shown to enhance detection and classification accuracy without compromising overall performance [3]. Specifically, the use of SLFH has been demonstrated to be effective, with optimal results compared to FLFH [3]. However, while prior studies have highlighted the efficacy of SLFH, they have not fully elucidated how to apply this technique within a specific model. Among the various Fuzzy hash types, existing research [2,3] has established Ssdeep hash as the most optimal. In light of these findings, this paper proposes a novel method that combines SLFH hash with the Fuzzy (SSdeep)-Imphash approach to analyze and classify ransomware samples.

2 Overview of Hashing Techniques in Malware Analysis

Import Hash (Imphash): The Import Hash technique quickly gauges similarity between two malware based on a small portion of a PE file called Imports

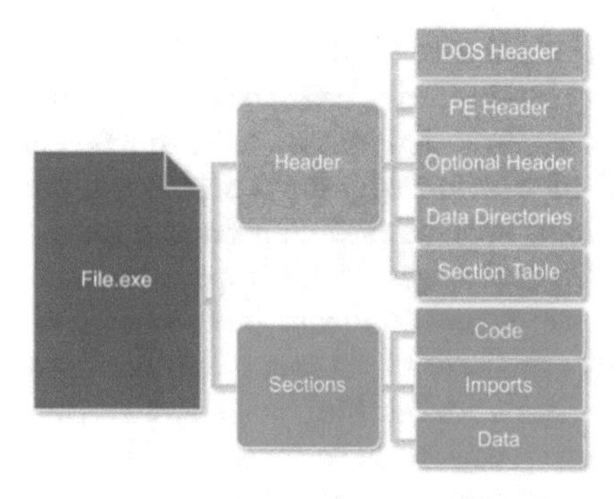

Fig. 1. The structure diagram of the PE file format.

(refer to Fig. 1). These represent function definitions bound into code by an end executable file. The Import Address Table (IAT) serves as the foundation for creating the Imphash of a PE program. Different orders of functions of Import Libraries produce distinct Imphash codes. Comparing Imphash results in a match or mismatch, unlike Fuzzy hash.

Fuzzy Hash: Cryptographic hash functions are inefficient in comparing and verifying malware because they only generate a true/false result to determine if two files are identical. While malware samples may not be identical, they often share similarities because they are developed based on a common code to create different variations of malware. Fuzzy hashing is a technique that divides a file into segments to calculate hash values, enabling the comparison of files with similar functions but distinct cryptographic hashes. Fuzzy hashing is valuable in analyzing and comparing test files and known malware, as it provides a degree of similarity between two malware samples [4–6].

Ssdeep Hash: Ssdeep is a Fuzzy hashing technique that differentiates between spam or junk emails and yields superior results compared to other Fuzzy hash algorithms [2,3]. Developed in 2006 by Kornblum, Ssdeep uses rolling hashing to generate a hash value or signature of an input file. Comparing the output hash values of two files, their similarity can be determined on a linear percentage scale. A value of 100 indicates that the files are almost or exactly the same, and a value of 0 indicates no similarity between them. Ssdeep is faster and more accurate than other Fuzzy Hash algorithms like Sdhash and Mv-hashB, making it widely utilized in malware analysis.

File Level Ssdeep Hash (FLSH): FLSH is the Ssdeep hash of a file that is named to distinguish it from the Ssdeep hash at other levels. The attackers can manipulate it by adding unnecessary sections to mimic benign software or making minor modifications to the file's content arrangement. When compared, these changes affect the similarity level, and relatively small modifications to the source code can reduce the similarity level to zero. Another issue arises at the file level, where the similarity level depends on the sequence and arrangement of the content in the file. PE files have a high-level structure, but malware authors can make minor modifications, such as inserting unused parts or moving parts out of their normal position. Such changes do not impact the loading and running of the software, but they do affect the similarity level when compared [3]. A study [7] revealed that relatively small modifications to the source code, such as swapping the sequence of instructions in the file, can significantly impact the similarity level, which can be reduced to zero in some cases. The study found that 99.8% of programs do not change in such a manner [3]. Malware authors also use another method to obscure executable files: compress them with a packaging utility like UPX or ASPACK.

Despite the drawbacks mentioned, comparing FLSH still produces faster and less resource-intensive results, as presented earlier. It operates with high accuracy when malware samples are not compressed. Therefore, the similarity level of FLSH is integrated into the model outlined in the proposed method.

Section Level Ssdeep Hash (SLSH): At the file level, the order of Sections can be manipulated by attackers to obscure and modify the FLSH hash of the file. This aims to make it more challenging for analysts to classify complex malware. There are many different sections in malware, such as: (i) *.text*: Contains instructions that CPU executes, and the only section contains code; (ii) *.rdata*: contains import (.idata) & export (.edata) information and stores other read-only data used by the program; (iii) *.data*: contains the program's global data, which can be accessed from anywhere in the program; (iv) *.rsrc*: includes resources like icons, images, lists, strings; (v) *.reloc*: contains information to relocate binaries.

At a deeper level within the PE file format structure, comparing sections can reduce the confusion caused by shuffling their order and modifying the FLSH. This shuffling is a common technique used by malicious authors, which can diminish the effectiveness of security software relying on FLSH for comparison. Comparing sections can help avoid this shuffling. Each suspected malware sample will have specific sections from which hash values can be extracted for analysis. The same holds for existing malware samples in a database. Extracting and storing all of the Ssdeep hash information for the sections of those samples in a new database will be a prerequisite for malware analysis based on SLSH.

3 The Proposed Method

3.1 The Combined Analysis Method of Imphash, File Level Ssdeep Hashing, and Section Level Ssdeep Hashing

Our method selects Ssdeep as the Fuzzy hashing technique due to its promising results. By sorting the hash types in order and comparing the extracted hash from the test sample with the pre-created hash database, this method provides the highly probable malware family of the test sample.

3.2 Preparing the Database

The step of collecting sample data for the proposed model is crucial in ensuring the reliability and accuracy of the analysis. The sources used to collect the malware samples are pre-classified sources, such as VirusShare, Bazaar, Hybrid-Analysis, and Any-run. The collected ransomware samples are then divided into directories based on their family names, and each sample is unpacked and named after the cryptographic hash for easy identification.

The *Pehash* tool, which is part of the PEV ecosystem toolset [10, 11], is used to extract necessary information from the samples, including SHA256 hash, Imphash hash, Ssdeep hash, accurate ransomware family name, a list of sections with names and Ssdeep hashes for each section of a sample. The extracted information is then exported to a CSV storage file divided into two sets: Test (*test.csv*) and Train (*train.csv*). The Test set accounts for 20% per ransomware family, while the remaining 80% is the Train set (refer to Table 1).

To conveniently compare the hashes of the sections of samples, the Train set is also extracted into another separate database (*sectionData.csv*). This database (a csv file) contains information about the sections taken from the samples of the Train set, including the ransomware family name, section name, SHA256 of the original sample, and Ssdeep hash for each section. This database is used to compare the Sections of the suspect sample. However, during the collection of these sections, some types of sections may have unfamiliar names that are rarely seen in the entire dataset. To address this issue, only section names that appear more than twenty times are considered for the analysis.

Table 1. The amount of malware that has been distributed.

Ransomware family	Number of Train set	Number of Test set	Total
Babuk	36	7	43
Cerber	99	20	119
Conti	66	11	77
Gandcrab	132	29	161
Locky	61	14	75
Ryuk	78	16	94
Stop	422	125	547
WannaCry	75	19	94

3.3 Predictive Model

Figure 2 illustrates the input process of the proposed model, where the suspected malware samples are used as inputs. The first step is to check whether the input file is in the PE format or not since Imphash only works with PE files. If the input file is not in PE format, it will be directed to the non-PE format stream

(stream 1 - dash line). Conversely, if the input file is in PE format, it will be directed to the PE format stream (stream 2 - straight line).

For stream 1, when the input file is not in PE format, only FLSH can be used. When comparing the FLSH hash of the sample with the pre-calculated hash database, in this case, the threshold will be set at 30%. This value is considered reasonable according to research [3]. If any value is greater than or equal to this threshold, the corresponding malware family with the FLSH hash in the database will be immediately returned as the analysis result. In the remaining case, if the suspected file is in PE format, the steps of comparing SLSH in the sections of the sample with the SLSH in the database will be performed, as described in Fig. 2.

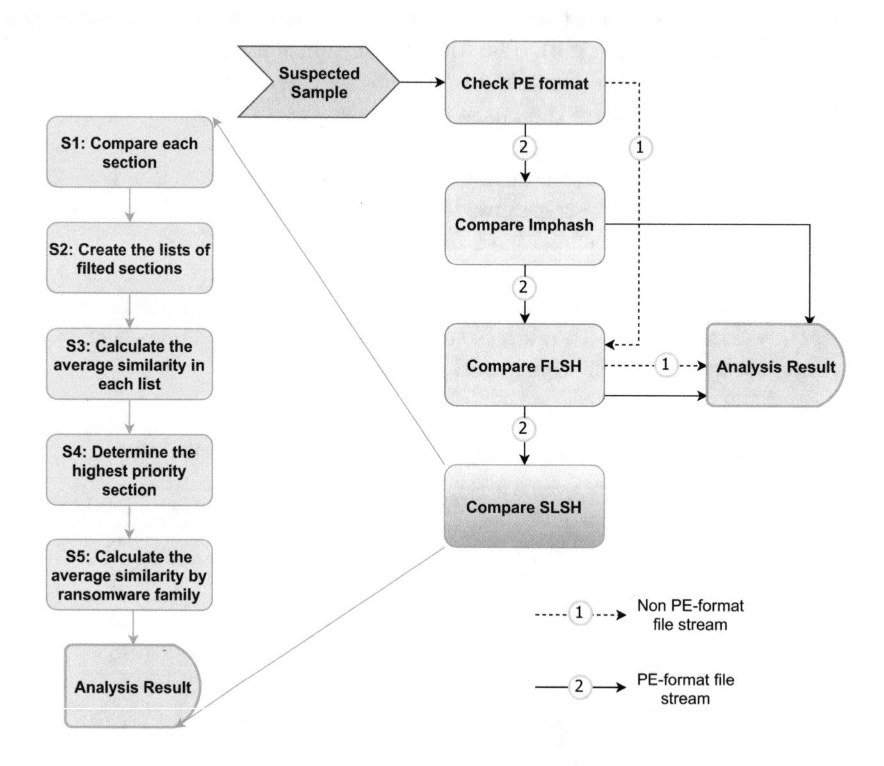

Fig. 2. The main operation flowchart of the proposed model

In the comparison steps of stream 2, if in *Compare Imphash* step, the hashes match, we can conclude immediately. If not, in *Compare FLSH* step, if the similarity level achieved exceeds 90%, the analysis result will be returned, analogous to stream 1. In contrast, for *Compare SLSH* step, a meticulous comparison of the suspected ransomware sample's sections with the system's secondary database is required. A comprehensive depiction of the SLSH comparison process will be elucidated in the SLSH comparison diagram (see the left part of Fig. 2).

The flowchart presented in Fig. 2 depicts the methodology employed. This diagram illustrates the step-by-step process of comparing the Ssdeep hash of the input sample with the database that contains pertinent information regarding the various Sections. The steps outlined in this diagram are critical components of the comparison process and will be expounded upon in the following paragraphs.

The SLSH comparison diagram in the left part of Fig. 2 outlines the five main steps involved in comparing Sections using Ssdeep hash at the section level and returning results. To explain these steps, we denote as follows:

- S_i : section i of suspected ransomware X, where $i = 1, 2, \ldots, n$
- h_{S_i} : SLSH value of section S_i
- S_{ij} : section i of jth ransomware in *sectionData.csv* file, where $j = 1, 2, \ldots, m$
- $h_{S_{ij}}$: SLSH value of section S_i of jth ransomware in *sectionData.csv* file
- $p_{i_{Xj}}$: similarity level value of section i of ransomware X and j^{th} ransomware
- L_{S_i} : a list of sections i of ransomware samples from *sectionDataset.csv* file, which have similarity level value $\geq 50\%$ compared with section i of suspected ransomware X
- $Avg_{L_{S_i}}$: the average of $p_{i_{Xj}}$ in each list L_{S_i}
- $Avg_{L_{L_S}}$: the average of $p_{i_{Xj}}$ values in a list L_{S_i} by ransomware family

Step 1: Compare S_i with S_{ij} by calculating $p_{i_{Xj}}$. If $p_{i_{Xj}} \geq 50\%$, add to the list of filtered section L_{S_i}. This list looks like $L_{S_i} = \{S_{ij}, S_{ij+2}, S_{ij+9} \ldots\}$. This list can include many different ransomware families.

Step 2: Create a set of lists of separate sections as described in **Step 1**.

Step 3: Calculate the average of $p_{i_{Xj}}$ in each list L_{S_i}, which is denoted by $Avg_{L_{S_i}}$.

Step 4: Compare the average values Avg_{L_i} of different lists in **Step 2** to determine which section will be the highest priority. The highest priority section will correspond to the L_{S_i} with the largest $Avg_{L_{S_i}}$ value. This section will influence the ransomware family decision for the output.

Step 5: Calculate the average similarity level $p_{i_{Xj}}$ by ransomware family in L_{S_i}. It means calculating the average of $p_{i_{Xj}}$ by each value of j in L_{S_i}. This value is denoted by $Avg_{jL_{S_i}}$. The family of ransomware with the largest $Avg_{jL_{S_i}}$ in the L_{S_i} under consideration will be returned in the analysis results.

3.4 Evaluation Criteria

The automated evaluation involved iterating through all samples in the Test set using the established model. The output results were then evaluated by comparing them to the accuracy results stored in the dataset and the predicted results returned by the model. To ensure the reliability of the evaluation, two different Test sets were collected and used for evaluation purposes.

The first Test set, consisting of data extracted from the original dataset, was utilized to assess the accuracy of ransomware classification. The accuracy measure was determined using the formula:

$$Acc = \frac{TP}{Total} \tag{1}$$

where *Acc* being the accuracy of the classification and *TP* (True Positive) [3] denotes the number of correct predictions made by the model that matched with the ransomware families stored in the Train set, and *Total* is the total number of samples tested. This evaluation aimed to assess the accuracy of classifying samples of known ransomware families stored in the database.

Specifically, this evaluation can demonstrate the accuracy of classifying samples identified as ransomware. However, this evaluation does not reflect the accuracy and errors when the input samples are benign files.

In order to evaluate the accuracy of the SLSH comparison method, it is necessary to identify and record the cases where SLSH is used. Then, the cases that utilize this method are extracted separately from the overall results, and their total count is denoted as *SectionTotal.* The number of correct predictions in the SLSH comparison cases is denoted as *SectionTP.* The accuracy of the SLSH comparison technique, denoted as *SectionAcc*, can be calculated using the following formula:

$$SectionAcc = \frac{SectionTP}{SectionTotal} \tag{2}$$

Next, to further evaluate the model's performance with respect to both malicious and benign samples, the size of the Test set was increased by combining the malicious samples from the previous Test set with benign software samples. At this point, the model's output can fall into one of four categories:

True Positive (TP): The model correctly classifies the ransomware family of the input file.

True Negative (TN): The model correctly identifies that the input file is not ransomware.

False Positive (FP): The model incorrectly classifies a benign file as ransomware.

False Negative (FN): The model incorrectly classifies a ransomware file as benign.

The parameters used to calculate accuracy are:

$$Precision = \frac{TP}{TP + FP} \tag{3}$$

$$Recall = \frac{TP}{TP + FN} \tag{4}$$

$$F1 = 2 \times \left(\frac{Precision \times Recall}{Precision + Recall}\right) \tag{5}$$

The *Precision* metric measures the accuracy of the model by calculating the ratio of the number of correctly predicted ransomware samples to the total number of samples predicted as ransomware. Similarly, *Recall* is calculated using a formula similar to *Precision* but includes the number of false negative (FN) predictions to evaluate the proportion of missed ransomware by the model [8,9]. The *F1 Score* is then calculated as a weighted harmonic mean of *Precision* and *Recall* to provide a more optimal evaluation of the model's performance.

4 Evaluations and Results

This section presents an analysis of the results obtained from the proposed model and a comparison with the Fuzzy-Imphash model described in previous sections. The evaluations and comments made on the results aim to demonstrate the improvements in the new method and highlight the weaknesses that have not been addressed.

4.1 Experiment with Test Set Containing only Ransomware

This section will present the results after testing the first Test set containing only ransomware. Table 2 illustrates a significant enhancement in accuracy when using the proposed model. Incorporating the SLSH comparison technique in the new model facilitates a more comprehensive comparison between the suspicious files and the database. Consequently, this enhancement in accuracy demonstrates that the previous model failed to detect numerous ransomware families during the testing phase.

Table 2. Comparison of accuracy between models

Evaluation value	Fuzzy-Imphash	Proposed Model
Total (sample)	241	241
TP (sample)	164	212
Acc (%)	68,05	87,97

Particularly for the new model, the implementation of sections must be evaluated for accuracy. Cases utilizing the SLSH comparison technique will be stored, and the accuracy will be calculated.

Table 3 presents the accuracy of the results obtained by applying the SLSH comparison technique, which is comparable to the overall accuracy achieved by the new model. Incorporating the SLSH comparison technique in the new model, originally based on the Fuzzy-Imphash method, highlights its efficacy in improving the model's performance. The high accuracy of the SLSH comparison technique underscores its potential as a valuable tool for ransomware classification and identification.

To provide a more comprehensive analysis, Table 4 presents the accuracy of each ransomware family in the entire dataset. The accuracy is computed using the same formula as the overall accuracy but filtered by ransomware families.

Table 3. Accuracy in cases using SLSH

Evaluation value	Proposed Model
Total (sample)	57
Correct classification (sample)	50
Incorrect classification (sample)	7
SectionAcc (%)	87.72

The tabulated accuracy results, as presented in Table 4, indicate a significant disparity in the accuracy rates among various ransomware families. The data reveals that the Stop ransomware family has experienced a remarkable increase in accuracy rates with the implementation of the proposed model. Furthermore, the Ryuk and Cerber families have exhibited higher levels of accuracy when subjected to the proposed model. Wannacry exhibits the highest accuracy level. This finding can be explained by two primary factors. Firstly, as all ransomware families share certain commonalities and are constructed using similar sources and platforms, their structural features are likely to be highly comparable, thus enabling accurate classification. Secondly, this family's low diversity and small sample size in the overall dataset may have contributed to the high accuracy rate. Conversely, the low accuracy of some families in the Fuzzy-Imphash model may be attributed to the fact that these ransomware samples were built on different platforms and architectures, resulting in greater variability in their structural features. However, we can also note that the accuracy for Wannacry is reduced when using the Proposed Model. Actually, there is no precise explanation for this phenomenon. We are working on it to improve the result in future works.

Table 4. Classification accuracy of ransomware families using Fuzzy-Imphash and Proposed Model with Test set containing only ransomware

Ransomware family	Fuzzy-Imphash	Proposed Model
STOP	53.6	90.4
RYUK	68.75	75
CERBER	70	80
LOCKY	78.57	78.57
CONTI	81.81	81.81
BABUK	85.71	85.71
GANDCRAB	93.1	93.1
WANNACRY	100	94.73

These findings provide evidence that the proposed model has demonstrated superior performance in accurately classifying ransomware within a dataset that solely comprises ransomware.

4.2 Experiment with Test Set Containing both Malicious and Benign Samples

The evaluation of the proposed models has been enhanced by including benign samples in the second set of tests. This approach offers a more comprehensive evaluation of the models when dealing with ransomware and benign samples. Moreover, intuitive parameters have been computed and tested for a more insightful analysis. The detailed results of the experiments are presented in Table 5.

Table 5. Comparison of Fuzzy-Imphash and Proposed Model with Test set containing ransomware and benign

Criteria	Fuzzy-Imphash	Proposed Model
TP (sample)	166	214
TN (sample)	243	212
FP (sample)	0	4
FN (sample)	74	14
Precision (%)	100	98.17
Recall (%)	69.16	91.85
F1 (%)	81.77	94.9

The results from Table 5 indicate a significant improvement in the Recall value of the proposed model, which suggests that it can effectively reduce the number of missed ransomware samples. In contrast, the Fuzzy-Imphash model has a lower Recall value, indicating that it misses many ransomware samples during the evaluation process. Moreover, the Precision value of both models is high, with the Fuzzy-Imphash model having a perfect Precision score due to the absence of false positives. The new method also has a high Precision rate and low false positives, indicating that the combination of Fuzzy and Import hash values can accurately identify ransomware.

Additionally, the F1-score is calculated based on the Precision and Recall values, and the new model outperforms the Fuzzy-Imphash model, demonstrating that it can produce better and more efficient results. These findings suggest that the proposed model is an improvement over the Fuzzy-Imphash model and can accurately identify and classify malicious code while minimizing false positives.

4.3 Advantages and Limitations of the Proposed Method

Advantages: The proposed model has been shown to improve the capability to identify the ransomware that may have been missed by the previous model. The results presented in the previous sections demonstrate that the application of this model produces more accurate outcomes.

The proposed model expedites the process of malware analysis and processing by returning immediate results if the ransomware matches the Imphash hash value or reaches a certain threshold of similarity compared to FLSH.

Using the SLSH comparison technique in the final step of the proposed model provides a more accurate and robust approach to handling ransomware samples with shuffled Section orders. When Sections are shuffled in this way, the similarity score produced by FLSH is often low, leading to incorrect predictions. In contrast, the SLSH comparison technique considers each Section's content, which provides a more comprehensive understanding of the malware sample. This approach improves the accuracy of the final results and is a significant advantage of the proposed model over the old model.

Limitations: Comparing many sections takes time. The SLSH comparison technique makes it slower, as it compares each section with the entire dataset. Storing sections can be challenging for real systems to meet processing time requirements.

The proposed model provides static processing and analysis but has limitations in fully demonstrating malware's behavior. The dynamic analysis provides a complete picture but needs more system resources and time. Packed or compressed malware reduces accuracy.

Updating the database and popular section list is crucial for the model's accuracy and efficiency. Outdated databases may fail to detect new malware types. Research and development are necessary to update the model in response to evolving malware threats.

The proposed model's section naming convention may not reflect their contents, leading to misclassification. A precise naming convention can improve the model's effectiveness.

5 Conclusions

In summary, the proposed model uses Fuzzy and Import hash values to analyze ransomware and SLSH comparison for better accuracy. The evaluation showed better performance compared to the Fuzzy-Imphash model. However, accuracy varied for specific ransomware families due to the limited sample size and similarities in the families. The inclusion of benign samples provided a more comprehensive evaluation. The model has limitations, including handling packed ransomware, and resource and time constraints in processing sections. Future research can incorporate Yara Rules [12] and a decompression stage to improve accuracy and handle packed malware. Expanding testing to other malware families can enhance the model's practicality.

References

1. Checkpoint, "Cyber Security Report 2021". https://www.checkpoint.com/pages/cyber-security-report-2021. Accessed 05 Jan 2023
2. Naik, N., Jenkins, P., Savage, N., Yang, L., Boongoen, T., Iam-On, N.: Fuzzy-import hashing: a static analysis technique for malware detection. Forensic Sci. Int. Digital Invest. **37**(301139), 1–13 (2021)
3. Shiel, I., O'Shaughnessy, S.: Improving file-level fuzzy hashes for malware variant classification. Digit. Investig. **28**, 88–94 (2019)
4. Gupta, S., Sharma, H., Kaur, S.: Malware characterization using windows API call sequences. In: Carlet, C., Hasan, M.A., Saraswat, V. (eds.) SPACE 2016. LNCS, vol. 10076, pp. 271–280. Springer, Cham (2016). https://doi.org/10.1007/978-3-319-49445-6_15
5. Microsoft, "Windows Devices," microsoft. https://news.microsoft.com/bythenumbers/en/windowsdevices. Accessed 05 Jan 2023
6. Breitinger, F., Baier, H.: A fuzzy hashing approach based on random sequences and hamming distance. In: ADFSL Conference on Digital Forensics, Security and Law, pp. 89–100 (2012)
7. Raff, E., Nicholas, C.: Lempel-Ziv Jaccard Distance, an effective alternative to ssdeep and sdhash. Digit. Investig. **24**, 34–49 (2018)
8. Kornblum, J.: Identifying almost identical files using context triggered piecewise hashing. Digit. Investig. **3**, 91–97 (2006)
9. Pagani, F., Dell'Amico, M., Balzarotti, D. : Beyond precision and recall: understanding uses (and misuses) of similarity hashes in binary analysis. In: Proceedings of the Eighth ACM Conference on Data and Application Security and Privacy, pp. 354–365 (2018)
10. Fernandes, E., Bezerra, F., Moraes, T.: Comparing PE pieces. https://pev.sourceforge.io/doc/manual/en_us/ch04s03.html. Accessed 05 Jan 2023
11. Fernandes, E., Bezerra, F., Moraes, T.: pev the PE file analysis toolkit. https://pev.sourceforge.io. Accessed 05 Jan 2023
12. Naik, N., Jenkins, P., Savage, N., et al.: Embedded YARA rules: strengthening YARA rules utilising fuzzy hashing and fuzzy rules for malware analysis. Complex Intell. Syst. **7**, 687–702 (2021)

Towards a New Multi-tasking Learning Approach for Human Fall Detection

Duc-Anh Nguyen[1], Cuong Pham[2], Rob Argent[3], Brian Caulfield[1], and Nhien-An Le-Khac[1](✉)

[1] University College Dublin, Dublin, Ireland
an.lekhac@ucd.ie

[2] Posts and Telecommunications Institute of Technology, Hanoi, Vietnam

[3] School of Pharmacy and Biomolecular Sciences, RCSI University of Medicine and Health Sciences, Dublin, Ireland

Abstract. Many fall detection systems are being used to provide real-time responses to fall occurrences. Automated fall detection is challenging because it requires near perfect accuracy to be clinically acceptable. Recent research has tried to improve the accuracy along with reducing the high rate of false positives. Nevertheless, there are still limitations in terms of having efficient learning approaches and proper datasets to train. To improve the accuracy, one approach is to include non-fall data from public datasets as negative examples to train the deep learning model. However, this approach could increase the imbalance of the training set. In this paper, we propose a multi-task deep learning model to tackle this problem. We divide datasets into multiple training sets for multiple tasks, and we prove this approach gives better results than a single-task model trained on all datasets. Many experiments are conducted to find the best combination of tasks for multi-task model training for fall detection.

Keywords: fall detection · multi-task learning · human activity recognition · data scarcity · deep learning

1 Introduction

Falls are the second greatest cause of unintentional injury deaths globally, with millions of falls each year requiring medical attention [1]. Risk factors in the elderly population include environmental considerations, comorbidities, medications and frailty [2]. Despite numerous interventions for falls prevention existing clinically including gait and balance retraining, education, medication reviews and environmental adaptations [1], it is not possible to prevent all falls. Therefore, the next step in management is early identification of the occurrence of a fall to access immediate assistance and minimise further injury.

Fall detection is done by capturing subject's movements using sensors, then an algorithm will determine if a fall has occurred. The algorithm can be trained on real or simulated fall data. There are many studies about fall detection using

N. T. Nguyen et al. (Eds.): CITA 2023, LNNS 734, pp. 50–61, 2023.
https://doi.org/10.1007/978-3-031-36886-8_5

different sensing strategies. A survey [3] categorised sensors into 3 groups, namely wearable sensors, visual sensors, ambient sensors. Another review [4] divided fall detection systems into 3 different groups, which are wearable inertial sensor-based, context-based, and radio frequency-based. Both papers also mentioned homogeneous fusion (sensors of the same type) and heterogeneous fusion (sensors of different types). Nevertheless, inertial sensors, especially accelerometer, are still the most popular choice given advantages including affordability, easy deployment and privacy preservation.

It is best that the training data is similar to real-world data. However, fall data collecting is very challenging given its unexpected nature. When collecting simulated data, the participants are aware of the falls and will prepare their postures to dampen the impact. Also, to ensure participants' safety, there are always soft and thick mats for them to fall on. This cannot be expected in real-world situations. As a result, there is always a gap between real and simulated data, which fall detection methods have to overcome. Good results on simulated data are not sufficient, and a reliable fall detection system should be evaluated on real-world data or simulated data based on real-world situations.

Fall detection is challenging because in real life, the acceptable threshold for performance is very high. If a system fails to detect a fall, the person may become stranded for hours with serious consequences. By contrast, if it produces regular false alarms, user acceptability will be impacted, there will be a loss of confidence in the system, and the device will not be used. A high sensitivity is hard to achieve given the challenges discussed in the previous paragraph. On the other hand, a high precision (low false alarm rate) is hard to achieve because there are infinite possibilities of how users act in the real world. Many actions are analogous to falls, resulting in a high false alarm rate. Adding non-fall data as negative examples to the training set could improve precision, especially since non-fall data is much easier to obtain than fall data. However, fall data usually occupies only a small proportion compared to non-fall data. By adding more negative data, the training set would be even more imbalanced. To tackle this, we propose a multi-task method, in which each task involves a different dataset and its own label list. Our contributions are listed as follows.

- We prove that multi-task learning can improve deep learning performance compared to regular single-task learning in fall detection when the amount of fall data is strictly limited. This provides a better balance between false alarm rate and sensitivity, which raise the opportunity for more improvement.
- We experiment with various settings to find the optimal combination of tasks to train the multi-task model for fall detection. The combination provides a good balance between tasks, and between activity classes.

2 Related Work

In this section, we review a range of fall detection studies and focus on wearable inertial sensors. Besides fall detection, human activity recognition methods that apply multi-task learning are also discussed.

2.1 Simulated Falls

[9] developed a threshold-based algorithm by observing patterns in real-world ADL data from the elderly and simulated fall data. Location data was used to define safe and dangerous zones. The method was tested on fall data simulated by experts and real ADL data from the elderly. However, the ADLs are confined to walking, going up the stairs, going down the stairs, standing up, and sitting down. The drawback of this method is the lack of flexibility as rules (e.g. threshold values, definitions of zones) are not the same for everyone. Nahian et al. [5] extracted 7700 features from raw accelerometer data. After data cleaning and feature selection, the feature set contained 39 or 5 features, depending on each experiment. The authors tested their feature sets using many machine learning models (Random Forest, SVM, Naive Bayes, etc.) on 3 public datasets (UR Fall, MobiFall v1, UP Fall). The experiments proved the usefulness of their proposed feature sets, while the best classifier varied for each experiment.

Researchers also use deep learning for fall detection. The authors of [6] proposed a Convolutional Neural Network (CNN) with 3 blocks. The model was tested on 3 datasets (UR Fall, Smartwatch, Notch). Their results proved that CNN was better than LSTM for fall detection, and data augmentation also played an important role. [8] used both accelerometer data and RGB images to train multi-modal deep learning models. Experiments were done on 2 public datasets (UR Fall and UP Fall) with CNNs and LSTM. They concluded that the best results were achieved with a CNNs solution. Whereas Recurrent networks are also used in other studies. The authors of [7] conducted experiments on 2 datasets, which were MobiAct v1 and Smartwatch. They compared multiple machine learning and deep learning models before concluding that GRU had the best performance in fall detection.

Though these methods achieved very good results, they were all tested only on datasets with simulated falls and simple activities of daily living (ADLs), which are not enough to represent real-world performance.

2.2 Real Falls

There are also studies that attempted to test methods on real-world fall data. The authors of [10] conducted their experiments on real-world fall data. The authors extracted 5 features and experimented with a range of machine learning models, the best one was SVM with 0.646 F1-score. [11] is a review paper summarising many studies that involved real-world fall data, and the following are some studies in the review. [12] evaluated a fall detection prototype in real-life usage. There was a total of 15500 hours of data collected from 16 older people, with 15 real falls. Their system achieved 0.8 sensitivity and a false alarm rate of 0.049 alarms per usage hour. [13] used 10 features and applied a synthetic oversampling technique to balance their dataset before training a Decision Tree to detect falls. They achieved 0.88 sensitivity and 0.87 specificity. [14] extracted 5 handcrafted features as inputs for an SVM model, and their method achieved a score of 0.9545 for both sensitivity and specificity. [15] developed a

Hidden Markov Model (HMM) based fall detection system and evaluated it on 2 datasets. The results were 0.981 precision and 0.992 recall on the simulated dataset, and 0.786 precision and 1.0 recall on the real-world dataset with 22 falls.

The above results are either not good or only achieved on a small test set. Also, many of them used real fall datasets not publicly available, so comparisons cannot be made. Though it is difficult to have a sufficient real fall dataset, more work is needed to close the gap between simulated and real-world data. There is no clinical or user specified acceptable threshold for false alarm rates, with few studies having explored the acceptability of fall detection technology with users. High false alarm rates have been shown to cause annoyance and potentially impacts the sustainability of adoption of such technologies.

2.3 Multi-task Learning

In deep learning, multi-task learning refers to a technique involving training a model to do more than 1 task simultaneously. Multi-task learning methods can be categorised into 2 groups based on their purposes as follows:

A model can be trained to do many equally important tasks. Learning one task may or may not improve the others. The authors of [16] trained a multi-task model to extract person and activity representations from human activity data. Two person representation vectors will be close to each other if the corresponding data is from the same person (the same for activity representation vectors). [17] used a multi-task model to classify human activity and estimate its intensity simultaneously to save cost as there was no need for 2 separate models. They reported that their model could improve activity classification accuracy while retaining intensity estimation accuracy. [18] jointly trained a model to classify simple and complex activities. For example, complex activities include relaxing, coffee time, cleanup, sandwich time, while simple activities are standing, walking, sitting, lying. Their experimental results showed that the model achieved good results on both tasks. [19] proposed to train activity segmentation and ergonomics risk assessment in 1 model using human pose data. They stated that their model outperformed single-task model due to the significance of activity type in the risk associated with a posture.

On the other hand, auxiliary tasks can also be included to improve the model's performance on the main task. [20] added 3 tasks (e.g. classification of body positions, genders, user identities) to the model to enhance the main task of activity classification. [21] proposed a multi-task self-supervised learning method that trained the model to classify types of transformation applied on unlabelled data. Then the model would be fine-tuned to classify human activities. In fact, this is more of a transfer learning method than a multi-task learning method as in its name.

2.4 Summary

There have been many studies on fall detection, however, there still exists a gap in fall detection performance between simulated and real-world data. We propose a multi-task learning approach to close this gap. The existing multi-task learning studies mostly focused on training many tasks on the same data to exploit mutual information of tasks, whereas none of them has investigated the effects of training auxiliary activity classification tasks on different data.

3 Multi-task Deep Neural Network for Fall Detection

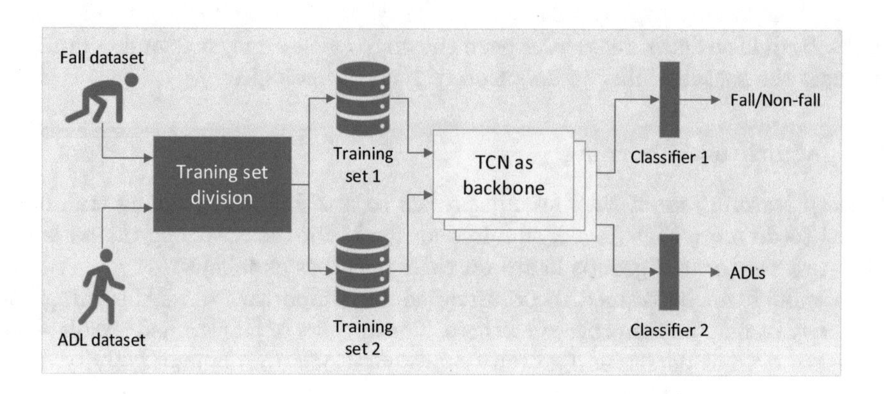

Fig. 1. Multi-task learning for fall detection

Given a small human fall dataset, our goal is to use an extra non-fall dataset as training data to improve fall detection performance and we propose a multi-task deep learning approach to achieve this objective. The proposed method allows us to add more negative examples (non-fall data) while maintaining the balance between precision and recall that opens up more potential improvements. Figure 1 describes the overall method. The implementation of our method is available online at https://github.com/nda97531/da_multitask.

3.1 Multiple Classifiers

The main idea in this proposal is to have multiple classifiers after one feature extractor. As mentioned in Sect. 1, any non-fall dataset could be added into the training set as negative examples. But instead of having one training set containing all the data, we treat each extra dataset as a separate task and have a classifier for it. In our particular case, the main task is to classify between fall and non-fall, and the others are classification tasks on different datasets. The way we divide data into multiple tasks is also a major contribution of this paper (*Training set division* in Fig. 1). Because this heavily depends on datasets, we will provide more details about this in Sect. 4.2.

By having multiple tasks, we can train the backbone (i.e. feature extractor) to extract useful information from diverse sources of data, while maintaining class balance for the classification step. Also, the backbone has to adapt to all classifiers, which have more than just fall and non-fall classes. It learns more aspects of the data and this could benefit the main task.

Each task has its own loss function and the final loss to optimise is the sum of all tasks' losses. This sum calculation is without any weight assigned for each component loss because the balance is already controlled by re-sampling (Sect. 3.3). In this paper, while all tasks are classification, the number of classes varies. For consistency in implementation, we use the cross entropy loss function for all tasks. This function works well for both binary classification and multi-class classification.

3.2 Temporal Convolutional Network as a Feature Extractor

We adopt the temporal neural network from our previous study as the backbone network. However, the paper cannot be cited because it is currently under review. This backbone consists of 2 parts as follows:

- Temporal Convolutional Network (TCN): We also used this backbone in [22], except that in this study, we keep the whole output sequence for the next layer instead of using only its last time step as it would retain more information. As a result, this TCN takes an input sequence and outputs a sequence of the same length after passing it through multiple residual blocks. In this network, each convolutional layer is followed by a batch normalisation layer, which is essential because we are dealing with data from different domains (different datasets).
- Temporal Attention Module (TAM): We adapt the Spatial Attention Module [23] for time-series data. It takes the output sequence from TCN and computes weights for each time step within the sequence. Each time step in the feature map is then multiplied with its corresponding weight.

Finally, the feature map is aggregated into a feature vector with a Global Average Pooling layer. With TAM, this can be considered as a weighted average of all time steps of the sequence.

3.3 Re-sampling for Class and Task Balance

To deal with the imbalance between fall and non-fall data, we apply re-sampling to balance both between classes and between classification tasks.

The class re-sampling is done by selecting data of each class evenly across all training batches. When selecting data to form a training batch, we pick a random class before getting a sample of that class. If the batch size is b, that selection will be repeated b times for every batch. After all samples of a class are already picked, we shuffle the sample list of that class and start picking again. By doing this, we ensure every data sample of a class is picked with the same frequency, while still having the randomness.

Similarly, task re-sampling is done by selecting the same number of samples from every task for each batch. For instance, if there are 2 tasks, each of them will have $b/2$ samples in every batch. For ease of implementation, the batch size is set to be divisible by the number of tasks.

3.4 Data Augmentation

In this paper, we apply only the rotation augmentation technique from [24] because other techniques do not resemble real-world fall situations. Accelerometer data is not human-readable, so we need to be mindful that strong augmentation may degrade the data.

In the real world and even in data collection, users may wear the sensors in different ways. For example, a smartwatch can be worn on the left hand, right hand, upside down, tightly, or loosely. A smartphone can be put in any pants pocket in any orientation. By rotating the triaxial accelerometer data, we try to simulate different sensor placements.

4 Experiments

In this section, we present details of our experiments. The experimental results prove that multi-task learning can outperform single-task learning in terms of fall detection. The results also suggest how a good combination of tasks should be.

4.1 Datasets

Table 1. List of used public datasets

Dataset	Classes	Dataset size	Participants
UR Fall [25]	Fall and Non-fall	30 falls and 40 non-fall sequences	5 participants
KFall [26]	15 falls and 21 ADLs	2729 sequences of ADLs (6.3 h) and 2346 falls	32 male participants (age 24.9 ± 3.7)
HAPT [27]	12 ADLs	6.2 h of data	30 participants (19–48 years old)
SFU-IMU [28]	Fall, ADL, and Near fall	240 sequences of ADLs 210 sequences of Falls 150 sequences of Near falls	10 participants (22–32 years old)

This study uses 4 datasets, in which 3 are training and validation set, and 1 is the test set. There are both fall and ADL-only datasets, details are shown in Table 1. In all of these datasets, participants wore the sensor on their waist or lower back, and accelerometer data includes gravity. We convert all data to the same frequency (50 Hz) and unit (g) before conducting experiments. Figure 2 shows randomly picked samples for fall and non-fall data from 4 datasets. Each sample is a 4-s window.

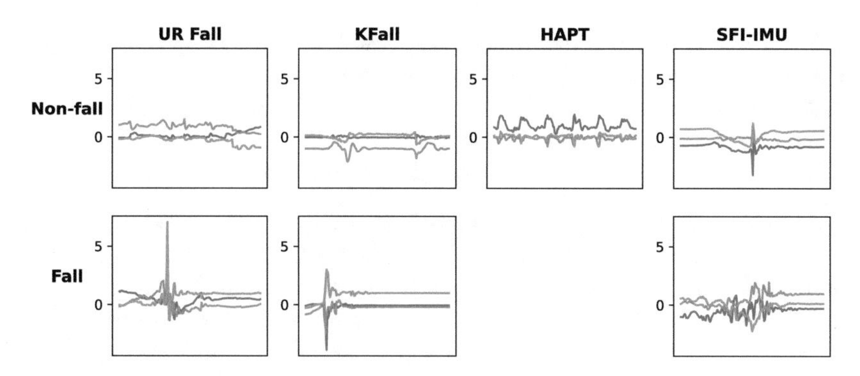

Fig. 2. Fall and non-fall 4-s samples of 4 datasets

We organise datasets for the experiments as depicted in Table 2. There are 2 training sets, in which D1 is treated as a base human fall dataset with 2 classes of fall and non-fall, while D2 is treated as an extra ADL dataset not containing any fall class. D2 is supposed to be a large non-fall training set with many ADLs classes. We use all ADLs data from the KFall dataset and not the HAPT dataset for this because its activities are mostly very short and dense. Thus it is not suitable for a classification model with a longer input window (window size will be described in Sect. 4.2). D1 consists of the HAPT dataset as non-fall data, fall data from the KFall dataset, and the UR Fall dataset, which has both classes but is relatively small. D1 is split into 2 equal halves, one as training data and the other is validation data.

While real fall datasets are hardly accessible, the SFU-IMU dataset is used as the test set because it was designed for realistic activities. The participants were trained and mimicked real falls captured on cameras.

Table 2. Experiment data

Name	Classes	Role	Components
D1	Fall and Non-fall	1/2 for training 1/2 for validation	UR Fall dataset HAPT dataset Fall data from KFall
D2	ADL classes of KFall	training	ADLs data from KFall
Test set	Fall and Non-fall	test	SFU-IMU dataset

4.2 Experiment Settings

We design experiments to show that multi-task learning can achieve better result than single-task learning with the same training set. Experiments are divided into 3 groups as below.

- **G1**: Two single-task learning settings. One has only D1 as the training set, the other combines D1 and D2 into a large training set.
- **G2**: Multi-task learning with the simplest task combination of D1 and D2. Each one is trained as a separate task.
- **G3**: Various combinations of D1 and D2 for 2 tasks of the multi-task model.

Details about the training sets composition are showed in Table 3, in which each row is an experiment. Column *Exp.* shows the experiment settings number. Within each cell of column *Dataset*, each line represents the training set of a task, and the column *Classes* shows the respective classes of every task. For example, in experiment G3.4, there are 2 tasks of the multi-task model. The first one is trained on D1 to classify between fall and non-fall, while the other is trained on D2 and all fall data of D1 to classify fall and ADL classes of D2 (D2 has 21 classes, so this is a 22-class classification problem).

Besides the training set composition, every other factor is identical for all experiments. The deep learning model, data augmentation, and data re-sampling are mentioned in Sect. 3. Data is fed to the model as 4-second windows. This is because as observed on the training set, we see that fall events are shorter than 4 s. The batch size is 16. For multi-task experiments, each of the 2 tasks has 8 samples in a batch. The model is trained for 40 epochs and the test results in Table 3 are obtained from the best model checkpoints on the validation set. The implementation of our experiments is available online at https://github.com/nda97531/da_multitask.

Table 3. Fall detection result on the whole SFU-IMU dataset

Exp.		Dataset	Classes	Precision	Recall	F1-score
G1	G1.1	D1	Fall and Non-fall	0.9382	0.7905	0.8577
	G1.2	D1+D2	Fall and Non-fall	0.9118	0.8492	0.8784
G2	G2.1	D1 D2	Fall and Non-fall D2 ADLs	0.8983	0.9222	0.9100
G3	G3.1	D1+D2 D1+D2	Fall and Non-fall Fall and Non-fall + D2 ADLs	0.8352	0.9667	0.8959
	G3.2	D1+D2 D1 fall +D2	Fall and Non-fall Fall + D2 ADLs	0.8393	0.9556	0.8930
	G3.3	D1+D2 D2	Fall and Non-fall D2 ADLs	0.8977	0.9254	0.9109
	G3.4	D1 D1 fall + D2	Fall and Non-fall Fall + D2 ADLs	0.9056	0.9381	0.9212
	G3.5	D1 D1 fall + D2	Fall and Non-fall Fall and Non-fall	0.9261	0.8524	0.8874
	G3.6	D1 D1 + D2	Fall and Non-fall Fall and Non-fall	0.9203	0.8762	0.8972
	G3.7	D1 D1 + D2	Fall and Non-fall Fall and Non-fall + D2 ADLs	0.8854	0.9444	0.9139

4.3 Evaluation and Discussion

Table 3 shows the results of 10 experiments. While G-mean (calculated from specificity and sensitivity) is another suitable metric, we use F1-score because both give similar insights in this case and F1-score is more common. Each test result in the table is the average of 3 runs. All the metrics are for binary classification with fall being the positive class. Overall, all multi-task experiments have better F1-scores than 2 single-task ones.

G1.1 is the basic case with a small fall dataset. After adding an ADL dataset (D2) to the training set, the single-task model's F1-score is improved from 0.8577 to 0.8784. Technically, when being trained with more negative data, the model would be biased towards the negative class, precision would increase and recall would decrease. But the opposite happens in G1.2 because of data re-sampling. We also try G1.2 without re-sampling and the result resembles that expectation.

With the same training data but divided into 2 tasks, all experiments in groups G2 and G3 give better results than G1. Experiments that duplicate the data too much like G3.1 and G3.2 tend to have bias towards falls and have low precision scores. Also, all models with an F1-score higher than 0.9 have 1 task trained to classify ADLs. For example, G3.4 and G3.5 have the exact same training set combination but with different classes. The model in G3.4 trained to classify fall and all ADLs has a nearly 4% higher F1-score than the model in G3.5. The same pattern also applies for G3.6 and G3.7. These experiments show that multi-task learning has better results than single-task learning, and ADLs classification does help achieving better fall detection performance.

5 Conclusion and Future Work

In this paper, we prove that multi-task learning can surpass single-task learning in fall detection. We conduct a set of experiments and analyse the results to find the best way to combine training sets and tasks. Some of our observations are as follows.

- With the exact same training data and configurations, dividing data into multiple tasks gives better results than a single task.
- Data re-sampling strongly affects precision and recall scores when adding more negative data.
- With the same training data and model, training the sub-task with all ADL classes improves results better than 2 classes.

From that, we consider there are some possible directions for future studies. Firstly, sampling strategies for multi-task models can be further improved. Also, we only conduct experiments with 2 tasks for a model in this paper. More diverse tasks added to the model could also be helpful. Furthermore, the choice of datasets for training data should also be investigated more.

References

1. World Health Organization. Falls (2021). https://www.who.int/news-room/fact-sheets/detail/falls
2. Rubenstein, L.Z.: Falls in older people: epidemiology, risk factors and strategies for prevention. Age Ageing **35**(Suppl. 2), ii37–ii41 (2006). https://doi.org/10.1093/ageing/afl084
3. Wang, X., Ellul, J., Azzopardi, G.: Elderly fall detection systems: a literature survey. Front. Robot. AI **7**, 74 (2020). https://doi.org/10.3389/frobt.2020.00071
4. Ren, L., Peng, Y.: Research of fall detection and fall prevention technologies: a systematic review. IEEE Access **7**, 77702–77722 (2019). https://doi.org/10.1109/ACCESS.2019.2922708
5. Nahian, M.J.A., et al.: Towards an accelerometer-based elderly fall detection system using cross-disciplinary time series features. IEEE Access **9**, 39413–39431 (2021). https://doi.org/10.1109/ACCESS.2021.3056441
6. Santos, G.L., Endo, P.T., de Carvalho Monteiro, K.H., da Silva Rocha, E., Silva, I., Lynn, T.: Accelerometer-based human fall detection using convolutional neural networks. Sensors **19**(7), 1644 (2019). https://doi.org/10.3390/s19071644
7. Xiaodan, W., Zheng, Y., Chu, C.-H., Cheng, L., Kim, J.: Applying deep learning technology for automatic fall detection using mobile sensors. Biomed. Sig. Process. Control **72**, 103355 (2022)
8. Galvão, Y.M., Ferreira, J., Albuquerque, V.A., Barros, P., Fernandes, B.J.T.: A multimodal approach using deep learning for fall detection. Exp. Syst. Appl. **168**, 114226 (2021)
9. Kostopoulos, P., Nunes, T., Salvi, K., Deriaz, M., Torrent, J.: F2D: a fall detection system tested with real data from daily life of elderly people. In: 2015 17th International Conference on E-health Networking, Application & Services (HealthCom), Boston, MA, USA, pp. 397–403 (2015). https://doi.org/10.1109/HealthCom.2015.7454533
10. Palmerini, L., Klenk, J., Becker, C., Chiari, L.: Accelerometer-based fall detection using machine learning: training and testing on real-world falls. Sensors **20**(22), 6479 (2020). https://doi.org/10.3390/s20226479
11. Broadley, R.W., Klenk, J., Thies, S.B., Kenney, L.P.J., Granat, M.H.: Methods for the real-world evaluation of fall detection technology: a scoping review. Sensors **18**(7), 2060 (2018). https://doi.org/10.3390/s18072060
12. Kangas, M., Korpelainen, R., Vikman, I., Nyberg, L., Jämsä, T.: Sensitivity and false alarm rate of a fall sensor in long-term fall detection in the elderly. Gerontology **61**(1), 61–68 (2015). Epub 2014 Aug 13. PMID: 25138139. https://doi.org/10.1159/000362720
13. Bourke, A.K., et al.: Fall detection algorithms for real-world falls harvested from lumbar sensors in the elderly population: a machine learning approach. In: 2016 Annual International Conference of the IEEE Engineering in Medicine and Biology Society, August 2016, pp. 3712–3715 (2016). PMID: 28269098. https://doi.org/10.1109/EMBC.2016.7591534
14. Chen, K.-H., Hsu, Y.-W., Yang, J.-J., Jaw, F.-S.: Enhanced characterization of an accelerometer-based fall detection algorithm using a repository. Instrum. Sci. Technol. **45** (2016). https://doi.org/10.1080/10739149.2016.1268155
15. Yu, S., Chen, H., Brown, R.A.: Hidden Markov model-based fall detection with motion sensor orientation calibration: a case for real-life home monitoring. IEEE J. Biomed. Health Inform. **22**(6), 1847–1853 (2018). https://doi.org/10.1109/JBHI.2017.2782079

16. Taoran Sheng and Manfred Huber. 2020. Weakly Supervised Multi-Task Representation Learning for Human Activity Analysis Using Wearables. Proc. ACM Interact. Mob. Wearable Ubiquitous Technol. **4**(2), 1–18 (2020). Article 57. https://doi.org/10.1145/3397330
17. Barut, O., Zhou, L., Luo, Y.: Multitask LSTM model for human activity recognition and intensity estimation using wearable sensor data. IEEE Internet Things J. **7**(9), 8760–8768 (2020). https://doi.org/10.1109/JIOT.2020.2996578
18. Peng, L., Chen, L., Ye, Z., Zhang, Y.: AROMA: a deep multi-task learning based simple and complex human activity recognition method using wearable sensors. Proc. ACM Interact. Mob. Wearable Ubiquitous Technol. **2**(2), 1–16 (2018). Article 74. https://doi.org/10.1145/3214277
19. Parsa, B., Banerjee, A.: A multi-task learning approach for human activity segmentation and ergonomics risk assessment. In: Proceedings of the IEEE/CVF Winter Conference on Applications of Computer Vision (WACV), pp. 2352–2362 (2021)
20. Li, Y., Zhang, S., Zhu, B., et al.: Accurate human activity recognition with multi-task learning. CCF Trans. Pervasive Comp. Interact. **2**, 288–298 (2020). https://doi.org/10.1007/s42486-020-00042-2
21. Saeed, A., Ozcelebi, T., Lukkien, J.: Multi-task self-supervised learning for human activity detection. Proc. ACM Interact. Mob. Wearable Ubiquitous Technol. **3**(2), 1–30 (2019). Article 61. https://doi.org/10.1145/3328932
22. Pham, C., Nguyen, L., Nguyen, A., et al.: Combining skeleton and accelerometer data for human fine-grained activity recognition and abnormal behaviour detection with deep temporal convolutional networks. Multimed. Tools Appl. **80**, 28919–28940 (2021). https://doi.org/10.1007/s11042-021-11058-w
23. Woo, S., Park, J., Lee, J., Kweon, I.S.: CBAM: convolutional block attention module. CoRR abs/1807.06521. arXiv: 1807.06521
24. Um, T.T., et al.: Data Augmentation of wearable sensor data for Parkinson's disease monitoring using convolutional neural networks. CoRR, abs/1706.00527 (2017)
25. Kwolek, B., Kepski, M.: Human fall detection on embedded platform using depth maps and wireless accelerometer. Comput. Meth. Program. Biomed. **117**(3), 489–501 (2014). ISSN 0169–2607
26. Yu, X., Jang, J., Xiong, S.: A large-scale open motion dataset (KFall) and benchmark algorithms for detecting pre-impact fall of the elderly using wearable inertial sensors. Front. Aging Neurosci. **13**, 1–14 (2021)
27. Reyes-Ortiz, J.-L., Oneto, L., SamÃ, A., Parra, X., Anguita, D.: Transition-aware human activity recognition using smartphones. Neurocomputing **171**, 754–767 (2016)
28. Aziz, O., Musngi, M., Park, E.J., et al.: A comparison of accuracy of fall detection algorithms (threshold-based vs. machine learning) using waist-mounted tri-axial accelerometer signals from a comprehensive set of falls and non-fall trials. Med. Biol. Eng. Comput. **55**, 45–55 (2017). https://doi.org/10.1007/s11517-016-1504-y

A Comparative Study of Wrapper Feature Selection Techniques in Software Fault Prediction

Nguyen Thanh Long[1,2], Ha Thi Minh Phuong[2], and Nguyen Thanh Binh[2](✉)

[1] The University of Danang - University of Science and Technology, Danang, Vietnam
ngthlo.doc@gmail.com

[2] Vietnam-Korea University of Information and Communication Technology, Danang, Vietnam
{htmphuong,ntbinh}@vku.udn.vn

Abstract. Software fault prediction aims to classify whether the module is defective or not-defective. In software systems, there are some software metrics may contain irrelevant or redundant information that leads to negative impact on the performance of the fault prediction model. Therefore, feature selection is an method that several studies have addressed to reduce computation time, improve prediction performance and provide a better understanding of data in machine learning. Additionally, the presence of imbalanced classes is one of the most challenge in software fault prediction. In this study, we examined the effectiveness of six different wrapper feature selection including Genetic Algorithm, Particle Swarm Optimization, Whale Optimization Algorithm, Cuckoo Search, Mayfly Algorithm and Binary Bat Algorithm for selecting the optimal subset of features. Then, we applied VanilaGAN to train the dataset with optimal features for handling the imbalanced problem. Subsequently, these generated training dataset and the testing dataset are fed to the machine learning techniques. Experimental validation has been done on five dataset collected from Promise repository and Precision, Recall, F1-score, and AUC are evaluation performance measurements.

Keywords: Software fault prediction · feature selection · Wrapper · VanillaGAN · dataset

1 Introduction

Software defect prediction (SDP) is a significant procedure in software engineering that aims to identify potential defects in software systems before they occur. The robust software predictive model predicts correctly bugs or defects before releasing a new software version. Therefore, developers can effectively plan the distribution of testing effort to reduce the overall cost of software development and improve the quality and reliability of the final product.

N. T. Nguyen et al. (Eds.): CITA 2023, LNNS 734, pp. 62–73, 2023.
https://doi.org/10.1007/978-3-031-36886-8_6

In recent studies, machine learning techniques have been exploited widely to construct a fault prediction model. These techniques use the historical defect dataset that has been collected from previous software projects to make predictions about the occurrence of future defects. The defect dataset consists of various software features (software metrics) e.g., Response for a Class, Depth of inheritance tree or Line of Code and labels which indicates defective or not for each module. During training, the predictive model learns the characteristics of software projects and makes predictions whether new module is faulty or not faulty [24]. Several researches state that the performance of software fault prediction models depends on software fault dataset [2,14]. However, there are two major challenges that the software defect prediction dataset faces the high dimensional features and imbalanced classes [22]. Additionally, some software metrics are irrelevant or redundant to fault-proneness modules. They are the most significant reasons that reduce the effectiveness of used machine learning techniques in SFP.

In SFP literature, feature selection is employed to address the problem of minimizing the redundant and irrelevant features that are not useful for prediction. The objective of feature selection is to generate a subset of optimal metrics from the input data to achieve better prediction performance [5]. Numerous feature selection techniques have been proposed in software fault prediction domain [10,19]. Feature selection techniques can be broadly categorized into three types, namely filter, wrapper and embedded techniques. The filter technique involves applying a statistical measure to each feature and selecting the top features based on their scores. Wrapper methods use a machine learning algorithm to evaluate the performance of a subset of features. The algorithm is trained using different subsets of features and the subset that yields the best performance is selected. Embedded methods incorporate feature selection into the learning algorithm itself. For example, some machine learning algorithms have built-in feature selection mechanisms, such as Lasso regression, decision trees, and random forests. According to Savina Colaco et al. [6], wrapper-based technique is the most popular use across all feature selection methods.

In this study, we examine the performance of different wrapper feature selection methods including Genetic Algorithm (GA), The Particle Swarm Optimization (PSO), Whale Optimization Algorithm (WOA), Cuckoo Search (CS), Mayfly Algorithm (MA) and Binary Bat Algorithm. Our main contribution is to evaluate the effectiveness of various wrapper feature selection techniques and compare the performance of classifiers in generating the optimal variables for the classification of software faults.

The remainder of this study is organized as follows: Sect. 2 presents the previous works related to current research. Used wrapper feature selection techniques are introduced in Sect. 3. Section 4 presents the proposed methodology followed by the experimental results in Sect. 5. Section 6 concludes the study with a discussion of our results and prospects for future work.

2 Related Work

Feature selection is one of the most significant tasks of preprocessing data in software fault prediction. The objective of feature selection is to select a subset of optimal features that excludes irrelevant or redundant features. Recently, several studies have proposed different feature selection techniques in SFP. Khoshgoftaar et al. [12] compared different filter feature-ranking selection techniques, namely information gain (IG), ReliefF, chi-squared (CS), symmetrical uncertainty (SU) and gainratio (GR) using 16 fault datasets. They also considered the signal-to-noise ratio (SNR) which is infrequently employed as software metric. Throughout the experimental results, the author suggested IG and SNR achieved the best classification performance across all datasets. However, Khoshgoftaar also recommended that although feature selection can improve the performance of fault prediction models but it has not yet addressed the imbalanced data problem. Wang [27] performed a study to address the problem of which feature elimination methods are stable in case of data change (the deletion or addition of features). They examined various filter-based and wrapper-based feature subset selection techniques using three real-world fault datasets. From the experimental results, they showed that the Correlation-Based Feature Selection (CFS) is the most stable method among different methods. Mohammad et al. [13] demonstrated the combination of feature selection and ensemble learning had a great performance of fault classification. They also concluded that greedy forward selection provided better performance than other methods and average probability ensemble which grouped seven devised classifiers outperformed conventional methods such as random forest and weighted SVMs. Wang [26] proposed a Cluster-based Hybrid Feature Selection that combined Filter and Wrapper methods to reduce the irrelevant and redundant information and enhance the performance the of predictive model. Their experimental results indicated that the Cluster-based Hybrid Feature Selection method obtained better performance compared to the other traditional methods in terms of accuracy evaluation measure. Savina Colaco et al. [6] presented a survey of different feature selection techniques specifically filter, wrapper and embedded methods. They found that wrapper methods indicated notable improvement in accuracy for the dataset containing larger features, hence wrapper methods are more popularly applied to obtain optimal features than other methods. Ghost et al. [10] proposed a hybrid of wrapper and filter feature selection based on ant colony optimization. They demonstrated their proposed method outperformed most of the modern methods used for feature selection.

3 Feature Selection

Feature selection is a high dimensionality reduction method by selecting the most important and relevant features, which is the smallest set from a large set of features in the given dataset. There are three types of feature selection methods including the Filter method, Wrapper method and Embedded method. In this study, Wrapper methods are adopted to select relevant features from the Promise dataset. Additionally, the features of the software fault dataset are defined as software metrics. Therefore, in this section, we concentrate on the software metrics and wrapper-based feature selection.

3.1 Software Metrics

Various kinds of features indicate the characteristic of software. Software metrics are used to evaluate various aspects of software development and software characteristics, such as software quality, complexity, performance, and maintainability. These metrics provide valuable insights into the development process and can help identify areas for improvement.

Software metrics can be broadly divided into four categories: Procedure, Object-Oriented, Hybrid and Miscellaneous Metrics [4,17]. Procedure software metrics typically focus on the internal characteristics of software, such as its size, complexity, and maintainability. Some examples of traditional software metrics are Line Of Code (LOC), McCabe [16] and Halstead Metrics [11]. Object-oriented (OO) metrics are applied to measure the properties of object oriented software such as Coupling between Objects (CBO), Depth of inheritance tree (DIT), Number of children (NOC), Response for a Class [1]. Software hybrid metrics are a set of software metrics that combine elements of traditional and object-oriented metrics to provide a more comprehensive view of software quality and complexity [15]. Hybrid metrics are designed to measure both the internal and external characteristics of software, such as its functionality, usability, and maintainability. Some common hybrid metrics [7,25] include Function point analysis (FPA), Maintainability index (MI). Miscellaneous metrics are process metrics [8], change metrics [18], etc. According to the research 's Ruchika Malhotra [15], the most commonly used metrics are recorded in the procedure and object-oriented metrics.

3.2 Genetic Algorithm

Genetic algorithm (GA) is an optimization algorithm that is inspired by natural selection. It is a population-based search algorithm that utilizes the concept of survival of the fittest. The new populations are produced by iterative use of genetic operators on individuals present in the population. The chromosome representation, selection, crossover, mutation, and fitness function computation are the key elements of GA.

3.3 Particle Swarm Optimization

Particle Swarm Optimization was proposed by Kennedy and Eberhart in 1995. This method is inspired by the habits of sharing information from finding food processes of a school of fish or flock of birds that move in a group. In a PSO algorithm, particles (which represent candidate solutions) are flying around, in a multi-dimensional search space. The positions of these particles are adjusted according to the particles' own memories and the best-fit particle of the neighboring particles [21].

3.4 Whale Optimization Algorithm

Whale optimization algorithm (WOA) is a nature-inspired meta-heuristic optimization algorithm that mimics the hunting behavior of humpback whales. The algorithm is inspired by the bubble-net hunting strategy. Foraging behavior of Humpback whales is called the bubble-net feeding method. Humpback whales prefer to hunt schools of krill or small fishes close to the surface. We can start WOA with a random initialization of the search agents and choose the best, first solution by calculating the fitness of each search agent. Next, the position is updated by basing on equations of Spiral Bubble-Net Feeding Behavior method and then search space constraints are checked and the best solution is updated if there is a better one found. Finally, the algorithm loop is terminated by the maximum number of iterations (default criterion) [3].

3.5 Cuckoo Search

Cuckoo search (CS) is a meta-heuristic algorithm inspired by the bird cuckoo, these are the "Brood parasites" birds. In a nest, each egg represents a solution and the cuckoo egg represents a new and good solution. The obtained solution is a new solution based on the existing one and the modification of some characteristics. In CS algorithm, there are three rules: 1) Each cuckoo lays one egg at a time, and dumps its egg in randomly chosen nest; 2) The best nests with high quality of eggs will carry over to the next generations; 3) The number of available host nests is fixed, and the egg laid by a cuckoo is discovered by the host bird with a probability $pa \in [0, 1]$. In this case, the host bird can either throw the egg away or abandon the nest, and build a completely new nest [28].

3.6 Mayfly Algorithm

Inspired from the flight behavior and the mating process of mayflies, the Mayfly Algorithm (MA) combines major advantages of swarm intelligence and evolutionary algorithms [29]. In MA, the mayfly swarms would be separated into male and female individuals. Specially, the male mayflies would always be strong and consequently, resulting in a better performance in optimization. And the individuals in MA would update their position according to their current position and velocity the same way as the individuals in PSO. All of the male mayflies and female mayflies would update their positions with the same measure. However, their velocity would be updated in different ways. The velocity of mayflies would be updated according to their current fitness values and the historical best fitness values due to the responsibility of carrying on exploration or exploitation procedures during iterations. The female mayflies would update their velocities with a different style, they would update their velocities based on the male mayflies they want to mate because of a short life cycle within seven days. All of the top half of female and male mayflies would be mated and given a pair of children for every one of them. Their offspring would be randomly evolved from their parents:

$$offsrping_1 = L * male + (1 - L) * female \quad (1)$$

$$offspring_2 = L * female + (1 - L) * male \quad (2)$$

where, L are random numbers in Gauss distribution [9].

3.7 Binary Bat Algorithm

Based on the bat's behaviour, an interesting meta-heuristic optimization technique called Bat Algorithm was developed. Such technique has been developed to behave as a band of bats tracking prey/foods using their capability of echolocation. In order to model this algorithm, some rules had been idealized, as follows: 1) All bats use echolocation to sense distance, and they also "know" the difference between food/prey and background barriers in some magical way; 2) A bat b_i fly randomly with velocity v_i at position x_i with a fixed frequency f_{min}, varying wavelength λ and loudness A_0 to search for prey. They can automatically adjust the wavelength (or frequency) of their emitted pulses and adjust the rate of pulse emission r $\in$ [0, 1], depending on the proximity of their target; 3) Although the loudness can vary in many ways, the loudness varies from a large (positive) was assumed that A_0 to a minimum constant value A_{min}. In feature selection, since the problem is to select or not a given feature, the bat's position is then represented by binary vectors restricting the new bat's position to only binary values using a sigmoid function (BBA) [20].

3.8 Feature Selection Details

In order to select features effectively, we taked advantage of feature selection module of sklearn library developed in Python that provides many machine learning algorithms. For the GA, we use 10 number of agents, cross and mutation probability are set to 0.7 and 0.1 correspondingly. For the rest, just one agents is used in selection process. Finally, $log_2(S)$ is set to max iterations of each method, where S is the number of samples.

4 Methodology

4.1 Proposed Approach

As we mentioned in introduction, software fault prediction is a vital procedure in software engineering and also depends on historical datasets that have been collected from previous software projects. And this process still has two major challenges including high dimensional data and imbalanced classes. Therefore we propose an approach to examine the effectiveness of six different Wrapper methods for selecting the subset of optimal features. Then the variant of generative adversarial network - VanillaGAN [23] is applied to generate synthetic instances of faulty modules to balance the ratio of faulty and non-faulty modules in the fault datasets.

The overall approach is shown in Fig. 1. In data preprocessing stage, we fill the missing values and apply Z-normalization for data normalization. Next, we apply six different wrapper methods, namely GA, PSO, WOA, CS, MA and BBA to select the most important and relevant features for reducing high dimension features. The datasets with optimal features are used to train the VanillaGAN network to overcome the imbalanced data. Finally, these training datsets, and testing datasets are used to train using machine learning techniques, e.g. K-Nearest Neighbors (KNN), Random Forest (RF), Decision Tree (DT), Naïve Bayes (NB) and Logistic Regression (LR). We use precision, recall, F1-score and AUC as measures to evaluate the performance of different wrapper-based feature selection methods.

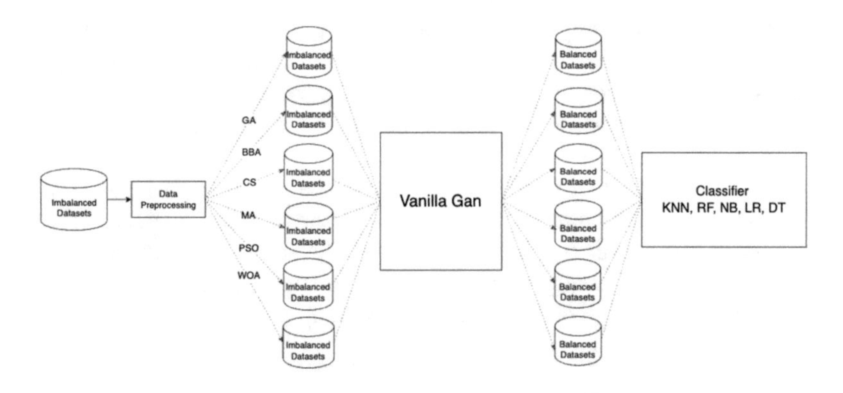

Fig. 1. Flow diagram of proposed approach

4.2 VanillaGAN for Handling Imbalanced Data

We use Vanilla Gan, which is the simplest version of GAN. The generator network lean the probability of the training set by mapping the noise z which is drawn independent and identically distributed (i.i.d) from $\mathbb{N}$ (0, 0.01) and added to both real and synthetic data to the probability distribution of training samples. Then, synthetic samples that are closer possible to real samples were generated by the generator network. It uses backpropagation to train both models. In Vanilla GAN, we use neural networks to approximate complex, high-dimensional distributions for both *Gen* and *Dis*. The discriminator training is done by minimizing its prediction error, whereas the generator is trained by maximizing the prediction error by the discriminator. This can be formalized as given follows:

$$min_{(\Theta_{Gen})} max_{(\Theta_{Dis})} (E_{x \sim p_D}[log Dis(x)] + E_{z \sim p_z}[log(1 - Dis(Gen(z)))]) \quad (3)$$

where p_D is the real data distribution, p_z is the prior distribution of the generative network, and ΘGen and ΘDis are the parameters of the generator (Gen) and discriminator (Dis). Given a strong discriminator, the generator's goal is achieved if p_D, the generator's distribution over x, equals p_z, the real distribution, which means that the Jensen-Shannon Divergence (JSD) is minimized [23].

4.3 Dataset

In this study, we used five sub-datasets collected from PROMISE repository which are collections of publicly available datasets. These resources are adopted to serve researchers in building predictive software models and the software engineering community at large. In terms of software fault prediction (SFP) problems, the datasets provide independent variables which are the different source code metrics such as line count of code (loc), coupling between objects (CBO), etc., and dependent variable in faulty and non-faulty with labels (0) and (1) correspondingly. The details of the used datasets are given in Table 1.

Table 1. Description of used software fault datasets

Dataset	Release	Instances	Metric	Faulty Instances	Imbalanced ratio (%)
PROMISE	CM1	505	40	48	9.504
	KC1	2107	21	325	15.424
	KC2	522	20	107	20.498
	PC1	1107	40	76	6.865
	JM1	10885	20	2106	19.347

4.4 Evaluation Measurement

Performance evaluation measures such as Precision, Recall, F1-score and AUC are used to evaluate how the wrapper feature selection algorithms affect the performance of prediction models. Precision is the ratio between the True Positive and all the Positive, meanwhile Recall is the measure of our model correctly identifying True Positive. In software fault prediction problems, Precision would be the measure of instances that our model correctly identifies as faulty labels out of all instances actually are faulty labels. For all instances which actually are faulty labels, Recall tells us how many instances our model correctly identifies as faulty labels. These metrics are defined by the following:

$$Precision(P) = \frac{TP}{TP + TF}, Recall(R) = \frac{TP}{TP + FN} \tag{4}$$

where TP = True positive, FP = False positive, FN = False negative. From two values of Precision and Recall, we can calculate additional metrics that are F1-score and AUC:

$$F1 - score = \frac{2 * P * R}{P + R} \tag{5}$$

The F1-score is the harmonic mean of the Precision and Recall. The highest possible value of an F-score is 1.0, indicating perfect precision and recall, and the lowest possible value is 0, if either precision or recall are zero. Besides that, AUC calculates the area under the receiver operating characteristics(ROC). This metric represents the probability that a random positive instance to the right of a random negative instance. The higher the probability is, the more perfect the prediction is.

5 Experimental Result

We use a separate training and testing dataset for each repetition of 10-time experiments; the final result is the average of the 10 iterations. The result has been reported in Table 2 and Table 3, including five machine learning techniques for the average result of five datasets of each Wrapper method with regard to used performance measures.

Table 2. Comparison results of presented Wrapper feature selection techniques combined with Vanilla GAN for all the used dataset.

Dataset	Performance Evaluation	GA+VanillaGAN					PSO+VanillaGAN					WOA + VanillaGAN					Without FS				
		KNN	RF	DT	NB	LR	KNN	RF	DT	NB	LR	KNN	RF	DT	NB	LR	KNN	RF	DT	NB	LR
CM1	Precision	0.808	0.790	0.804	0.750	**0.859**	0.806	0.813	0.81	0.805	**0.833**	0.821	0.782	0.808	0.803	0.829	0.792	0.780	0.801	0.749	0.800
	Recall	0.726	0.829	0.811	0.734	0.786	0.740	0.860	0.849	0.863	0.686	0.754	0.860	0.837	**0.866**	0.706	0.707	0.825	0.811	0.803	0.876
	F1-score	0.758	0.808	0.806	0.741	0.820	0.768	**0.832**	0.827	0.831	0.738	0.783	0.819	0.819	**0.833**	0.752	0.747	0.801	0.805	0.775	0.826
	AUC	0.617	0.688	0.646	0.509	0.690	0.689	**0.740**	0.645	0.524	0.687	0.675	0.729	0.617	0.543	0.720	0.608	0.683	0.637	0.506	0.686
KC1	Precision	0.813	0.814	0.793	0.831	0.824	0.822	0.829	0.818	**0.837**	0.821	0.825	0.832	0.818	0.834	0.824	0.810	0.819	0.816	0.815	0.801
	Recall	0.809	0.833	0.828	0.778	0.670	0.831	**0.855**	0.847	0.657	0.666	0.836	**0.854**	0.852	0.706	0.697	0.802	0.830	0.819	0.697	0.656
	F1-score	0.811	0.818	0.804	0.797	0.713	0.825	0.817	0.819	0.701	0.710	**0.828**	0.816	0.821	0.741	0.735	0.805	0.810	0.813	0.751	0.701
	AUC	0.726	0.789	0.769	0.783	0.755	0.746	**0.817**	0.795	0.769	0.744	0.741	0.804	0.790	0.780	0.766	0.715	0.785	0.763	0.764	0.799
KC2	Precision	0.869	0.876	0.882	0.879	**0.884**	**0.881**	0.879	0.870	0.871	0.876	0.874	0.879	0.876	**0.884**	0.872	0.866	0.863	0.866	0.872	0.870
	Recall	0.841	0.875	0.862	0.745	0.693	0.866	**0.924**	0.913	0.900	0.789	0.866	0.922	0.913	0.907	0.793	0.817	0.884	0.857	0.879	0.769
	F1-score	0.853	0.875	0.870	0.793	0.761	0.872	**0.894**	0.889	0.884	0.826	0.869	**0.894**	0.892	0.893	0.827	0.840	0.870	0.861	0.875	0.816
	AUC	0.719	0.76	0.721	0.707	0.707	0.739	**0.785**	0.715	0.672	0.658	0.729	**0.783**	0.727	0.671	0.635	0.710	0.759	0.701	0.67	0.637
PC1	Precision	0.817	0.818	0.788	0.819	0.828	0.797	0.811	0.801	0.820	0.824	0.814	0.808	0.798	0.806	0.832	0.812	0.807	0.782	0.81	0.819
	Recall	0.798	0.813	0.792	**0.828**	0.755	0.783	0.825	0.819	0.823	0.749	0.798	0.825	0.817	0.817	0.796	0.727	0.810	0.803	0.82	0.722
	F1-score	0.804	0.811	0.784	**0.816**	0.774	0.789	0.811	0.789	0.818	0.769	0.800	0.806	0.797	0.808	0.807	0.767	0.808	0.792	0.814	0.767
	AUC	0.772	0.804	0.738	0.817	0.820	0.760	0.826	0.817	0.818	0.817	0.788	0.806	0.798	0.772	0.819	0.770	0.767	0.737	0.74	0.809
JM1	Precision	0.747	0.766	0.758	**0.768**	0.759	0.747	0.762	0.757	0.764	0.750	0.742	0.749	0.747	0.762	0.761	0.738	0.747	0.742	0.756	0.740
	Recall	0.768	**0.809**	0.806	0.796	0.704	0.774	0.807	0.805	0.785	0.555	0.768	0.770	0.776	0.796	0.591	0.766	0.760	0.777	0.785	0.511
	F1-score	0.756	0.760	0.755	0.777	0.725	0.758	0.756	0.754	0.771	0.595	0.753	0.753	0.751	0.771	0.631	0.751	0.759	0.753	0.770	0.541
	AUC	0.658	**0.708**	0.702	0.663	0.668	0.659	0.708	0.700	0.674	0.643	0.651	0.69	0.684	0.68	0.674	0.674	0.679	0.681	0.642	0.638
Average results																					
	Precision	0.811	0.813	0.805	0.809	0.831	0.811	**0.819**	0.811	0.819	0.821	0.815	0.810	0.809	0.818	0.824	0.804	0.803	0.801	0.800	0.806
	Recall	0.788	0.832	0.820	0.776	0.722	0.799	**0.854**	0.847	0.806	0.689	0.804	0.846	0.839	0.818	0.717	0.764	0.822	0.813	0.797	0.707
	F1-score	0.796	0.814	0.804	0.785	0.759	0.802	**0.822**	0.816	0.801	0.728	0.807	0.818	0.816	0.809	0.750	0.782	0.810	0.805	0.797	0.730
	AUC	0.698	0.750	0.715	0.696	0.728	0.719	**0.775**	0.734	0.691	0.710	0.717	0.762	0.723	0.689	0.723	0.695	0.735	0.704	0.664	0.714

FS: Feature Selection

For the Precision measure, values are almost higher than 0.80 for all the techniques except JM1 dataset, the highest value is 0.884 for the PC1 dataset in GA and WOA method, and the lowest value is 0.732 for the JM1 dataset in MA method. For the Recall measure, values almost higher than 0.80, PSO and CS obtained the highest value of 0.924 for PC1 and the lowest value is 0.591 for JM1 dataset in WOA method. For the F1-score, BBA reached the highest value of 0.895 for the PC1 dataset, PSO and CS reached a value of 0.894, which is also good, and the lowest value is 0.40 for the CM1 dataset. In terms of AUC

Table 3. Comparison results of presented Wrapper feature selection techniques combined with Vanilla GAN for all the used dataset (cont).

Dataset	Performance Evaluation	MA+VanillaGAN					CS+VanillaGAN					BBA+ VanillaGAN					Without FS				
		KNN	RF	DT	NB	LR	KNN	RF	DT	NB	LR	KNN	RF	DT	NB	LR	KNN	RF	DT	NB	LR
CM1	Precision	0.814	0.809	0.804	0.787	**0.844**	0.819	0.818	0.817	0.785	**0.847**	0.82	0.794	0.798	0.789	0.836	0.792	0.780	0.801	0.749	0.800
	Recall	0.737	0.866	0.843	0.854	0.726	0.731	0.874	0.846	0.854	0.723	0.771	0.854	0.834	**0.851**	0.72	0.707	0.825	0.811	0.803	0.876
	F1-score	0.769	0.83	0.82	0.818	0.768	0.766	**0.839**	0.828	0.816	0.766	0.792	0.82	0.813	**0.817**	0.763	0.747	0.801	0.805	0.775	0.826
	AUC	0.676	0.708	0.644	0.542	0.727	0.683	**0.744**	0.671	0.568	0.715	0.686	0.727	0.623	0.538	0.689	0.608	0.683	0.637	0.506	0.686
KC1	Precision	0.818	0.84	0.825	0.828	0.824	0.829	0.823	0.815	**0.833**	0.819	0.817	0.834	0.817	0.813	0.824	0.810	0.819	0.816	0.815	0.801
	Recall	0.829	0.859	0.853	0.721	0.688	0.829	**0.853**	0.848	0.666	0.673	0.827	**0.858**	0.849	0.663	0.673	0.802	0.830	0.819	0.697	0.656
	F1-score	0.822	0.827	0.82	0.752	0.728	0.818	0.814	0.813	0.709	0.715	**0.821**	0.827	0.818	0.694	0.716	0.805	0.810	0.813	0.751	0.701
	AUC	0.742	0.807	0.782	0.782	0.771	0.742	**0.805**	0.779	0.776	0.747	0.743	0.812	0.784	0.767	0.757	0.715	0.785	0.763	0.764	0.799
KC2	Precision	0.876	0.879	0.871	0.879	**0.878**	**0.87**	0.879	0.87	0.878	0.874	0.882	0.877	0.87	**0.879**	0.879	0.866	0.863	0.866	0.872	0.870
	Recall	0.859	0.922	0.914	0.904	0.789	0.854	**0.924**	0.913	0.897	0.779	0.868	0.924	0.909	0.897	0.786	0.817	0.884	0.857	0.879	0.769
	F1-score	0.867	0.894	0.89	0.89	0.827	0.861	**0.894**	0.889	0.887	0.818	0.874	**0.895**	0.887	0.886	0.824	0.840	0.870	0.861	0.875	0.816
	AUC	0.717	0.789	0.716	0.665	0.659	0.716	**0.79**	0.72	0.675	0.652	0.736	**0.801**	0.719	0.669	0.652	0.710	0.759	0.701	0.67	0.637
PC1	Precision	0.797	0.807	0.782	0.817	0.823	0.783	0.824	0.829	0.826	0.83	0.827	0.815	0.785	0.812	0.828	0.812	0.807	0.782	0.81	0.819
	Recall	0.774	0.823	0.794	**0.825**	0.747	0.758	0.819	0.825	0.832	0.766	0.809	0.826	0.806	0.821	0.787	0.727	0.810	0.803	0.82	0.722
	F1-score	0.782	0.811	0.785	**0.815**	0.768	0.767	0.818	0.824	0.822	0.784	0.815	0.816	0.792	0.81	0.798	0.767	0.808	0.792	0.814	0.767
	AUC	0.75	0.787	0.735	0.815	0.815	0.747	0.805	0.783	0.826	0.819	0.806	0.808	0.754	0.781	0.812	0.770	0.767	0.737	0.74	0.809
JM1	Precision	0.747	0.742	0.732	**0.756**	0.757	0.76	0.76	0.757	0.767	0.761	0.76	0.764	0.764	0.767	0.745	0.738	0.747	0.742	0.756	0.740
	Recall	0.778	**0.757**	0.762	0.777	0.593	0.787	0.806	0.803	0.793	0.725	0.784	0.8	0.801	0.803	0.604	0.766	0.760	0.777	0.785	0.511
	F1-score	0.756	0.745	0.741	0.764	0.63	0.769	0.758	0.759	0.776	0.74	0.769	0.765	0.76	0.774	0.642	0.751	0.759	0.753	0.770	0.541
	AUC	0.654	**0.679**	0.668	0.655	0.669	0.666	0.72	0.709	0.663	0.672	0.686	0.709	0.703	0.682	0.651	0.674	0.679	0.681	0.642	0.638
Average results																					
	Precision	0.81	0.815	0.803	0.813	0.825	0.812	**0.821**	0.818	0.818	0.826	0.821	0.817	0.807	0.812	0.822	0.804	0.803	0.801	0.800	0.806
	Recall	0.795	0.845	0.833	0.816	0.709	0.792	**0.855**	0.847	0.808	0.733	0.812	0.852	0.84	0.807	0.714	0.764	0.822	0.813	0.797	0.707
	F1-score	0.799	0.821	0.811	0.808	0.744	0.796	**0.825**	0.823	0.802	0.765	0.814	0.825	0.814	0.796	0.749	0.782	0.810	0.805	0.797	0.730
	AUC	0.708	0.754	0.709	0.692	0.728	0.711	**0.773**	0.732	0.702	0.721	0.731	0.771	0.717	0.687	0.712	0.695	0.735	0.704	0.664	0.714

FS: Feature Selection

measure, for most of the cases, values are greater than 0.700 in CM1, PC1, KC1 and KC2. The highest value of AUC is 0.826 for KC2 in PSO method and the lowest value is 0.509 for CM1. Additionally, the performance of the classifiers based on wrapper feature selection methods was better than when no feature selection methods are applied.

From the average results, it can be observed that PSO with Vanilla GAN and Cuckoo Search with VanillaGAN significantly produced highlighted values, which are higher than others for most of the cases, followed by Mayfly Algorithm with Vanilla GAN. Moreover, it is clear that the RF technique produces the best performance for most cases among the used wrapper methods for evaluation measures.

6 Conclusion

The performance of software fault prediction (SFP) is affected by two major factors that are high dimensional features and imbalanced classes. One of the common ways to overcome these problem can be a hybrid feature selection method combined with data sampling that generate balanced dataset with optimal features.

For the high dimension reduction, this paper presented six Wrapper methods that are Genetic Algorithm, the Particle Swarm Optimization, Whale Optimization Algorithm, Cuckoo Search, Mayfly Algorithm and Binary Bat Algorithm to select important and relevant features from Promise dataset. Then the Vanilla GAN was trained with the collected dataset to generate synthetic instances of minority class to handle imbalanced data. Afterward, five machine learning techniques were applied on balanced data in SFP. Experiment results show us a significant achievement of combination of wrapper feature selection and GAN network for SFP, specifically the Particle Swarm Optimization with Vanilla GAN and Cuckoo Search with VanillaGAN produced a good performance on most of the dataset, followed by Mayfly Algorithm with Vanilla GAN.

In the future work, we intend to investigate on ensemble learning for feature selection and several variations of GAN for handing the imbalanced dataset in SFP.

References

1. Aggarwal, K., Singh, Y., Kaur, A., Malhotra, R.: Empirical study of object-oriented metrics. J. Object Technol. **5**(8), 149–173 (2006)
2. Arisholm, E., Briand, L.C., Johannessen, E.B.: A systematic and comprehensive investigation of methods to build and evaluate fault prediction models. J. Syst. Softw. **83**(1), 2–17 (2010)
3. Brodzicki, A., Piekarski, M.J.K.J.: The whale optimization algorithm approach for deep neural networks. Sensors **21**, 8003 (2021)
4. Caglayan, B., Tosun, A., Miranskyy, A., Bener, A., Ruffolo, N.: Usage of multiple prediction models based on defect categories. In: Proceedings of the 6th International Conference on Predictive Models in Software Engineering, pp. 1–9 (2010)
5. Chandrashekar, G., Sahin, F.: A survey on feature selection methods. Comput. Electr. Eng. **40**(1), 16–28 (2014)
6. Colaco, S., Kumar, S., Tamang, A., Biju, V.G.: A review on feature selection algorithms. In: Emerging Research in Computing, Information, Communication and Applications, ERCICA 2018, vol. 2 pp. 133–153 (2019)
7. De Carvalho, A.B., Pozo, A., Vergilio, S.R.: A symbolic fault-prediction model based on multiobjective particle swarm optimization. J. Syst. Softw. **83**(5), 868–882 (2010)
8. Elish, K.O., Elish, M.O.: Predicting defect-prone software modules using support vector machines. J. Syst. Softw. **81**(5), 649–660 (2008)
9. Gao, Z.M., Zhao, J., Li, S.R., Hu, Y.R.: The improved mayfly optimization algorithm. J. Phys. Conf. Ser. **1684**, 012077 (2020). https://doi.org/10.1088/1742-6596/1684/1/012077
10. Ghosh, M., Guha, R., Sarkar, R., Abraham, A.: A wrapper-filter feature selection technique based on ant colony optimization. Neural Comput. Appl. **32**, 7839–7857 (2020)
11. Halstead, M.H.: Elements of Software Science. Operating and Programming Systems Series. Elsevier Science Inc. (1977)
12. Khoshgoftaar, T.M., Gao, K., Napolitano, A.: An empirical study of feature ranking techniques for software quality prediction. Int. J. Softw. Eng. Knowl. Eng. **22**, 161–183 (2012)
13. Laradji, I.H., Alshayeb, M., Ghouti, L.: Software defect prediction using ensemble learning on selected features. Inf. Softw. Technol. **58**, 388–402 (2015)
14. Lessmann, S., Baesens, B., Mues, C., Pietsch, S.: Benchmarking classification models for software defect prediction: a proposed framework and novel findings. IEEE Trans. Softw. Eng. **34**(4), 485–496 (2008)

15. Malhotra, R.: A systematic review of machine learning techniques for software fault prediction. Appl. Soft Comput. **27**, 504–518 (2015)
16. McCabe, T.J.: A complexity measure. IEEE Trans. Softw. Eng. **4**, 308–320 (1976)
17. Meiliana, Karim, S., Warnars, H.L.H.S., Gaol, F.L., Abdurachman, E., Soewito, B.: Software metrics for fault prediction using machine learning approaches: a literature review with promise repository dataset. In: 2017 IEEE International Conference on Cybernetics and Computational Intelligence (CyberneticsCom), pp. 19–23 (2017). https://doi.org/10.1109/CYBERNETICSCOM.2017.8311708
18. Moser, R., Pedrycz, W., Succi, G.: A comparative analysis of the efficiency of change metrics and static code attributes for defect prediction. In: Proceedings of the 30th International Conference on Software Engineering, pp. 181–190 (2008)
19. Moslehi, F., Haeri, A.: A novel hybrid wrapper-filter approach based on genetic algorithm, particle swarm optimization for feature subset selection. J. Ambient. Intell. Humaniz. Comput. **11**, 1105–1127 (2020)
20. Nakamura, R., Pereira, L., Costa, K., Rodrigues, D., Papa, J., Yang, X.S.: BBA: a Binary Bat Algorithm for feature selection, pp. 291–297, August 2012. https://doi.org/10.1109/SIBGRAPI.2012.47
21. Ovat, F., Anyandi, A.J.: The particle swarm optimization (PSO) algorithm application - a review. Glob. J. Eng. Technol. Adv. **3**, 001–006 (2020)
22. Pandey, S.K., Mishra, R.B., Tripathi, A.K.: Machine learning based methods for software fault prediction: a survey. Exp. Syst. Appl. **172**, 114595 (2021)
23. Rathore, S.S., Chouhan, S.S., Jain, D.K., Vachhani, A.G.: Generative oversampling methods for handling imbalanced data in software fault prediction. IEEE Trans. Reliab. **71**(2), 747–762 (2022)
24. Shepperd, M., Song, Q., Sun, Z., Mair, C.: Data quality: some comments on the NASA software defect datasets. IEEE Trans. Softw. Eng. **39**(9), 1208–1215 (2013)
25. Turhan, B., Kocak, G., Bener, A.: Software defect prediction using call graph based ranking (CGBR) framework. In: 2008 34th Euromicro Conference Software Engineering and Advanced Applications, pp. 191–198. IEEE (2008)
26. Wang, F., Ai, J., Zou, Z.: A cluster-based hybrid feature selection method for defect prediction. In: 2019 IEEE 19th International Conference on Software Quality, Reliability and Security (QRS), pp. 1–9. IEEE (2019)
27. Wang, H., Khoshgoftaar, T.M., Napolitano, A.: Stability of filter-and wrapper-based software metric selection techniques. In: Proceedings of the 2014 IEEE 15th International Conference on Information Reuse and Integration, IEEE IRI 2014, pp. 309–314. IEEE (2014)
28. Yang, X.S., Deb, S.: Cuckoo search via lévy flights. In: 2009 World Congress on Nature & Biologically Inspired Computing (NaBIC), pp. 210–214 (2009)
29. Zervoudakis, K., Tsafarakis, S.: A mayfly optimization algorithm. Comput. Ind. Eng. **145**, 106559 (2020)

Incorporating Natural Language-Based and Sequence-Based Features to Predict Protein Sumoylation Sites

Thi-Xuan Tran[1], Van-Nui Nguyen[2(✉)], and Nguyen Quoc Khanh Le[3,4]

[1] University of Economics and Business Administration, Thai Nguyen, Vietnam
tranxuantbhd@tueba.edu.vn

[2] University of Information and Communication Technology, Thai Nguyen, Vietnam
nvnui@ictu.edu.vn

[3] Professional Master Program in Artificial Intelligence in Medicine, Taipei Medical University, New Taipei, Taiwan
khanhlee@tmu.edu.tw

[4] Research Center for Artificial Intelligence in Medicine, Taipei Medical University, New Taipei, Taiwan

Abstract. The incidence of thyroid cancer and breast cancer is increasing every year, and the specific pathogenesis is unclear. Post-translational modifications are an important regulatory mechanism that affects the function of almost all proteins. They are essential for a diverse and well-functioning proteome and can integrate metabolism with physiological and pathological processes. In recent years, post-translational modifications have become a research hotspot, with methylation, phosphorylation, acetylation and succinylation being the main focus. SUMOylated proteins are predominantly localized in the nucleus, and SUMO regulates nuclear processes, including cell cycle control and DNA repair. SUMOylated proteins are predominantly localized in the nucleus, and SUMO regulates nuclear processes, including cell cycle control and DNA repair. SUMOylation has been increasingly implicated in cancer, Alzheimer's, and Parkinson's diseases. Therefore, identification and characterization SUMOylation sites are essential for determining modification-specific proteomics. This study aims to propose a novel schema for predicting protein SUMOylation sites based on the incorporation of natural language features (Word2Vec) and sequence-based features. In addition, the novel model, called RSX_SUMO, is proposed for the prediction of protein SUMOylation sites. Our experiments reveal that the performance of RSX_SUMO model achieves the highest performance in both five-fold cross-validation and independent testing, obtain the performance on independent testing with acccuracy at 88.6% and MCC value of 0.743. In addition, the comparison with several existing prediction models show that our proposed model outperforms and obtains the highest performance. We hope that our findings would provide effective suggestions and be a great helpful for researchers related to their related studies.

Keywords: SUMOylation sites prediction · Machine learning · Word2Vec · Random forest · XGBoost · SVM

N. T. Nguyen et al. (Eds.): CITA 2023, LNNS 734, pp. 74–88, 2023.
https://doi.org/10.1007/978-3-031-36886-8_7

1 Introduction

Post-translational modificataions (PTMs), such as methylation, acetylation, glycosylation, ubiquitination, succinylation, and phosphorylation, are chemical modifications that play a critical role in the fuctional diversity and complexity levels of promotes following the protein biosynthesis. They regulate localization, activity and interactions with other cellular molecules in most biological processes. Small ubiquitin-like modifier (SUMO) proteins are a type of PTM that play an important role in subcellular transport, transcription, DNA repair and signal transduction [1–3]. The schematical structure of a SUMO protein is displayed in Fig. 1 (Fig. 1A: Representing the backbone of the protein as a ribbon, highlighting secondary structure; N-terminus in blue, C-terminus in red. Figure 1B: Representing atoms as spheres, showing the shape of the protein). Commonly, SUMO protein has four isoforms, including: SUMO-1, SUMO-2, SUMO-3, and SUMO-4.

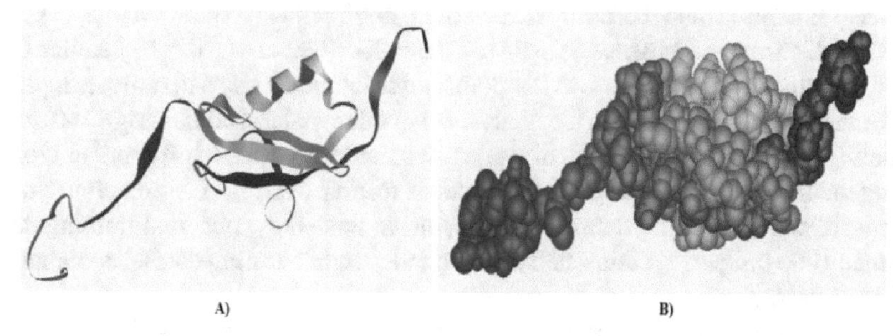

Fig. 1. Schematical structure of a human SUMO protein (Generated by iMol based on PDB ID = 1A5R)

Recent researches have indicated that sumoylation can promote the binding ability of proteins, and that some proteins, such as claspin, rely on sumoylation for their binding function. Many diseases and disorders, such as Alzheimer's and Parkinson's diseases, have been found to be closely related with sumoylation [4–6]. Therefore, identification of sumoylation sites in proteins is important not only for an in-depth understanding of many important biological processes but also for developing effective drugs.

Although purification strategies have greatly contributed to the study of SUMOylation, they have disadvantages such as high laboriousness and cost. In contrast, computational approaches for predicting SUMOylation sites have attracted considerable attention because of their convenience and efficiency. In recent years, the amount of interests in the prediction of protein based on computational approaches has been increasing rapidly, such as: SUMOsp [7], pSumo-CD [8], SUMOgo [9], SumSec [10], HseSUMO [11], SUMO-Forest [12], C-iSUMO [13], and iSUMOK-PseAAC [14], GPS-SUMO [15], SUMOhydro [16]. These approaches can be clustered into two groups: structure-based, which includes three classifiers, and sequence-based, which covers the remainder. In the structured-based group, HseSUMO relies on four half-sphere exposure-based features, SumSec combines bootstrap resampling, C4.5 decision tree, and majority voting with the two types of structural features predicted by SPIDER2, and C-ISUMO makes use of the AdaBoost algorithm combined with the sine and cosine of backbone torsion angles and accessible surface area predicted by SPIDER2. However, considering the importance of the topic and the urgent need for more powerful computational tools in this area, further efforts aiming at predicting protein sumoylation sites are definitely needed.

Continue with previous works [17–24], we are motivated to propose a novel schema for predicting protein SUMOylation sites based on the incorporation of natural language features (Word2Vec) and sequence-based features. Various features, including individual features and hybrid features, have been investiagated for the construction of predictive model serving for predicting protein SUMOylation sites.

2 Methodology

2.1 Data Preparation and Pre-processing

Experimentally verified SUMOylation sites are collected from various different open resources and published literatures, including dbPTM3.0 [25], SUMOsp [7], GPS-SUMO [26], pSumo-CD [8], seeSUMO [27], and SUMOhydro [16]. The details of these datasets are displayed in Table 1. After doing some technical steps to remove duplicated or redundant proteins, we obtain the final non-redundant dataset containing 1160 uniques proteins. To prepare for independent testing, we randomly select 160 proteins from the non-redundant dataset to serve as independent testing dataset. The remaining data is then considered as training dataset. As a result, in this study, our final training dataset contained 1000 uniques proteins, and the final independent testing dataset contained 160 uniques proteins.

In this work, we focus on the sequence-based characterization of SUMOylation sites with substrate specificities. Therefore, to generate the positive data (SUMO-data), we use a window length of 2n + 1 to extract sequence fragments that center at the experimentally verified SUMOylated lysine (K) residue, as well as containing n upstream and n downstream flanking amino acids. The Given a number of experimentally verified SUMOylated proteins, the sequence fragments containing window length of 2n + 1 amino acids and centering at lysine residue without the annotation of SUMOylation were regarded as the negative training data (non-SUMO data). Based on previous studies [18–23, 28] and our preliminary evaluation of various window sizes, the window size of 13 (n = 6) is found to be optimal in the identification of SUMOylation sites. Therefore, the training dataset consisted of 1820 positive training sequences and 37222 negative training sequences. However, as some negative data (non-SUMO data) may be identical to positive data (SUMO-data) the predictive model's performance may be overestimated in both the training and testing datasets.. To prevent this, we applied the CD-HIT program [29] to remove homologous data. After filtered out sequences with 40% sequence identity, the training dataset contained 745 positive training sequences and 2458 negative training sequences.

In this study, an independent testing dataset is randomly selected from the non-redundant dataset, which contained 160 proteins. The positive and negative independent testing dataset are constructed using the same approach as applied to the training dataset. The program cd-hit-2d, which applies a sequence identity cut-off at high similarity, is used again to remove the data redundancy between the independent testing dataset and the training dataset. As a result, the independent testing dataset contains 117 positive

Table 1. Data statistics of experimentally verified SUMOylation sites

Resources	SUMOylated proteins	Sumosites
dbPTM 3.0	434	1029
SUMOsp	144	239
seeSUMO	247	377
GPS-SUMO	545	983
JASSA	505	877
pSUMO-CD	510	755
SUMOhydro	238	394
Total:	**2623**	**4654**
Combined NR data	**1160**	**2109**
Training dataset	**1000**	**1820**
Testing dataset	**160**	**289**

and 1260 negative data. As negative data is often much larger than the positive dataset in both the training and testing sets, data imbalance needs to be avoided. To address this issue, negative data is randomly selected in a 1:2 ratio to the positive dataset to form the training and testing datasets. Finally, the final training and testing datasets in this study are displayed in Table 2.

Table 2. Training dataset and testing dataset to be used in this study

	Positive sites	Negative sites	Total
Training dataset	745	1490	2235
Testing dataset	117	234	351

2.2 Feature Extraction and Encoding

In order to construct the predictive models for the identification of SUMOylation sites, we have investigated not only the four common features (AAINDEX, CKSAAGP: Composition of k-Spaced Amino Acid Group Pairs, BLOSUM62, Word2Vec), but also the hybrid features that are generated by incoprating two or more single features above. The detail of features and their size are displayed in Table 3. For the first three single features (AAINDEX, BLOSUM62 and CKSAAGP), we exploit [30] to generate corresponding binary vectors to be used for the model learning.

Word2Vec Feature: Similar to human language, protein sequences can be naturally represented as strings of letters. The protein alphabet consists of 20 common amino acids.

Furthermore, like natural language, naturally evolved proteins are typically composed of reused modular elements exhibiting slight variations that can be rearranged and assembled in a hierarchical fashion. By this analogy, common protein motifs and domains, which are the basic functional buiding blocks of proteins, are akin to words, phrases and sentences in human language. Word2Vec understands and vectorizes the meaning of words in a document based on the hypothesis that words with similar meanings in a given context exhibit close istance [31]. Both the learning algorithms exhibit input, projection, and output layers although their output derivation processes are different. The input layer receives $W_n = (W_{(t-2)}, W_{(t-1)}, \ldots, W_{(t+1)}, W_{(t+2)})$ as arguments, where W_n denotes words. The projection layer corresponds to an array of multidimensional vectors and stores the sum of several vectors. The output layer corresponds to the layer that outputs the results of the vectors from the projection layer. The basic principle of CBOW involves predicting when a certain word appears via analyzing neighboring words. The projection layer of CBOW projects all words at the same position, and thus, the vectors of all words maintain an average and share the positions of all words. The structure of CBOW exhibits the advantage of uniformly organizing the information distributed in the dataset. Conversely, the Skip-gram exhibits a structure for predicting vectors of other words from one word. The basic principle of Skip-gram involves predicting other words that appear around a certain word. The projection layer of the Skip-gram predicts neighboring words around the word inserted into the input layer. The structure of the Skip-gram exhibits the advantage of vectorizing when new words appear. Based on the study by Mikolov, CBOW is faster and better suited when compared to Skip-gram when the data size is large, and Skipgram exhibits a better performance when compared to CBOW while learning new words. However, other studies that compare the performance of CBOW and Skip-gram state that the performance of Skip-gram exceeds that of CBOW [32, 33].

Table 3. Feature vectors and their size used in the study

Feature	Size
Single features:	
AAINDEX	6903
CKSAAGP	150
BLOSUM62	260
Word2Vec (Skip-gram)	100
Hybrid features:	
AAINDEX_CKSAAGP	7053
AAINDEX_BLOSUM62	7163

(continued)

Table 3. (*continued*)

Feature	Size
AAINDEX_Word2Vec	7003
CKSAAGP_BLOSUM62	410
CKSAAGP_Word2Vec	250
BLOSUM62_Word2Vec	360
AAINDEX_CKSAAGP_BLOSUM62	7313
AAINDEX_CKSAAGP_Word2Vec	7153
CKSAAGP_BLOSUM62_Word2Vec	510
AAINDEX+CKSAAGP+BLOSUM62+Word2Vec	7413

2.3 Model Construction, Learning and Evaluation

Machine learning (ML) is a multidisciplinary field that covers computer science, probability theory, statistics, approximation theory, and complex algorithms. Its theories and methods have been widely used to solve complex problems in engineering applications. ML uses data to "train" and learns how to complete tasks from the data through various algorithms. These algorithms attempt to mine hidden information from a large amount of historical data and use them for regression or classification. ML algorithms include Support Vector Machine (SVM), Naive Bayes (NB), K- nearest neighbor (KNN), Decision tree (DT), Random forest (RF), and Adaptive network fuzzy inference system (ANFIS), Gradient Boosting, eXtreme Gradient Boosting (XGBoost). As the most important branch of ML, deep learning (DL) has developed rapidly in recent years and has gradually become a research hotspot in ML.

To construct the predictive model for the identification of protein SUMOylation, we have investigated various machine learning algorithms and features. An initial survey on the performance of models generated by these algorithms has been conducted to determine the optimal algorithms. In this research, the three well-known machine algorithms, including the Random forest (RF), eXtreme gradient boosting (XGBoost) and Suport Vector Machine (SVM), are appeared to be optimal, so they are used to generate for our final model serving for the prediction of protein SUMOylation sites (Fig. 2).

In order to evaluate the performance of the predictive models, the 5-fold cross-validation approach has been performed to assess the classifying power of the ML models. The following measurements are commonly used to evaluate the performance of the constructed models: Sensitivity (SEN), Specificity (SPE), Accuracy (ACC), Matthew's Correlation Coefficient (MCC), Recall, and F1-score.

$$\text{Sensitivity}(SEN) = \frac{TP}{TP + FN};\ \text{Specificity}(SPE) = \frac{TN}{TN + FP};\ \text{Accuracy}(ACC) = \frac{TP}{TP + FN}$$

$$\text{Matthew corelation coeficient }(MCC) = \frac{(TP \times TN) - (FN \times FP)}{\sqrt{(TP + FN) \times (TN + FP)(TP + FP)(TN + FN)}}$$

$$Recall = \frac{TP}{TP + FN}$$

$$Precision = \frac{TP}{TP + FP}$$

$$F1 - Score = 2\frac{Precision * Recall}{Precision + Recall}$$

Here, TP, TN, FP and FN represent the numbers of true positives, true negatives, false positives and false negatives, respectively.

After running 5-fold cross-validation process, the constructed model with the highest values of MCC and accuracy were selected as the optimal model for identifying potential phage virion protein. Moreover, an independent testing approach was carried out to evaluate the ability of selected model in the real case.

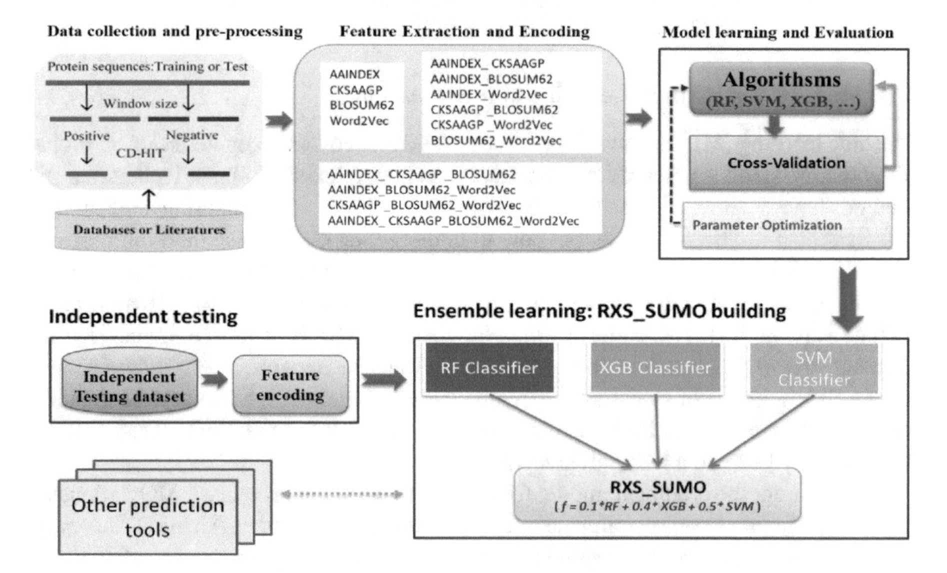

Fig. 2. The flowchart in our study.

3 Results and Discussion

3.1 Impact of Amino Acid Composition and Single Features

To examine the position-specific amino acid compostion for SUMOylation sites, WebLogo [34] is applied to generate the graphical sequence logo for the relative frequency of the corresponding amino acid at postions surrounding SUMOylation sites. Using WebLogo, the flanking sequences of substrate sites (at position 0) could be graphically visualized in the entropy plots of the sequence logo. Through the identified motif, we can easily observe the conservation of the amino acids around the SUMOylation sites. The identified motif is subsequently evaluated for their ability to distinguish SUMOylation from non-SUMOylation using five-fold cross-validation.

The investigation of differences between the amino acids composition surrounding SUMOylation and those of non-SUMOylation shows that the overall trends are similar with slight variations. As shown in Fig. 3 (a), prominent amino acid residues include Glutamate (E), Valine (V), Isoleucine (I) and Leucine (L); whereas Tryptophan (W), Tyrosine (Y), and Phenylalanine (F) are three of the least significant amino acid residues. The sequence logo displays the most enriched residues surrounding the SUMOylation (Lysine K). As shown in Fig. 3(b), it also visualizes that the most conserved amino acid residues including Glutamate (E), Valine (V), Isoleucine (I), Leucine (L), Lysine (K) and Serine (S). In addition, TwoSampleLogo [35] is used to visualize the difference between SUMO-sites and non-SUMO sites. As displayed in Fig. 3(c), The enriched residues appears to be Glutamate (E), Valine (V), Leucine (L) and Isoleucine (I); while the depleted amino acid residues includes Valine (V), Leucine (L), and Tyrosine (Y).

In this research, the four well-known single features (AAINDEX, CKSAAGP, BLOSUM62 and Word2Vec) have been investigated to learn the predictive models for the

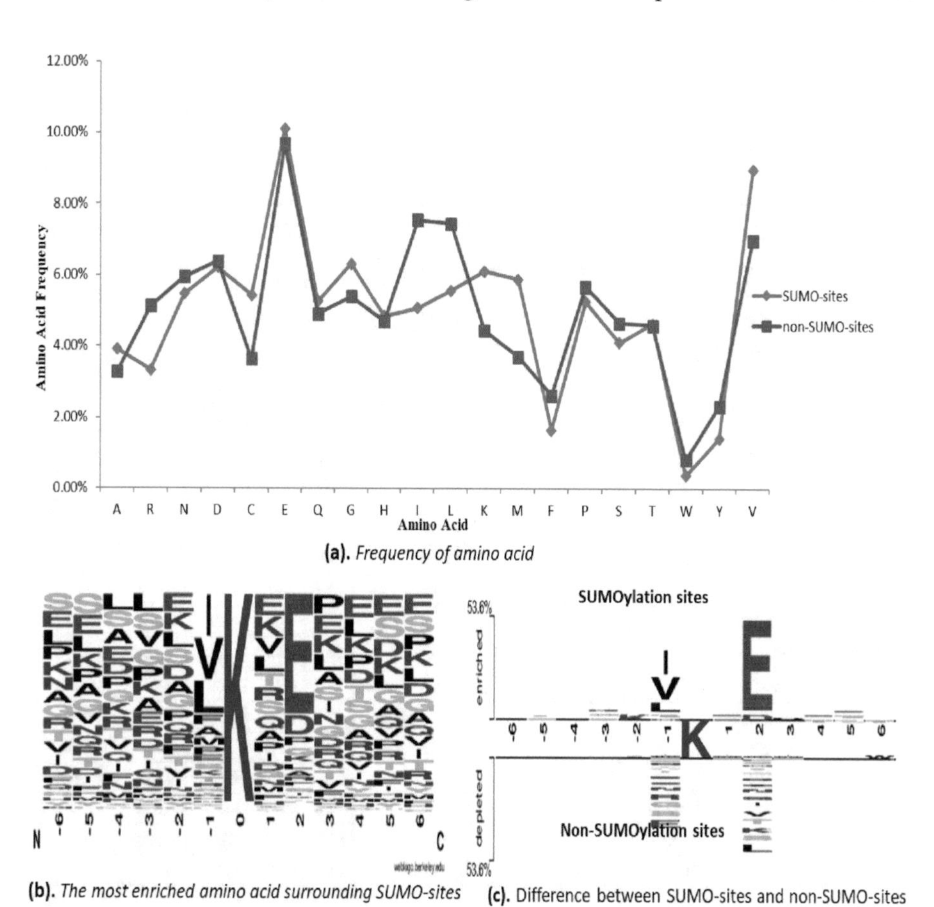

(a). *Frequency of amino acid*

(b). *The most enriched amino acid surrounding SUMO-sites* **(c).** Difference between SUMO-sites and non-SUMO-sites

Fig. 3. Frequency of the amino acid composition surrounding the SUMOylation sites.

prediction of protein SUMOylation sites. The performance of the models generated based on these features is displayed in detail in Tables 4 and 5.

3.2 Impact of Hybrid Features

It is straightforward and very beneficial to combine two or more different approaches in machine learning to exploit advantages from them. In this research, hybrid features are built from the incorporation of two or more single features in order to form new features for the investigation. As a consequence, the hybrid features are found to be the most effective in predicting protein SUMOylation sites. The performance of the predictive models when evaluated with hybrid features using five-fold cross-validation and independent testing are shown in Tables 4 and 5. Fortunately, the hybrid feature of *AAINDEX_CKSAAGP_BLOSUM62_Word2Vec* has been demonstrated to generate the best model with all three algorithms (RF, XGBoost and SVM), getting the highest performance.

Table 4. Performance of the predictive models evaluated by using five-fold cross-validation

Algorithm	Feature	ACC (%)	SEN (%)	SPE (%)	MCC	Recall	Precision	F1-score
Random Forest	AAINDEX	81.9	79.8	82.6	0.577	0.798	0.611	0.692
	CKGAAP	78.5	75.7	79.4	0.492	0.757	0.523	0.619
	BLOSUM62	78.5	75.2	79.5	0.493	0.752	0.530	0.622
	Word2Vec	77.6	73.3	78.9	0.470	0.733	0.517	0.606
	AAINDEX_CKSAAGP	82.8	82.1	83.0	0.599	0.821	0.617	0.705
	AAINDEX_BLOSUM62	83.0	82.9	83.0	0.604	0.829	0.617	0.708
	AAINDEX_Word2Vec	82.6	81.4	82.9	0.593	0.814	0.617	0.702
	CKSAAGP_BLOSUM62	81.7	78.6	82.7	0.572	0.786	0.617	0.692
	CKSAAGP_Word2Vec	78.7	76.5	79.4	0.498	0.765	0.523	0.622
	BLOSUM62_Word2Vec	78.3	74.5	79.5	0.487	0.745	0.530	0.620
	AAINDEX_CKSAAGP_BLOSUM62	83.0	84.1	82.6	0.604	0.841	0.604	0.703
	AAINDEX_CKSAAGP_Word2Vec	83.0	83.5	82.8	0.604	0.835	0.611	0.705
	CKSAAGP_BLOSUM62_Word2Vec	83.2	84.9	82.7	0.610	0.849	0.604	0.706
	AAINDEX_CKSAAGP_BLOSUM62_Word2Vec	**83.9**	**85.3**	**83.4**	**0.626**	**0.853**	**0.624**	**0.721**
XGBoost	AAINDEX	80.1	73.8	82.6	0.538	0.738	0.624	0.676
	CKGAAP	75.2	65.6	78.8	0.419	0.656	0.537	0.590
	BLOSUM62	72.5	58.0	80.7	0.395	0.580	0.631	0.605
	Word2Vec	70.0	54.7	78.7	0.339	0.547	0.591	0.568
	AAINDEX_CKSAAGP	81.2	76.4	83.0	0.563	0.764	0.631	0.691
	AAINDEX_BLOSUM62	80.3	74.4	82.6	0.543	0.744	0.624	0.679
	AAINDEX_Word2Vec	80.9	76.2	82.7	0.556	0.762	0.623	0.685
	CKSAAGP_BLOSUM62	79.9	75.2	81.5	0.529	0.752	0.591	0.662

(*continued*)

Table 4. (*continued*)

Algorithm	Feature	ACC (%)	SEN (%)	SPE (%)	MCC	Recall	Precision	F1-score
	CKSAAGP_Word2Vec	77.2	70.8	79.3	0.462	0.708	0.537	0.611
	BLOSUM62_Word2Vec	76.3	66.9	80.0	0.449	0.669	0.570	0.616
	AAINDEX_CKSAAGP_BLOSUM62	80.5	74.6	82.9	0.548	0.746	0.631	0.684
	AAINDEX_CKSAAGP_Word2Vec	81.4	76.6	83.3	0.569	0.766	0.638	0.696
	CKSAAGP_BLOSUM62_Word2Vec	81.0	77.1	82.4	0.556	0.771	0.611	0.682
	AAINDEX_CKSAAGP_BLOSUM62_Word2Vec	**81.9**	**77.4**	**83.6**	**0.579**	**0.774**	**0.644**	**0.703**
SVM	AAINDEX	78.7	73.3	80.7	0.502	0.733	0.570	0.642
	CKGAAP	78.1	72.2	80.1	0.485	0.722	0.557	0.629
	BLOSUM62	78.7	72.5	81.0	0.503	0.725	0.584	0.647
	Word2Vec	79.0	72.7	81.3	0.509	0.727	0.591	0.652
	AAINDEX_CKSAAGP	79.0	73.9	80.7	0.507	0.739	0.570	0.644
	AAINDEX_BLOSUM62	79.2	73.7	81.2	0.513	0.737	0.584	0.652
	AAINDEX_Word2Vec	79.0	72.7	81.3	0.509	0.727	0.591	0.652
	CKSAAGP_BLOSUM62	79.2	73.3	81.3	0.514	0.733	0.591	0.654
	CKSAAGP_Word2Vec	79.2	74.4	81.2	0.518	0.744	0.584	0.654
	BLOSUM62_Word2Vec	79.4	73.9	81.4	0.519	0.739	0.591	0.657
	AAINDEX_CKSAAGP_BLOSUM62	79.6	74.2	81.7	0.525	0.742	0.597	0.662
	AAINDEX_CKSAAGP_Word2Vec	79.6	74.6	81.5	0.524	0.746	0.591	0.659
	CKSAAGP_BLOSUM62_Word2Vec	79.6	74.2	81.7	0.525	0.742	0.597	0.662
	AAINDEX_CKSAAGP_BLOSUM62_Word2Vec	**81.7**	**79.6**	**82.3**	**0.571**	**0.796**	**0.604**	**0.687**

Table 5. Performance of the predictive models evaluated by using Independent testing

Algorithm	Feature	ACC (%)	SEN (%)	SPE (%)	MCC	Recall	Precision	F1-score
Random Forest	AAINDEX	84.6	76.0	89.1	0.657	0.760	0.786	0.773
	CKGAAP	83.5	75.2	87.6	0.628	0.752	0.752	0.752
	BLOSUM62	82.6	72.6	88.1	0.615	0.726	0.769	0.747
	Word2Vec	71.2	58.9	75.5	0.318	0.589	0.453	0.512
	AAINDEX_CKSAAGP	85.8	79.1	89.0	0.678	0.791	0.778	0.784
	AAINDEX_BLOSUM62	85.8	79.6	88.7	0.677	0.796	0.769	0.783
	AAINDEX_Word2Vec	86.3	80.5	89.1	0.690	0.805	0.778	0.791
	CKSAAGP_BLOSUM62	86.3	80.5	89.1	0.690	0.805	0.778	0.791
	CKSAAGP_Word2Vec	86.3	85.6	86.6	0.685	0.856	0.709	0.776
	BLOSUM62_Word2Vec	85.2	79.8	87.6	0.662	0.798	0.744	0.770
	AAINDEX_CKSAAGP_BLOSUM62	86.9	82.0	89.2	0.702	0.820	0.778	0.798
	AAINDEX_CKSAAGP_Word2Vec	87.2	82.1	89.5	0.709	0.821	0.786	0.803

(*continued*)

Table 5. (*continued*)

Algorithm	Feature	ACC (%)	SEN (%)	SPE (%)	MCC	Recall	Precision	F1-score
	CKSAAGP_BLOSUM62_Word2Vec	86.9	81.4	89.5	0.703	0.814	0.786	0.800
	AAINDEX_CKSAAGP_BLOSUM62_Word2Vec	**87.5**	**82.9**	**89.6**	**0.715**	**0.829**	**0.786**	**0.807**
XGBoost	AAINDEX	84.0	75.2	88.7	0.644	0.752	0.778	0.765
	CKGAAP	83.8	73.4	89.7	0.645	0.734	0.803	0.767
	BLOSUM62	84.6	76.0	89.1	0.657	0.760	0.786	0.773
	Word2Vec	72.9	62.2	76.6	0.360	0.622	0.479	0.541
	AAINDEX_CKSAAGP	86.0	79.3	89.4	0.685	0.793	0.786	0.790
	AAINDEX_BLOSUM62	85.2	78.3	88.6	0.665	0.783	0.769	0.776
	AAINDEX_Word2Vec	85.2	77.8	88.9	0.667	0.778	0.778	0.778
	CKSAAGP_BLOSUM62	85.8	78.6	89.3	0.679	0.786	0.786	0.786
	CKSAAGP_Word2Vec	85.2	79.3	87.9	0.663	0.793	0.752	0.772
	BLOSUM62_Word2Vec	86.3	81.1	88.8	0.689	0.811	0.769	0.789
	AAINDEX_CKSAAGP_BLOSUM62	86.7	82.0	89.2	0.702	0.820	0.778	0.798
	AAINDEX_CKSAAGP_Word2Vec	86.6	84.3	87.6	0.692	0.843	0.735	0.785
	CKSAAGP_BLOSUM62_Word2Vec	86.6	80.7	89.5	0.697	0.807	0.786	0.797
	AAINDEX_CKSAAGP_BLOSUM62_Word2Vec	**87.7**	**83.0**	**90.0**	**0.722**	**0.830**	**0.795**	**0.812**
SVM	AAINDEX	84.6	77.9	87.8	0.651	0.779	0.752	0.765
	CKGAAP	84.9	79.6	87.2	0.655	0.796	0.735	0.764
	BLOSUM62	85.5	78.0	89.3	0.674	0.780	0.786	0.783
	Word2Vec	71.8	62.5	74.2	0.314	0.625	0.385	0.476
	AAINDEX_CKSAAGP	85.8	80.2	88.3	0.676	0.802	0.761	0.781
	AAINDEX_BLOSUM62	85.8	79.6	88.7	0.677	0.796	0.769	0.783
	AAINDEX_Word2Vec	84.9	78.1	88.2	0.658	0.781	0.761	0.771
	CKSAAGP_BLOSUM62	85.2	78.3	88.6	0.665	0.783	0.769	0.776
	CKSAAGP_Word2Vec	85.8	81.9	87.4	0.673	0.819	0.735	0.775
	BLOSUM62_Word2Vec	86.3	79.5	89.7	0.692	0.795	0.795	0.795
	AAINDEX_CKSAAGP_BLOSUM62	86.0	80.4	88.7	0.683	0.804	0.769	0.786
	AAINDEX_CKSAAGP_Word2Vec	86.9	83.2	88.5	0.700	0.832	0.761	0.795
	CKSAAGP_BLOSUM62_Word2Vec	87.2	82.7	89.2	0.708	0.827	0.778	0.802
	AAINDEX_CKSAAGP_BLOSUM62_Word2Vec	**87.3**	**82.3**	**89.9**	**0.716**	**0.823**	**0.795**	**0.809**

3.3 RSX_SUMO Model Construction

In an attempt to further improve the performance for the identification of protein SUMOylation sites, we have exploited the idea of ensemble learning to construct the RSX_SUMO model.

In this study, the RSX_SUMO model is formulated by incorporating the models generating by using the three well-known algorithms (including RF, XGBoost and SVM). The detail information of RSX_SUMO model is as below:

+ Feature to be used: the hybrid feature of *AAINDEX_CKSAAGP_BLOSUM62_Word2Vec*.
+ Algorithm: The incorporation of RF, XGBoost and SVM algorithm.

+ Objective function: $f = aX + bY + cZ$ (where X, Y, Z are the prediction results of RF, XGBoost, SVM algorithms, respectively. Variables a, b, c are weight of RF, XGBoost, SVM algorithms, respectively; X, Y, Z, a, b, c in range of [0; 1]; a + b + c = 1).

After some technical steps to optimize variable of weights, we found that the RSX_SUMO model would perform the best and obtain the highest performance with the weights on variables a, b, c at values of 0.1, 0.4, 0.5 respectively. In other word, the optimal objective function of RSX_SUMO model is $f = 0.1X + 0.4Y + 0.5Z$.

As displayed in Tables 6 and 7, the performance of RSX_SUMO model achieves the highest performance in both five-fold cross-validation and independent testing, obtain the performance on independent testing with acccuracy at 88.6% and MCC value of 0.743.

Table 6. Performance of the RSX_SUMO model evaluating by using five-fold cross-validation

Algorithm	ACC (%)	SEN (%)	SPE (%)	MCC	Recall	Precision	F1-score
RF	83.9	85.3	83.4	0.626	0.853	0.624	0.721
XGB	81.9	77.4	83.6	0.579	0.774	0.644	0.703
SVM	81.7	79.6	82.3	0.571	0.796	0.604	0.687
RSX_SUMO	**84.1**	**86.1**	**83.5**	**0.632**	**0.861**	**0.624**	**0.724**

Table 7. Performance of the RSX_SUMO model evaluating by using *Independent testing*

Algorithm	ACC (%)	SEN (%)	SPE (%)	MCC	Recall	Precision	F1-score
RF	87.5	81.7	90.2	0.717	0.803	0.796	0.883
XGB	87.7	83.0	90.0	0.722	0.830	0.795	0.812
SVM	87.3	82.3	89.9	0.716	0.823	0.795	0.809
RSX_SUMO	**88.6**	**83.5**	**91.1**	**0.743**	**0.725**	**0.780**	**0.795**

3.4 Comparison with Other Predictors

As mentioned previously, in order to assess the ability and practicality of the predictive model, it is necessary to compare the performance of our proposed model with similar existing predictors. GPS-SUMO 2.0 is the up-to-date prediction tool of SUMOylation sites, that is developed based on GPS-SUMO, SUMOsp and SUMOsp 2.0. Therefore, in this study, we select 2 similar existing predictors (GPS-SUMO 2.0 and seeSUMO 2.0) to make comparison with our proposed model, using same independent testing. Fortunately, as displayed in Table 8, our proposed model outperforms than SUMOsp2.0 and seeSUMO2.0, achieves highest accurracy of 88.6% and highest value of MCC at 0.743. The result demonstrates the ability of our proposed model in the prediction of protein SUMOylation sites.

Table 8. Performance of the proposed model comparison with other previous predictors

	Threshold	TP	FP	TN	FN	ACC(%)	SEN(%)	SPE(%)	MCC
GPS-SUMO 2.0	Low	84	33	231	11	87.74	88.42	87.50	0.715
	Medium	77	40	208	34	79.39	69.37	83.87	0.525
	High	84	33	231	11	87.74	88.42	87.50	0.715
seeSUMO 2.0	Low	82	35	225	17	85.52	82.83	86.54	0.661
	Medium	76	41	200	42	76.88	64.41	82.99	0.475
	High	79	38	221	21	83.57	79.00	85.33	0.615
RXS_SUMO		96	21	215	19	88.60	83.48	0.911	0.743

4 Conclusion

Protein SUMOylation is a post-translational modification that plays critical roles in many cellular processes, including DNA replication, signaling, and trafficking. This modification is found in all eukaryotic cells. Inhibition of SUMOylation has been extensively investigated to suppress the activity of oncogenic Ras protein to achieve antitumor activity. SUMOylation inhibitors are being investigated as potential therapeutics against several diseases, including cancers, progeria, aging, parasitic diseases, bacterial and viral infections. In this study, we propose a novel schema for predicting protein SUMOylation sites based on the incorporation of natural language features (Word2Vec) and sequence-based features. The experimental result shows that the hybrid feature of *AAINDEX_CKSAAGP_BLOSUM62_Word2Vec*, which combines natural language features and sequence-based features, is optimal, leading to the generation of the predictive model with the highest performance. Besides, the novel model, called RSX_SUMO, has been proposed for the prediction of protein SUMOylation sites. We hope that the idea that incorporates natural language features and sequence-based features for the prediction of protein SUMOylation sites could provide an effective suggestion for researchers related to the problem of SUMOylation characterization for their research.

Acknowledgment. The authors sincerely thank to TUEBA for partly financial supported this research under the TNU-level project ID: ĐH2023-TN08-05.

References

1. Geiss-Friedlander, R., Melchior, F.: Concepts in sumoylation: a decade on. Nat. Rev. Mol. Cell Biol. **8**(12), 947–956 (2007)
2. Hay, R.T.: SUMO: a history of modification. Mol. Cell **18**(1), 1–12 (2005)
3. Müller, S., et al.: SUMO, ubiquitin's mysterious cousin. Nat. Rev. Mol. Cell Biol. **2**(3), 202–210 (2001)
4. Marmor-Kollet, H.S., et al.: Spatiotemporal proteomic analysis of stress granule disassembly using APEX reveals regulation by SUMOylation and links to ALS pathogenesis. Mol. Cell. **80**, 15 (2020)

5. Princz, A.T.: N. SUMOylation in neurodegenerative diseases. Gerontology **66**, 8 (2020)
6. Seeler, J.S.B., Nacerddine, K.O., Dejean, A.: SUMO, the three Rs and cancer. Curr. Top. Microbiol. Immunol. **313**, 22 (2007)
7. Ren, J., et al.: Systematic study of protein sumoylation: development of a site-specific predictor of SUMOsp 2.0. Proteomics **9**(12), 3409–3412 (2009)
8. Jia, J., et al.: pSumo-CD: predicting sumoylation sites in proteins with covariance discriminant algorithm by incorporating sequence-coupled effects into general PseAAC. Bioinformatics **32**(20), 3133–3141 (2016)
9. Chang, C.C., et al.: SUMOgo: prediction of sumoylation sites on lysines by motif screening models and the effects of various post-translational modifications. Sci. Rep. **8**(1), 15512 (2018)
10. Dehzangi, A., et al.: SumSec: accurate prediction of sumoylation sites using predicted secondary structure. Molecules **23**(12) (2018)
11. Sharma, A., et al.: HseSUMO: sumoylation site prediction using half-sphere exposures of amino acids residues. BMC Gen. **19**(Suppl. 9), 982 (2019)
12. Qian, Y., et al.: SUMO-Forest: a cascade forest based method for the prediction of SUMOylation sites on imbalanced data. Gene **741**, 144536 (2020)
13. Lopez, Y., Dehzangi, A., Reddy, H.M., Sharma, A.: C-iSUMO: a sumoylation site predictor that incorporates intrinsic characteristics of amino acid sequences. Comput. Biol. Chem. **87** (2020)
14. Khan, Y.D., et al.: iSUMOK-PseAAC: prediction of lysine sumoylation sites using statistical moments and Chou's PseAAC. PeerJ **9**, e11581 (2021)
15. Zhao, Q., et al.: GPS-SUMO: a tool for the prediction of sumoylation sites and SUMO-interaction motifs. Nucleic Acids Res.. **42**(Web Server issue), W325–W330 (2014)
16. Chen, Y.Z., Chen, Z., Gong, Y.A., Ying, G.: SUMOhydro: a novel method for the prediction of sumoylation sites based on hydrophobic properties. PloS One **7**(6), e39195 (2012)
17. Nguyen, V.-N., Nguyen, H.-M., Tran, T.-X.: An approach by exploiting support vector machine to characterize and identify protein SUMOylation sites. JASSA. **505**, 877 (2012)
18. Nguyen, V.-N., et al.: Characterization and identification of ubiquitin conjugation sites with E3 ligase recognition specificities. BMC Bioinform. BioMed Central (2015)
19. Nguyen, V.-N., et al.: A new scheme to characterize and identify protein ubiquitination sites. IEEE/ACM Trans. Comput. Biol. Bioinform. **14**(2), 393–403 (2016)
20. Bui, V.-M., Nguyen, V.-N.: The prediction of Succinylation site in protein by analyzing amino acid composition. In: Akagi, M., Nguyen, T.T., Vu, D.T., Phung, T.N., Huynh, V.N. (eds.) Advances in Information and Communication Technology. ICTA 2016. Advances in Intelligent Systems and Computing, vol. 538, pp. 633–642. Springer, Cham (2017). https://doi.org/10.1007/978-3-319-49073-1_67
21. Le, N.Q.K., Ho, Q.T., Ou, Y.Y.: Incorporating deep learning with convolutional neural networks and position specific scoring matrices for identifying electron transport proteins. J. Comput. Chem. **38**(23), 2000–2006 (2017)
22. *Nguyen, V.-N., et al. A new schema to identify S-farnesyl cysteine prenylation sites with substrate motifs. in Advances in Information and Communication Technology: Proceedings of the International Conference, ICTA 2016. 2017. Springer*
23. Le, N.Q.K., et al.: Identification of Clathrin proteins by incorporating hyperparameter optimization in deep learning and PSSM profiles. Comput. Methods Programs Biomed. **177**, 81–88 (2019)
24. Nguyen, V.-N., et al.: Exploiting two-layer support vector machine to predict protein sumoylation sites. In: Fujita, H., Nguyen, D., Vu, N., Banh, T., Puta, H. (eds.) Advances in Engineering Research and Application. ICERA 2018. Lecture Notes in Networks and Systems, vol. 63, pp. 324–332. Springer, Cham. https://doi.org/10.1007/978-3-030-04792-4_43

25. Lu, C.T., et al.: DbPTM 3.0: an informative resource for investigating substrate site specificity and functional association of protein post-translational modifications. Nucleic Acids Res. **41**(Database issue), D295–305 (2013)
26. Beauclair, G., et al.: JASSA: a comprehensive tool for prediction of SUMOylation sites and SIMs. Bioinformatics **31**(21), 3483–3491 (2015)
27. Teng, S., Luo, H., Wang, L.: Predicting protein sumoylation sites from sequence features. Amino Acids **43**, 447–455 (2012)
28. Ho Thanh Lam, L., et al.: Machine learning model for identifying antioxidant proteins using features calculated from primary sequences. Biology **9**(10), 325 (2020)
29. Huang, Y., et al.: CD-HIT Suite: a web server for clustering and comparing biological sequences. Bioinformatics **26**(5), 680–682 (2010)
30. Chen, Z., et al.: iFeature: a Python package and web server for features extraction and selection from protein and peptide sequences. Bioinformatics **34**(14), 2499–2502 (2018)
31. Sahlgren, M.: The distributional hypothesis. Ital. J. Disabil. Stud. **20**, 33–53 (2008)
32. Chiu, B., et al.: How to train good word embeddings for biomedical NLP. In: Proceedings of the 15th Workshop on Biomedical Natural Language Processing (2016)
33. Lai, S., et al.: How to generate a good word embedding. IEEE Intell. Syst. **31**(6), 5–14 (2016)
34. Crooks, G.E., et al.: WebLogo: a sequence logo generator. Genome Res. **14**(6), 1188–1190 (2004)
35. Vacic, V., Iakoucheva, L.M., Radivojac, P.: Two Sample Logo: a graphical representation of the differences between two sets of sequence alignments. Bioinformatics **22**(12), 1536–1537 (2006)

A Model for Alliance Partner Selection Based on GM (1, 1) and DEA Frameworks - Case of Vietnamese Coffee Industry

Ngoc-Thang Nguyen[1], Van-Thanh Phan[2](✉), Thi Ai Nhi Duong[1], Thanh-Ha Le[1], Thao-Vy Pham[1], Nghiem Hong Ngoc Bich Pham[1], and Thanh-Giang Kieu[3]

[1] Tay Nguyen University, 567 Le Duan Street, Buon Ma Thuot, Đăk Lăk, Vietnam
{nnthang,dtanhi,ltha,ptvy,pnhnbich}@ttn.edu.vn

[2] Quang Binh University, 312 Ly Thuong Kiet Street, Đong Hoi, Quang Binh, Vietnam
thanhkem2710@gmail.com

[3] Buon Don Ethnic Minority Boarding High School, Buon Ma Thuot, Đăk Lăk, Vietnam

Abstract. Due to the fierce market competition, the alliance between companies aiming to increase operational efficiency and competitiveness has become the inevitable trend. However, not all of the alliance strategies are 100% successful in reality. The most important question for the company leader is: How to find out the best partner when an alliance? In order to deal with this problem, this study proposes a systematic approach to find out the best partner in the alliance process based on the Grey prediction models and Data Envelopment Analysis (DEA). To illustrate above approach, 07 coffee trading companies with full data in Đăk Lăk province, Vietnam are used as Decision-Making Units (DMUs). The empirical results show that 721 Coffee one member company (DMU_7) has become the best partner for Thang Loi Coffee Joint Stock Company in the alliance. In the future direction, this proposed approach will be extended and applied in this or many fields by considering lots of different factors or using different methodologies to deal with practical scenarios.

Keywords: Grey prediction · Data envelopment analysis · Strategic alliances · Coffee trading companies

1 Introduction

Nowadays, there is a large number of coffee trading companies in the Vietnamese market. Besides the big domestic coffee companies like Trung Nguyen, Vinacafe, many foreign companies are investing or doing business in this field. Among them must be mentioned NesCafé. In addition, there are a lot of new local companies that have been established beside the existing companies. That reason why it increases the competitive pressure among companies in the coffee trading market. Moreover, the regulatory agencies failed to take appropriate measures and lead to a serious oversupply in the market, serious vicious competition, a decline in business efficiency, operating loss and the malaise of the entire industry.

N. T. Nguyen et al. (Eds.): CITA 2023, LNNS 734, pp. 89–101, 2023.
https://doi.org/10.1007/978-3-031-36886-8_8

In 2020, the Covid-19 outbreak disrupted the coffee supply chain, decreased consumer demand for coffee in several regions, and significantly impacted coffee business' earnings [1]. Face the above difficulties, there are many solutions to increase operational efficiency and enhance competitiveness. Some enterprises have enough financial resources try to change their whole business model or manufacturing methods by using contemporary technologies or investing in machinery and equipment. For some of the insufficient financial resources companies, they try to change a part of the business model or find a partner for financial support. Some companies are bankrupt.

To help the companies exist in the market and increase operational efficiency as well as competitiveness, Recently, Strategic alliance (SA) is one of the efficient solutions of enterprises used by managers because It helps the company easy of market entry, shared risks, shared knowledge and expertise and synergy and competitive advantage [2]. However, It's not easy to find out a suitable partner in alliance strategies in reality. This issue has become an important task for the leadership of the company. In order to deal with this problem, this study proposes a systematic approach for helping managers find out the best partner in the alliance process based on the Grey prediction models and Data Envelopment Analysis (DEA). To illustrate above approach, 07 coffee trading companies with complete data in Đăk Lăk province, Vietnam was used. Input and output variables were selected based on the Pearson correlation in assumption of DEA. Grey forecasting model is used to forecast the company's performance in 2023. Then, the super-SBM I-V model was used to evaluate the past, current and future performance of these DMUs. For strategic alliance, the Thang Loi Coffee Joint Stock Company (DMU_3) was selected as the target company to illustrate the formation of the right partner. The result showed that the 721 Coffee one member company (DMU_7) is the best partner for the Thang Loi Coffee Joint Stock Company.

The remainder of the paper is organized as follows. Section 2 provides the literature review and details of methodologies of grey model and super-SBM I-V model. Section 3 proposed a systematic approach. The research results are shown in Sect. 4. The last section presents the conclusion and suggestions for future research.

2 Related Works

In recent years, many researchers have combined the grey model with DEA to evaluate the performance of alliance strategy. For instant, Wang and Wang [13] used the DEA model to evaluate the performance of Merger & Acquisition in High-technology in Taiwan. Wang and Phan based on the grey forecasting method (GM) and data envelopment analysis (DEA) proposed an effective model to help top managers finding out the best partner for strategic alliances in banking [14]. The results indicate that an effective model can help enterprises to find the best partners of cooperation. But the previous study just used the simple model of grey forecasting model and DEA model. In this study, the GM (1, 1) and supper-SBM–I-V as a foundation to propose a systematic approach to identify the best partners in alliance strategy in order to improve the efficiency of the organization.

2.1 Strategic Alliance

There are several definitions about strategic alliance. Dussauge and Garrette [3] defined that "an alliance is a cooperative agreement or association between two or more independent enterprises, which will manage one specific project, with a determined duration, for which they will be together to improve their competencies. It is constituted to allow its partners to pool resources and coordinate efforts in order to achieve results that neither could obtain by acting alone. The key parameters surrounding alliances are opportunism, necessity and speed". At the same time, Faulkner [4] also defined that "A strategic alliance is a particular mode of inter-organizational relationship in which the partners make substantial investments in developing a long-term collaborative effort, and common orientation". Another shorter definition is "Strategic alliances are voluntary arrangements between firms involving exchange, sharing, or co-development of products, technologies, or services" [5].

In this research, the strategic alliance is defined as cooperation between two or more independent enterprises aiming to increase the efficient operation, enhance the competitiveness, ease of market entry, shared risks, shared knowledge and expertise.

2.2 Grey Forecasting Model

Grey system theory was developed by Prof. Deng in early 1982 [6], recently many scientists and scholars using this method to deal with uncertainty problems under discrete data and incomplete information in the real case. In this paper also presented the advantages of grey theory compared with statistical models under two aspects. The first one is grey theory requires a small sample of data to estimate the behavior and the second is dealing with problems under incomplete information. In addition, this model is easy for the user use to establish modeling and requires a small sequence data. Therefore, the spread and influence of it has rapidly expanded over time and making it an effective tool for forecasting in a variety of fields, including finance and economics [7], the energy industry [8] and the coffee industry [9]. One of models in grey forecasting series is GM (1, 1) model. Its was constructed based on the first - order differential equation. The detailed procedure of GM (1, 1) was show as follows:

Step 1: Assume that $x^{(0)}$ denotes a non-negative sequence of raw data as

$$x^{(0)} = (x^{(0)}(1), x^{(0)}(2), ...x^{(0)}(n)), n \geq 4 \tag{1}$$

where n is length of sequence data

Step 2: Set up a new sequence $X^{(1)}$ by one- time AGO aim to smooth the randomness of primitive sequence as:

$$X^{(1)} = (x^{(1)}(1), x^{(1)}(2), ...x^{(1)}(n)), n \geq 4 \tag{2}$$

where

$$x^{(1)}(1) = x^{(0)}(1) \text{ and } x^{(1)}(k) = \sum_{i=1}^{k} x^{(0)}(i), k = 2, 3, ..., n. \tag{3}$$

To calculate the mean generating sequence $z^{(1)}$ of $x^{(1)}$ we use function bellows

$$z^{(1)}(k) = 0.5x^{(1)}((k) + (k-1)), k = 2, 3, ..., n. \tag{4}$$

And we get the mean sequence is

$$z^{(1)} = (z^{(1)}(1), (z^{(1)}(2), ..., (z^{(1)}(n)) \tag{5}$$

Step 3: Establish the GM (1, 1) model by first order differential equation for $x^{(1)}(k)$ as:

$$\frac{dx^{(1)}(k)}{d(k)} + ax^{(1)}(k) = b \tag{6}$$

The solution, also known as time response function, of above equation is given by:

$$\hat{x}^{(1)}(k) = \left[x^{(0)}(1) - \frac{b}{a}\right]e^{-a(k-1)} + \frac{b}{a} \tag{7}$$

where $x^{(1)}(k)$ denotes the prediction x at time point k.

To calculate the coefficients $[a, b]^T$ we use the Ordinary Least Squares (OLS) method:

$$[a, b]^T = (B^T B)^{-1} B^T Y \tag{8}$$

In that

$$Y = \begin{bmatrix} x^{(0)}(2) \\ x^{(0)}(3) \\ \dots\dots \\ \dots\dots \\ x^{(0)}(n) \end{bmatrix} \text{ and } B = \begin{bmatrix} -z^{(1)}(2) & 1 \\ -z^{(1)}(3) & 1 \\ \dots\dots & . \\ \dots\dots & . \\ -z^{(1)}(n) & 1 \end{bmatrix}$$

where: Y is called data series, B is called data matrix, and $[a, b]^T$ is called parameter series.

Step 4: we use Inverse AGO (IAGO) to find predicted values of primitive sequence:

$$\hat{x}^{(0)}(k) = \left[x^{(0)}(1) - \frac{b}{a}\right]e^{-a(k-1)}(1 - e^a) \tag{9}$$

Therefore, the fitted and predicted sequence is given $\hat{x}^{(0)}$ as:

Step 5: $\hat{x}^{(0)} = (\hat{x}^{(0)}(1), \hat{x}^{(0)}(2), ...\hat{x}^{(0)}(n)$ where:

$$\hat{x}^{(0)}(1) = x^{(0)}(1) \tag{10}$$

$\hat{x}^{(0)}(1), \hat{x}^{(0)}(2), ...\hat{x}^{(0)}(n)$ are called the GM (1, 1) fitted sequence while $\hat{x}^{(0)}(n+1), \hat{x}^{(0)}(n+2), ...$ are called the GM (1, 1) forecast values.

2.3 Data Envelopment Analysis

Related to data envelopment analysis (DEA), it is a non-parametric method, was developed by Charnes, Cooper and Rhodes (1978) based on the original idea of Farrell (1957) to measure the efficiency score of decision making units [10]. Because of the non-parametric approaches have the benefit of not assuming a particular functional form/shape for the frontier; however they do not provide a general relationship (equation) relating output and input. Therefore, DEA has become a widely used analytical tool for measuring and evaluating performance of organizations in various areas [11]. Nowadays, in the DEA have more than 181 models, but each model focuses on the difference aspect. In this study, a DEA model "Slacks-based measure of super-efficiency" (super SBM) was used. This model was developed on "Slacks-based measure of efficiency" (SBM) introduced by Tone (2001) [12]. In this model with n DMUs with the input and output matrices $X = \left(x_{ij}\right) \in R^{m\times n}$ and $Y = \left(Y_{ij}\right) \in R^{s\times n}$, respectively. λ is a non-negative vector in R^n. The vectors $S^- \in R^m$ and $S^+ \in R^s$ indicate the input excess and output shortfall respectively. SBM model in fractional form is as follows (Tone 2001):

$$\min \rho = \frac{1-\frac{1}{\mathrm{m}}\sum_{i=1}^{m} s_i^-/x_{i0}}{1+\frac{1}{s}\sum_{i=1}^{s} s_i^-/y_{i0}} \tag{11}$$

Subject to $x_0 = X\lambda + s^-$,

$$y_0 = Y\lambda - s^+,$$

$$\lambda \geq 0,\ \mathrm{s}^- \geq 0,\ \mathrm{s}^+ \geq 0.$$

Let an optimal solution for SBM be $(p^*, \lambda^*, s^{-*}, s^{+*})$. A DMU (x_0, y_0) is SBM-efficient, if $p^* = 1$. This condition is equivalent to $S^{-*} = 0$ and $S^{+*} = 0$, no input excesses and no output shortfalls in any optimal solution. SBM is non-radial and deals with input/output slacks directly. The SBM returns and efficiency measure between 0 and 1.

The best performers have the full efficient status denoted by unity. In Tones research also discriminated these efficient DMUs and ranked the efficient DMUs by super SBM model. Assuming that the DMU (x_0, y_0) is SBM-efficient, $p^* = 1$, super SBM model is as follows:

$$\min \delta = \frac{\frac{1}{\mathrm{m}}\sum_{i=1}^{m} \overline{x}_i/x_{i0}}{\frac{1}{s}\sum_{r=1}^{s} \overline{y_r}/y_{r0}} \tag{12}$$

Subject to $\overline{x} \geq \sum\limits_{j=1,\neq 0}^{n} \lambda_j x_j$,

$$\overline{y} \leq \sum_{j=1,\neq 0}^{n} \lambda_j x_j,$$

$$\overline{y} \geq x_0 \text{ and } \overline{y} \leq y_0,$$

$$\overline{y}\,\overline{y} \geq y_0,\ \lambda \geq 0.$$

As in many DEA models, it is crucial to consider how to deal with negative outputs in the evaluation of efficiency in SBM models too. However, negative data should have their duly role in measuring efficiency, hence a new scheme was introduced in DEA-Solver pro 4.1 Manuel, and the scheme was changed as follows:

Let us suppose $y_{ro} \leq 0$. it is defined $\overline{y}_r^+$ and y_{-r}^+ by

$$\overline{y}_r^+ = \max_{j=1,\ldots,n} \{y_{rj} | y_{rj} > 0\}, \tag{13}$$

$$\overline{y}_r^+ = \min_{j=1,\ldots,n} \{y_{rj} | y_{rj} > 0\}. \tag{14}$$

If the output r has no positive elements, then it is defined as $\overline{y}_r^+ = y_{-r}^+ = 1$. The term is replaced s_r^+/y_{r0} in the objective function in the following way. The value y_{r0} is never changed in the constraints.

(1) $\overline{y}_r^+ = y_{-r}^+ = 1$, the term is replaced by

$$s_r^+ / \frac{y_{-r}^+(\overline{y}_r^+ - y_{-r}^+)}{\overline{y}_r^+ - y_{r0}} \tag{15}$$

$$s_r^+ / \frac{(y_{-r}^+)^2}{B(\overline{y}_r^+ - y_{r0})}, \tag{16}$$

where B is a large positive number, (in DEA-Solver $B = 100$).

In any case, the denominator is positive and strictly less than y_{-r}^+. Furthermore, it is inverse proportion to the distance $\overline{y}_r^+ - y_{r0}$. This scheme, therefore, concerns the magnitude of the non-positive output positively. The score obtained is units invariant, i.e., it is independent of the units of measurement used.

3 Proposal Research

Awarded of the power and usability of GM (1, 1) and DEA in the real case, this paper proposes a systematic approach for helping manager find out the best alliance partner. The process of alliance partner selection was carried out through the following steps.

Step 1. Collect the DMU

- Searching coffee business enterprises in Đăk Lăk province
- Collect history data on candidates' companies.

Step 2. Inputs/Output selection

- DEA assumptions checking
- Pearson coefficient checking

Step 3. Grey prediction

- Based on the data during 2020–2023 to forecast input and output variables
- MAPE indicator to test the forecasting performance

Step 4: Analysis the performance of all DMUs before alliance by Super SBM-I-V.
Use the super - SBM–I-V model to analyze the performance of all DMUs
Step 5. Form virtual alliance and analysis performance of all DMU after alliance by Super SBM-I-V

Based on the definition of our research already mentioned in the literature review, we set up the virtual companies by summing their factors results together and then using the super - SBM–I-V model to evaluate the performance of all DMUs before and after the alliance. Finally, we make decisions based on the ranking and efficiency score.

Step 6. Partner selection
Based on the rank and the efficiency score, we divided into three groups (1) good group: win - win situation when applying alliance "both of them increasing the efficiency score". (2) normal group: win - lost situation when applying alliance "only one DMU increasing performance". (3) bad group: "Lost - lost situation "both of them inefficiency".

4 Case Study

4.1 DMU Collection

After survey and collected all enterprises related to business coffee in Đăk Lăk province, the totally is 13 companies are operating in the market. However, some companies have not published financial statements or just established only one or two years does not meet the requirements of this study. So only 07 companies with fully data are selected to be our DMUs to assess the performance. In order to convenience for presentation in the manuscript, the name of all companies is coded in Table 1.

4.2 Inputs and Outputs Selection

The DEA requires that the relationship between input and output factors must isotonicity". Hence, after using Pearson correlation analysis the outputs in this study are total revenue $Y(_1)$ and the profit after tax (Y_2). The inputs are fixed asset (X_1), equity (X_2) and number of staff (X_3). Table 2 and 3 show the results of Pearson correlation coefficients obtained from the year 2020 to 2022 and 2023, respectively.

Table 1. Code of companies

DMU_S	Name of company
DMU_1	Buon Ho Coffee Company
DMU_2	Cu Pul Coffee two member limited liability company
DMU_3	Thang Loi Coffee Joint Stock Company
DMU_4	Robanme Coffee limited company
DMU_5	Duc Lap Coffee company
DMU_6	Dak Man VietNam limited
DMU_7	721 Coffee one member company

Table 2. Pearson correlation coefficients inputs and outputs in 2020–2022

2020	X_1	X_2	X_3	Y_1	Y_2
X_1	1	0.935105	0.916639	0.685013	0.419510
X_2	0.935105	1	0.838620	0.695738	0.326586
X_3	0.916639	0.838620	1	0.548458	0.673741
Y_1	0.685013	0.695738	0.548458	1	0.126076
Y_2	0.419510	0.3265864	0.673741	0.126076	1
2021	X_1	X_2	X_3	Y_1	Y_2
X_1	1	0.842251	0.985304	0.621757	0.682861
X_2	0.842251	1	0.822172	0.798965	0.541510
X_3	0.985304	0.822172	1	0.581854	0.726212
Y_1	0.621757	0.798965	0.581854	1	0.137436
Y_2	0.682861	0.541510	0.726212	0.137436	1
2022	X_1	X_2	X_3	Y_1	Y_2
X_1	1	0.719567	0.958436	0.556471	0.676066
X_2	0.719567	1	0.811176	0.812817	0.433101
X_3	0.958436	0.811176	1	0.591018	0.740682
Y_1	0.556471	0.812817	0.591018	1	0.290429
Y_2	0.676066	0.433101	0.740682	0.290429	1

The results in Table 2 and 3 indicated that the relationship between input and output variables fits the assumption of DEA (isotonicity) and high degree relationship in each year. Therefore, the inputs and outputs selection is suitable in this research.

After the input and output variables selection, All information and data sets of all DMUs in 2020 was shown in Table 4.

Table 3. Pearson correlation coefficients inputs and outputs in the year of 2023

2023	X_1	X_2	X_3	Y_1	Y_2
X_1	1	0.441935	0.79575	0.301306	0.507553
X_2	0.441935	1	0.78601	0.849927	0.511894
X_3	0.79575	0.78601	1	0.617964	0.788155
Y_1	0.301306	0.849927	0.617964	1	0.37576
Y_2	0.507553	0.511894	0.788155	0.37576	1

Table 4. Original data of all DMU_S in the year of 2020

DMUs	Inputs (VND)			Outputs (VND)	
	Total Asset	Equity	Number of staff	Total revenue	Profit after tax
DMU_1	13436122623	25643562326	98	25814635770	2507779479
DMU_2	5267549600	10734493239	29	1743798018	−6847804307
DMU_3	40167537648	128096247913	123	253962573600	1596247913
DMU_4	555717339	1004916912	34	382678876	48449990
DMU_5	40050964428	104173503742	120	1062646546000	224371043
DMU_6	39050988928	112173573542	113	1948423457753	244334543
DMU_7	23691006971	18634460346	86	41788913895	1102661411

4.3 Forecast the Performance of All DMU by GM (1, 1) Model.

According to the previous data during the period time 2020- 2022, we use Microsoft Excel based on the algorithm of GM (1, 1) to forecasts the input and output variables in the year 2023. The results of the forecast data was shown in the Table 5.

Table 5. Forecast data of all DMU in the year of 2023 (Unit: VND)

DMU 2023	Total Asset	Equity	Number of staff	Total revenue	Profit after tax
DMU_1	66339190097	25801415818	103	22111946206	404136206
DMU_2	1592814683	879108137	27	3465374299	−5592233503
DMU_3	36497780173	139637587828	124	614081778572	2659890731
DMU_4	3069764190	711429732	35	1866785023	75376458
DMU_5	32359752067	100929807161	121	1727153055871	2969219248
DMU_6	44100334479	167048588887	114	2238960307508	274964349
DMU_7	28587313489	19120864361	87	55419407555	1354077311

This study uses the MAPE (Means absolute percentage error) to evaluate the forecast model's accuracy. Table 6 shows the result of MAPE for all DMU.

Table 6. Average MAPE error of DMUs

DMUs	Average MAPE
DMU_1	7.97%
DMU_2	11.83%
DMU_3	2.83%
DMU_4	11.09%
DMU_5	5.08%
DMU_6	1.40%
DMU_7	1.29%
Average MAPE of 07 DMU_s	5.93%

The MAPE of forecast result show in Table 6 is very low less than 10% which confirms that GM (1, 1) model is fit model in this case study. Therefore, the forecast results in Table 6 have high reliability.

4.4 Analysis Performance Before Alliance

In this step, the Thang Loi Coffee Joint Stock Company (DMU_3) is selected as a target company that is supposed to form a partnership with other companies. In order to compare the performance of DMU_3 before and after alliance. The super- SBM–I-V model was chosen to evaluate the performance of all companies because it has more advantages than traditional DEA model [11], Tone et al. [12] show that super - SBM model could rank extremely DEA efficient DMUs, is able to eliminate the drawback of earlier model. Table 7 shows the efficiency score and the ranking of the 07 DMUs the year 2023.

Table 7. Efficiency score and ranking of 07 DMUs from the year of 2023

Rank	DMU	Score
1	DMU_4	10.669447
2	DMU_7	1.4510607
3	DMU_2	1.4133597
3	DMU_6	1
5	DMU_5	1
6	**DMU_3**	**0.7829562**
7	DMU_1	0.2778048

Table 7 shows that DMU_4 is the best performance with an efficiency score (10.669447), while the DMU_3 performed poorly due to being ranked to 6. For further investigation the performance of DMU_3 in the past and recently compare with other DMUs, we run the software again. All the ranking and efficiency score was shown in Table 8.

Table 8. Efficiency score and ranking of 07 DMUs from the time period 2020–2022

2020			2021			2022		
Rank	DMU	Score	Rank	DMU	Score	Rank	DMU	Score
1	DMU_4	14.760161	1	DMU_4	11.462365	1	DMU_4	11.368421
2	DMU_2	1.0581018	2	DMU_7	1.1176094	2	DMU_7	1.8331801
3	DMU_6	1	3	DMU_2	1.0719401	3	DMU_2	1.0721442
3	DMU_1	1	4	$\mathbf{DMU_3}$	1	4	DMU_6	1
5	DMU_5	0.6005291	4	DMU_6	1	4	DMU_5	1
6	DMU_7	0.5738959	6	DMU_5	0.8260251	6	DMU_1	0.6055984
7	$\mathbf{DMU_3}$	0.4132502	7	DMU_1	0.5218013	7	$\mathbf{DMU_3}$	0.5600184

Table 8 shows that about two thirds of the DMUs performed efficiency as their efficiency score greater than 1, this is indicated that they are efficient in the time period 2020–2022. In this table also indicated that the DMU_3 once again show a poor ranking number (ranked to 7 in two years) in the time period 2020–2022, which implies that this company requires a change on its current status.

4.5 Analysis the Performance After Alliance

In order to help the company of DMU_3 in increasing the performance in the future. We make alliance with others DMUs, A totals have 13 scenarios (07 of them are individual DMUs and 06 of them are combinations of DMU_3) are initiated for comparison. Table 9 shows the efficiency score and ranking of these scenarios, after applying a strategic alliance based on the data of the year 2023.

The results in the Table 9 suggested that DMU_3 should make alliance with DMU_5, DMU_6 and DMU_7 to increase the performance. Especially, the alliance with DMU_7 is the best partner for DMU_3 as this alliance can improve the ranking of the DMU_3 from 9 to 5. Furthermore when apply alliance with DMU_7 also make the DMU_7 increasing the performance, this issue is easy happened in reality. By contrast, DMU_3 can make alliance with DMU_5 and DMU_6 but this issue is difficult happened in reality because when the alliance with DMU_3, DMU_5 and DMU_6 will be get a risk.

Table 9. Efficiency score and ranking of different strategic alliance scenarios in 2023

Rank	DMU	Score	Good/normal/bad partnership
1	DMU_4	5.1634949	
2	DMU_2	1.314759	
3	DMU_6	1.241477	
4	DMU_5	1.1530123	
5	$\mathbf{DMU_3+DMU_7}$	1	Good
5	DMU_3+DMU_6	1	Normal
5	DMU_3+DMU_5	1	Normal
5	$\mathbf{DMU_7}$	1	
9	$\mathbf{DMU_3}$	**0.8156404**	
10	DMU_3+DMU_4	0.8064652	Bad
11	DMU_3+DMU_1	0.6558259	Bad
12	DMU_3+DMU_2	0.5298331	Bad
13	DMU_1	0.4545397	

5 Conclusions

This research provides a hybrid approach by combing GM (1,1) and Super SBM–I-V model for finding out the right partner in alliance under assessing business performance of DMUs with three inputs and two outputs variable. Experimental results found that DMU_7 will be become the best partner of DMU_3 for increasing the efficient score. These results are good sound for helping organization in alliance partner selection. Because of the lack of in information of some non-listed companies and time limitation so, we cannot list them on our decision-making unit in this study. We are also planning to solve the prediction task taking into account the prediction values generated by different sources such as more inputs and outputs variables and different DEA models. In this case there may appear inconsistency among the sources and the methods for making consensus and integration can be used for determining the best prediction value [15–17].

References

1. Duc, N.T., Nguyen, L.V., Do, M.H., et al.: The impact of COVID-19 pandemic on coffee exportation and Vietnamese farmers' livelihood. Agriculture **11**, 79 (2021)
2. Soares, B.: The use of strategic alliances as an instrument for rapid growth by New Zealand based quested companies, United New Zealand School of Business Dissertations and Theses (2007)
3. Dussauge, P., Garrette, B.: Determinants of success in international strategic alliances: evidence from the global aerospace industry. J. Int. Bus. Stud. **26**(3), 505–530 (1995)
4. Faulkner, D.: International Strategic Alliances: Cooperating to Compete, McGraw-Hill Book Company (1995)

5. Gulati, R.: Alliance and networks. Strateg. Manag. J. **19**, 293–317 (1998)
6. Deng, J.L.: Introduction to grey system theory. J. Grey Syst. **1**, 1–24 (1989)
7. Wang, C.N., Phan, V.-T.: Enhancing the accurate of grey prediction for GDP growth rate in Vietnam. In: 2014 International Symposium on Computer, Consumer and Control (IS3C). IEEE (2014)
8. Nguyen, N.T., Phan, V.T., Malara, Z.: Nonlinear grey Bernoulli model based on Fourier transformation and its application in forecasting the electricity consumption in Vietnam. J. Intell. Fuzzy Syst. **37**(6), 7631–7641 (2019)
9. Nguyen, N.T., Phan, V.T., Nguyen, V.Đ., Le, T.H., Pham, T.V.: Forecasting the coffee consumption demand in Vietnam based on grey forecasting model. Vietnam J. Comput. Sci. **9**(3), 245–259 (2022)
10. Website of DEA. http://www.emp.pdx.edu/dea/homedea.html
11. Cooper, W.W., Seiford, L.M., Zhu, J.: Data envelopment analysis: history, models, and interpretations. In: Cooper, W., Seiford, L.M., Zhu, J. (eds.) Handbook on Data Envelopment Analysis. ISOR, vol. 164, pp. 1–39. Springer, Boston (2011). https://doi.org/10.1007/978-1-4419-6151-8_1
12. Tone, K.: A slacks-based measure of efficiency in data envelopment analysis. Eur. J. Oper. Res. **30**(3), 498–5091 (2001)
13. Wang, C.N., Wang, C.-H.: A DEA application model for merger & acquisition in high-tech businesses. In: Proceedings of IEEE International Engineering Management Conference (2005)
14. Phan, V.T., Wang, C.N.: Model for alliance partners selection based on the grey model and DEA application-case by Vietnamese bank industry. Indones. J. Electr. Eng. Comput. Sci. **12** (2014)
15. Nguyen, N.T., Sobecki, J.: Using Consensus methods to construct adaptive interfaces in multimodal web-based systems. J. Univ. Access Inf. Soc. **2**(4), 342–358 (2003)
16. Duong, T.H., Nguyen, N.T., Jo, G.S.: A hybrid method for integrating multiple ontologies cybernetics and systems **40**(2), 123–145 (2009)
17. Sliwko, L., Nguyen, N.T.: Using multi-agent systems and consensus methods for information retrieval in internet. Int. J. Intell. Inf. Database Syst. **1**(2), 181–198 (2007)

Car Detector Based on YOLOv5 for Parking Management

Duy-Linh Nguyen, Xuan-Thuy Vo, Adri Priadana, and Kang-Hyun Jo(✉)

Department of Electrical, Electronic and Computer Engineering, University of Ulsan, Ulsan 44610, South Korea
{ndlinh301,priadana}@mail.ulsan.ac.kr, xthuy@islab.ulsan.ac.kr, acejo@ulsan.ac.kr

Abstract. Nowadays, YOLOv5 is one of the most widely used object detection network architectures in real-time systems for traffic management and regulation. To develop a parking management tool, this paper proposes a car detection network based on redesigning the YOLOv5 network architecture. This research focuses on network parameter optimization using lightweight modules from EfficientNet and PP-LCNet architectures. The proposed network is trained and evaluated on two benchmark datasets which are the Car Parking Lot Dataset and the Pontifical Catholic University of Parana+ Dataset and reported on mAP@0.5 and mAP@0.5:0.95 measurement units. As a result, this network achieves the best performances at 95.8 % and 97.4 % of mAP@0.5 on the Car Parking Lot Dataset and the Pontifical Catholic University of Parana+ Dataset, respectively.

Keywords: Convolutional neural network (CNN) · EfficientNet · PP-LCNet · Parking management · YOLOv5

1 Introduction

Along with the rapid development of modern and smart cities, the number of vehicles in general and cars in particular has also increased in both quantity and type. According to a report by the Statista website [15], there are currently about one and a half million cars in the world and it is predicted that in 2023, the number of cars sold will reach nearly 69.9 million. This number will increase further in the coming years. Therefore, the management and development of tools to support parking lots are essential. To construct smart parking lots, researchers propose many methods based on geomagnetic [25], ultrasonic [16], infrared [2], and wireless techniques [21]. These approaches mainly rely on the operation of sensors designed and installed in the parking lot. Although these designs achieve high accuracy, they require large investment, labor, and maintenance costs, especially when deployed in large-scale parking lots. Exploiting the benefits of convolutional neural networks (CNNs) in the field of computer vision,

N. T. Nguyen et al. (Eds.): CITA 2023, LNNS 734, pp. 102–113, 2023.
https://doi.org/10.1007/978-3-031-36886-8_9

several researchers have designed networks to detect empty or occupied parking spaces using conventional cameras with quite good accuracy [5,12,13]. Following that trend, this paper proposes a car detector to support smart parking management. This work explores lightweight network architectures and redesigned modules inside of the YOLOv5 network to balance network parameters, detection accuracy, and computational complexity. It ensures deployment in real-time systems with the lowest deployment cost. The main contributions of this paper are shown below:

1 - Proposes an improved YOLOv5 architecture for car detection that can be applied to parking management and other related fields of computer vision.
2 - The proposed detector performs better than other detectors on the Car Parking Lot Dataset and the Pontifical Catholic University of Parana+ Dataset.

The distribution of the remaining parts in the paper is as follows: Sect. 2 presents the car detection-based methods. Section 3 explains the proposed architecture in detail. Section 4 introduces the experimental setup and analyzes the experimental results. Section 5 summarizes the issue and future work orientation.

2 Related Works

2.1 Traditional Machine Learning-Based Methods

The car detection process of traditional machine learning-based techniques is divided into two stages, manual feature extraction and classification. First, feature extractors generate feature vectors using classical methods such as Scale-invariant Feature Transform (SIFT), Histograms of Oriented Gradients (HOG), and Haar-like features [18,19,22]. Then, the feature vectors go through classifiers like the Support Vector Machine (SVM) and Adaboost [6,14] to obtain the target classification result. The traditional feature extraction methods rely heavily on prior knowledge. However, in the practical application, there are many objective confounding factors including weather, exposure, distortion, etc. Therefore, the applicability of these techniques on real-time systems is limited due to low accuracy.

2.2 CNN-Based Methods

Parking lot images obtained from drones or overhead cameras contain many small-sized cars. In order to detect these objects well, many studies have focused on the small object detection topic using a combination of CNN and traditional methods or one-stage detectors. The authors in [1,3,24] fuse the modern CNNs and SVM networks to achieve high spatial resolution in vehicle count detection and counting. Research in [11] develops a network based on the YOLOv3 network architecture in which the backbone network is combined between ResNet and DarkNet to solve object vision in drone images. The work in [10] proposes

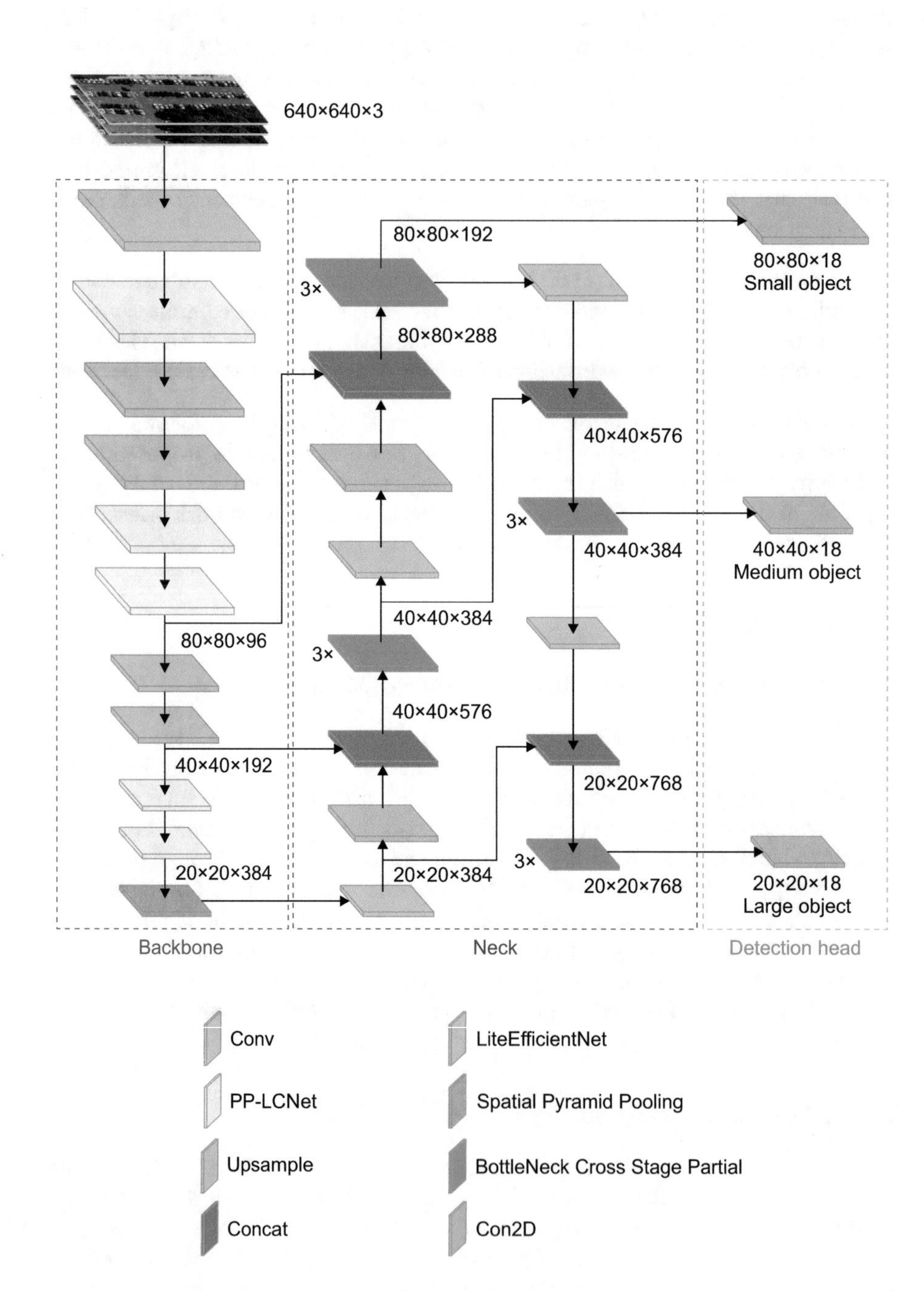

Fig. 1. The architecture of proposed car detector.

a new feature-matching method and a spatial context analysis for pedestrian-vehicle discrimination. An improved YOLOv5 network architecture is designed by [7] for vehicle detection and classification in Unmanned Aerial Vehicle (UAV) imagery and [23] for real-world imagery. Another study in [20] provides a one-stage detector (SF-SSD) with a new spatial cognition algorithm for car detection in UAV imagery. The advantage of modern machine learning methods is high detection and classification accuracy, especially for small-sized objects. However, they require the network to have a high-level feature extraction and fusion, and a certain complexity to ensure operation in real-world conditions.

3 Methodology

The proposed car detection network is shown in Fig. 1. This network is an improved YOLOv5 architecture [9] including three main parts: backbone, neck, and detection head.

3.1 Proposed Network Architecture

Basically, the structure of the proposed network follows the design of the YOLOv5 network architecture with many changes inside the backbone and neck modules. Specifically, the Focus module is replaced by a simple block called Conv. This block is constructed with a standard convolution layer (Con2D) with kernel size of 1×1 followed by a batch normalization (BN) and a ReLU activation function as shown in Fig. 2 (a). Subsequent blocks in the backbone module are also redesigned based on inspiration from lightweight network architectures such as PP-LCNet [4] and EfficientNet [17]. The design of the PP-LCNet (PP-LC) layer is described in detail in Fig. 3 (a). It consists of a depthwise convolution layer

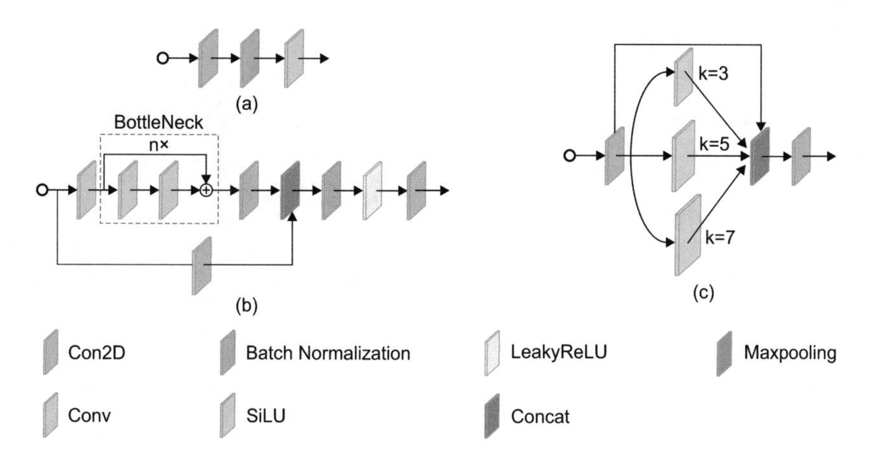

Fig. 2. The architecture of Conv (a), BottleNeck Cross Stage Partial (b), and Spatial Pyramid Pooling (c) blocks.

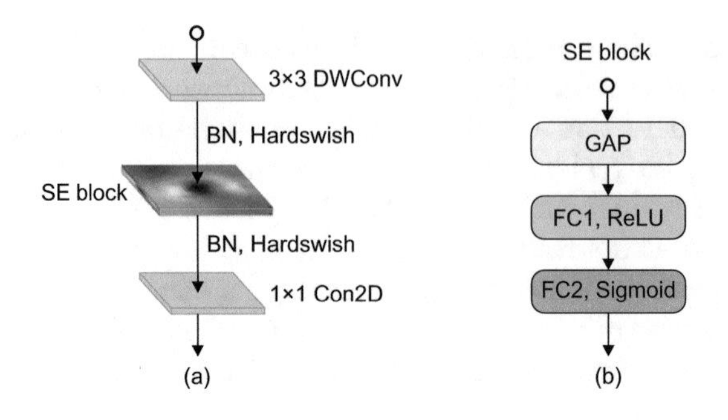

Fig. 3. The architecture of PP-LCNet (a) and SE (b) blocks.

(3×3 DWConv), an attention block (SE block), and ends with a standard convolution layer (1×1 Con2D). In between these layers, the BN and the hardswish activation function are used. The SE block is an attention mechanism based on a global average pooling (GAP) layer, a fully connected layer (FC1) followed by a rectified linear unit activation function (ReLU), and a second fully connected layer (FC2) followed by a sigmoid activation function as Fig. 3 (b). This method uses lightweight convolution layers that save a lot of network parameters. In addition, the attention mechanism helps the network focus on learning important information about the object on each feature map level. The next block is LiteEfficientNet (LE). This block is very simple and is divided into two types corresponding to two stride levels (stride = 1 or stride = 2). In the first type with stride = 2, the LiteEfficientNet block uses an extended convolution layer (1×1 Con2D), a depth-wise convolution layer (3×3 DWConv), and ends with a project convolution layer (1×1 Con2D). For the second type with stride = 1, the LiteEfficientNet block is exactly designed the same as the first type and added a skip connection to merge the current and original feature maps with

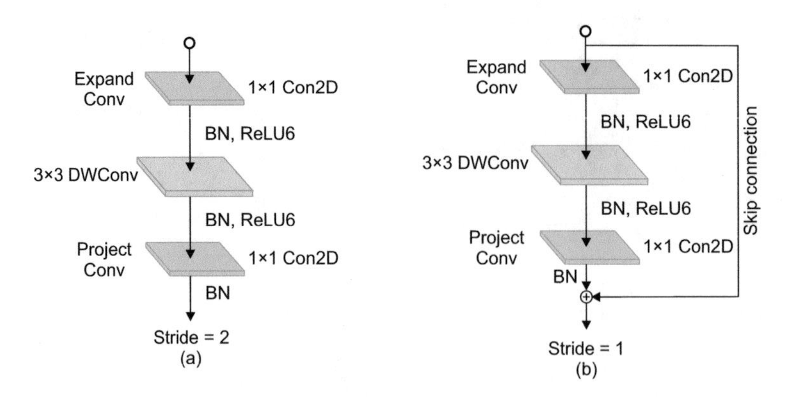

Fig. 4. The two types of LiteEfficientNet (LE) architecture, stride = 2 (a) and stride = 1 (b)

the addition operation. This block extracts the feature maps on the channel dimension. The combined use of PP-LCNet and LiteEfficientNet blocks ensures that feature extraction is both spatial and channel dimensions of each feature map level. The detail of the LiteEfficientNet block is shown in Fig. 4. The last block in the backbone module is the Spatial Pyramid Pooling (SPP) block. This work re-applies the architecture of SPP in the YOLOv5 as Fig. 2 (c). However, to minimize the network parameters, the max pooling kernel sizes are reduced from 5×5, 9×9, and 13×13 to 3×3, 5×5, and 7×7, respectively.

The neck module in the proposed network utilizes the Path Aggregation Network (PAN) architecture following the original YOLOv5. This module combines the current feature maps with previous feature maps by concatenation operations. It generates the output with three multi-scale feature maps that are enriched information. These serve as three inputs for the detection heads.

The detection head module also leverages the construction of three detection heads from the YOLOv5. Three feature map scales of the PAN neck go through three convolution operations to conduct prediction on three object scales: small, medium, and large. Each detection head uses three anchor sizes that describe in Table 1.

Table 1. Detection heads and anchors sizes.

Heads	Input	Anchor sizes	Ouput	Object
1	$80 \times 80 \times 129$	(10, 13), (16, 30), (33, 23)	$80 \times 80 \times 18$	Small
2	$40 \times 40 \times 384$	(30, 61), (62, 45), (59, 119)	$40 \times 40 \times 18$	Medium
3	$20 \times 20 \times 768$	(116, 90), (156, 198), (373, 326)	$20 \times 20 \times 18$	Large

3.2 Loss Function

The definition of the loss function is shown as follows:

$$\mathcal{L} = \lambda_{box}\mathcal{L}_{box} + \lambda_{obj}\mathcal{L}_{obj} + \lambda_{cls}\mathcal{L}_{cls}, \tag{1}$$

where $\mathcal{L}_{box}$ uses CIoU loss to compute the bounding box regression. The object confidence score loss $\mathcal{L}_{obj}$ and the classification loss $\mathcal{L}_{cls}$ using Binary Cross Entropy loss to calculate. λ_{box}, λ_{obj}, and λ_{cls} are balancing parameters.

4 Experiments

4.1 Datasets

The proposed network is trained and evaluated on two benchmark datasets, the Car Parking Lot Dataset (CarPK) and the Pontifical Catholic University of Parana+ Dataset (PUCPR+) [8]. The CarPK dataset contains 89,777 cars collected from the Phantom 3 Professional drone. The images were taken from four parking lots with an approximate height of 40 m. The CarPK dataset is divided into 988 images for training and 459 images for validation phases. The PUCPR+ dataset is selected from a part of the PUCPR dataset consisting of 16,456 cars. The PUCPR+ dataset provides 100 images for training and 25 images for validation. These are image datasets for car counting in different parking lots. The cars in the image are annotated by bounding boxes with top-left and bottom-right angles and stored as text files (*.txt files). To accommodate the training and evaluation processes, this experiment converts the entire format of the annotation files to the YOLOv5 format.

4.2 Experimental Setup

The proposed network is conducted on the Pytorch framework and the Python programming language. This network is trained on a Testla V100 32GB GPU and evaluated on a GeForce GTX 1080Ti 11GB GPU. The optimizer is Adam optimization. The learning rate is initialized at 10^3 and ends at 10^5. The momentum set at 0.8 and then increased to 0.937. The training process goes through 300 epochs with a batch size of 64. The balance parameters are set as follows: λ_{box}=0.05, λ_{obj}=1, and λ_{cls}=0.5. To increase training scenarios and avoid the over-fitting issue, this experiment applies data augmentation methods such as mosaic, translate, scale, and flip. For the inference process, other arguments are set like an image size of 1024×1024, a batch size of 32, a confidence threshold = 0.5, and an IoU threshold = 0.5. The speed results are reported in milliseconds (ms).

4.3 Experimental Results

The performance of the proposed network is evaluated lying on the comparison results with the retrained networks from scratch and the recent research on the two above benchmark datasets. Specifically, this work conducts the training and evaluation of the proposed network and the four versions of YOLOv5 architectures (l, m, s, n). Then, it compares the results obtained with the results in [7,20] on the CarPK dataset and the results in [20] on the PUCPR+ dataset. As a result, the proposed network achieves 95.8% of mean Average Precision with an IoU threshold of 0.5 (mAP@0.5) and 63.1% of mAP with ten IoU thresholds from 0.5 to 0.95 (mAP@0.5:0.95). This result shows the superior ability of the proposed network compared to other networks. While the speed (inferent time) is only 1.7 ms higher than retrained YOLOv5m network, nearly 1.5 times lower than the retrained YOLOv5l network, and quite lower than other experiments in [7] from 2.3 (YOLOv5m) to 7.9 (YOLOv5m) times. Besides, the weight of the network (22.7 MB) and the computational complexity (23.9 GFLOPs) are only half of the retrained YOLOv5m architecture. The comparison results on the CarPK validation set are presented in Table 2. For the PUCPR+ dataset, the proposed network achieves 97.4% of mAP@0.5 and 58.0% of mAP@0.5:0.95. This result is outstanding compared to other competitors and is only 0.3% of

Table 2. Comparison result of proposed car detection network with other networks and retrained YOLOv5 on CarPK validation set. The symbol "∗" denotes the retrained networks. N/A means not-available values.

Models	Parameter	Weight (MB)	GFLOPs	mAP@0.5	mAP@0.5:0.95	Inf. time (ms)
YOLOv5l*	46,631,350	93.7	114.2	95.3	62.3	26.4
YOLOv5m*	21,056,406	42.4	50.4	94.4	61.5	15.9
YOLOv5s*	7,022,326	14.3	15.8	95.6	62.7	8.7
YOLOv5n*	1,765,270	3.7	4.2	93.9	57.8	6.3
YOLOv5x [7]	N/A	167.0	205.0	94.5	57.9	138.2
YOLOv5l [7]	N/A	90.6	108.0	95.0	59.2	72.1
YOLOv5m [7]	N/A	41.1	48.0	94.6	57.8	40.4
Modified YOLOv5 [7]	N/A	44.0	57.7	94.9	61.1	50.5
SSD [20]	N/A	N/A	N/A	68.7	N/A	N/A
YOLO9000 [20]	N/A	N/A	N/A	20.9	N/A	N/A
YOLOv3 [20]	N/A	N/A	N/A	85.3	N/A	N/A
YOLOv4 [20]	N/A	N/A	N/A	87.81	N/A	N/A
SA+CF+CRT [20]	N/A	N/A	N/A	89.8	N/A	N/A
SF-SSD [20]	N/A	N/A	N/A	90.1	N/A	N/A
Our	**11,188,534**	**22.7**	**23.9**	**95.8**	**63.1**	**17.6**

mAP@0.5 and 2.5% of mAP@0.5:095 lower than retrained YOLOv5m, respectively. However, the proposed network has a speed of 17.9 ms, which is only slightly higher than the retrained YOLOv5m network (2.3 ms ↑) and lower than the retrained YOLOv5l network (4.5 ms ↓). The comparison results are shown in Table 3 and several qualitative results are shown in Fig. 5.

From the mentioned results, the proposed network has a balance in performance, speed, and network parameters. Therefore, it can be implemented in parking management systems on low-computing and embedded devices. However, the process of testing this network also revealed some disadvantages. Since

Table 3. Comparison result of proposed car detection network with other networks and retrained YOLOv5 on PUCPR+ validation set. The symbol " ∗ " denotes the retrained networks. N/A means not-available values.

Models	Parameter	Weight (MB)	GFLOPs	mAP@0.5	mAP@0.5:0.95	Inf. time (ms)
YOLOv5l*	46,631,350	93.7	114.2	96.4	53.8	22.4
YOLOv5m*	21,056,406	42.4	50.4	97.7	60.5	15.6
YOLOv5s*	7,022,326	14.3	15.8	84.6	38.9	7.4
YOLOv5n*	1,765,270	3.7	4.2	89.7	41.6	5.9
SSD [20]	N/A	N/A	N/A	32.6	N/A	N/A
YOLO9000 [20]	N/A	N/A	N/A	12.3	N/A	N/A
YOLOv3 [20]	N/A	N/A	N/A	95.0	N/A	N/A
YOLOv4 [20]	N/A	N/A	N/A	94.1	N/A	N/A
SA+CF+CRT [20]	N/A	N/A	N/A	92.9	N/A	N/A
SF-SSD [20]	N/A	N/A	N/A	90.8	N/A	N/A
Our	**11,188,534**	**22.7**	**23.9**	**97.4**	**58.0**	**17.9**

the car detection network is mainly based on the signal obtained from the drone-view or floor-view camera, it is influenced by a number of environmental factors, including illumination, weather, car density, occlusion, shadow, objects similarity, and the distance from the camera to the cars. Several mistaken cases are listed in Fig. 5 with yellow circles.

Fig. 5. The qualitative results and several mistakes of the proposed network on the validation set of the CarPK and PUCPR+ datasets with IoU threshold = 0.5 and confidence score = 0.5. Yellow circles denote the wrong detection areas. (Color figure online)

4.4 Ablation Study

The experiment conducted several ablation studies to inspect the importance of each block in the proposed backbones. The blocks are replaced in turn, trained on the CarPK training set, and evaluated on the CarPK validation set as shown in Table 4. The results from this table show that the PP-LCNet block increases the network performance at mAP@ 0.5 (1.1% $\uparrow$) but decreased in mAP@0.5:0.95 (0.8% $\downarrow$) when compared to the LiteEfficientNet block. Combining these two blocks gives a perfect result along with the starting Conv and the ending SPP blocks. Besides, it also shows the superiority of the SPP block (0.4% $\uparrow$ of mAP@0.5 and mAP@0.5:0.95) over the SPPF block when they generate the same GFLOPs and network parameters.

Table 4. Ablation studies with different types of backbones on the CarPK validation set.

Blocks	Proposed backbones			
Conv	✓	✓	✓	✓
PP-LCNet		✓	✓	✓
LiteEfficientNet	✓		✓	✓
SPPF			✓	
SPP	✓	✓		✓
Parameter	10,728,766	9,780,850	11,188,534	11,188,534
Weight (MB)	21.9	19.9	22.7	22.7
GFLOPs	20.8	18.5	23.9	23.9
mAP@0.5	95.1	94.3	95.4	**95.8**
mAP@0.5:0.95	58.2	59.3	62.7	**63.1**

5 Conclusion

This paper introduces an improved YOLOv5 architecture for car detection in parking management systems. The proposed network contains three main modules: backbone, neck, and detection head. The backbone module is redesigned using lightweight architectures: PP-LCNet and LiteEfficientNet. The network achieves 95.8 % of mAP@0.5 and 63.1 % of mAP@0.5:0.95 and better performance results when compared to recent works. The optimization of network parameters, speed, and detection accuracy provides the ability to deploy on real-time systems. In the future, the neck and detection head modules will be developed to detect smaller vehicles and implement on larger datasets.

Acknowledgement. This result was supported by the "Regional Innovation Strategy (RIS)" through the National Research Foundation of Korea(NRF) funded by the Ministry of Education(MOE)(2021RIS-003).

References

1. Ammour, N., Alhichri, H., Bazi, Y., Benjdira, B., Alajlan, N., Zuair, M.: Deep learning approach for car detection in uav imagery. Remote Sensing **9**, 1–15 (2017). https://doi.org/10.3390/rs9040312
2. Chen, H.C., Huang, C.J., Lu, K.H.: Design of a non-processor obu device for parking system based on infrared communication. In: 2017 IEEE International Conference on Consumer Electronics - Taiwan (ICCE-TW), pp. 297–298 (2017). https://doi.org/10.1109/ICCE-China.2017.7991113
3. Chen, S., Zhang, S., Shang, J., Chen, B., Zheng, N.: Brain-inspired cognitive model with attention for self-driving cars. IEEE Trans. Cognitive Developm. Syst. **11**(1), 13–25 (2019). https://doi.org/10.1109/TCDS.2017.2717451
4. Cui, C., et al.: Pp-lcnet: A lightweight CPU convolutional neural network. CoRR abs/ arXiv: 2109.15099 (2021)
5. Ding, X., Yang, R.: Vehicle and parking space detection based on improved yolo network model. J. Phys: Conf. Ser. **1325**, 012084 (2019). https://doi.org/10.1088/1742-6596/1325/1/012084
6. Freund, Y., Schapire, R.E.: A desicion-theoretic generalization of on-line learning and an application to boosting. In: Vitányi, P. (ed.) EuroCOLT 1995. LNCS, vol. 904, pp. 23–37. Springer, Heidelberg (1995). https://doi.org/10.1007/3-540-59119-2_166
7. Hamzenejadi, M.H., Mohseni, H.: Real-time vehicle detection and classification in uav imagery using improved yolov5. In: 2022 12th International Conference on Computer and Knowledge Engineering (ICCKE), pp. 231–236 (2022). https://doi.org/10.1109/ICCKE57176.2022.9960099
8. Hsieh, M., Lin, Y., Hsu, W.H.: Drone-based object counting by spatially regularized regional proposal network. CoRR abs/ arXiv: 1707.05972 (2017)
9. Jocher, G., et al.: ultralytics/yolov5: v3.1 - Bug Fixes and Performance Improvements (Oct 2020). https://doi.org/10.5281/zenodo.4154370
10. Liang, X., Zhang, J., Zhuo, L., Li, Y., Tian, Q.: Small object detection in unmanned aerial vehicle images using feature fusion and scaling-based single shot detector with spatial context analysis. IEEE Trans. Circ. Syst. Video Technol, 1758–1770 (2019)
11. Liu, M., Wang, X., Zhou, A., Fu, X., Ma, Y., Piao, C.: Uav-yolo: Small object detection on unmanned aerial vehicle perspective. Sensors **20**(8) (2020). https://doi.org/10.3390/s20082238, https://www.mdpi.com/1424-8220/20/8/2238
12. Martín Nieto, R., García-Martín, A., Hauptmann, A.G., Martínez, J.M.: Automatic vacant parking places management system using multicamera vehicle detection. IEEE Trans. Intell. Trans. Syst. **20**(3), 1069–1080 (2019). https://doi.org/10.1109/TITS.2018.2838128
13. Mettupally, S.N.R., Menon, V.: A smart eco-system for parking detection using deep learning and big data analytics. In: 2019 SoutheastCon, pp. 1–4 (2019). https://doi.org/10.1109/SoutheastCon42311.2019.9020502
14. Mitra, V., Wang, C.J., Banerjee, S.: Text classification: A least square support vector machine approach. Appli. Soft Comput. **7**, 908–914 (2007). https://doi.org/10.1016/j.asoc.2006.04.002
15. Scotiabank: Number of cars sold worldwide from 2010 to 2022, with a 2023 forecast (in million units). https://www.statista.com/statistics/200002/internationalcar-sales-since-1990/, note = (Accessed 01 Jan 2023)

16. Shao, Y., Chen, P., Tongtong, C.: A grid projection method based on ultrasonic sensor for parking space detection, pp. 3378–3381 (July 2018). https://doi.org/10.1109/IGARSS.2018.8519022
17. Tan, M., Le, Q.V.: Efficientnet: Rethinking model scaling for convolutional neural networks. CoRR abs/ arXiv: 1905.11946 (2019)
18. Viola, P., Jones, M.: Rapid object detection using a boosted cascade of simple features. In: Proceedings of the 2001 IEEE Computer Society Conference on Computer Vision and Pattern Recognition, CVPR 2001. vol. 1 (2001). https://doi.org/10.1109/CVPR.2001.990517
19. Xu, Z., Huang, W., Wang, Y.: Multi-class vehicle detection in surveillance video based on deep learning. J. Comput. Appli. **39**(3), 700 (2019)
20. Yu, J., Gao, H., Sun, J., Zhou, D., Ju, Z.: Spatial cognition-driven deep learning for car detection in unmanned aerial vehicle imagery. IEEE Trans. Cognitive Develop. Syst. **14**(4), 1574–1583 (2022). https://doi.org/10.1109/TCDS.2021.3124764
21. Yuan, C., Qian, L.: Design of intelligent parking lot system based on wireless network. In: 2017 29th Chinese Control And Decision Conference (CCDC), pp. 3596–3601 (2017). https://doi.org/10.1109/CCDC.2017.7979129
22. Zhang, S., Wang, X.: Human detection and object tracking based on histograms of oriented gradients. In: 2013 Ninth International Conference on Natural Computation (ICNC), pp. 1349–1353 (2013). https://doi.org/10.1109/ICNC.2013.6818189
23. Zhang, Y., Guo, Z., Wu, J., Tian, Y., Tang, H., Guo, X.: Real-time vehicle detection based on improved yolo v5. Sustainability **14**(19) (2022). https://doi.org/10.3390/su141912274, https://www.mdpi.com/2071-1050/14/19/12274
24. Zhao, F., Kong, Q., Zeng, Y., Xu, B.: A brain-inspired visual fear responses model for UAV emergent obstacle dodging. IEEE Trans. Cognit. Develop. Syst. **12**(1), 124–132 (2020). https://doi.org/10.1109/TCDS.2019.2939024
25. Zhou, F., Li, Q.: Parking guidance system based on zigbee and geomagnetic sensor technology. In: 2014 13th International Symposium on Distributed Computing and Applications to Business, Engineering and Science, pp. 268–271 (2014). https://doi.org/10.1109/DCABES.2014.58

Deep Learning-Based Approach for Automatic Detection of Malaria in Peripheral Blood Smear Images

Vu-Thu-Nguyet Pham[1], Quang-Chung Nguyen[1], Quang-Vu Nguyen[1](✉), and Huu-Hung Huynh[2]

[1] The University of Danang - Vietnam-Korea University of Information and Communication Technology, Danang, Vietnam
{pvtnguyet.19it1,nqchung.19it1,nqvu}@vku.udn.vn
[2] The University of Danang - University of Science and Technology, Danang, Vietnam
hhhung@dut.udn.vn

Abstract. Malaria is a deadly disease that affects millions of people around the world every year. An accurate and timely diagnosis of malaria is essential for effective treatment and control of the disease. In this study, we propose a deep learning-based approach for automatic detection of malaria in peripheral blood smear images. Our approach consists of two stages: object detection & binary classification using Faster R-CNN, and multi-class classification using EfficientNetv2-L with SVM as the head. We evaluate the performance of our approach using the mean average precision at IoU = 0.5 (mAP@0.5) metric. Our approach achieves an overall performance of 88.7%, demonstrating the potential of deep learning-based approaches for accurate and efficient detection of malaria in peripheral blood smear images. Our study has several implications for the field of malaria diagnosis and treatment. The use of deep learning-based approaches for malaria detection could significantly improve the accuracy and speed of diagnosis, leading to earlier and more effective treatment of the disease.

Keywords: Deep Learning · Bioinformatics · Parasite Detection

1 Introduction

Malaria is a disease caused by protozoan parasites transmitted through infected female Anopheles mosquitoes. Although it is extremely dangerous, malaria is a curable and preventable disease if it is diagnosed at an early stage. Microscopic thick and thin blood smear examinations are the most reliable and commonly used method to detect different types of malaria [1].

However, this method has some disadvantages, including the need for extensive human intervention, which can lead to late and erroneous diagnosis. Automatic parasite counting is a potential solution to this problem, as it provides a more reliable and standardized interpretation of blood films and reduces the workload of malaria field workers.

N. T. Nguyen et al. (Eds.): CITA 2023, LNNS 734, pp. 114–125, 2023.
https://doi.org/10.1007/978-3-031-36886-8_10

This study proposes an automated malaria screening approach and discusses previous works on identifying and classifying malaria parasites. The study includes sections on some related researches, materials and methods, experimental results, and conclusion.

2 Related Works

Formerly, a majority of image analysis-based computer-aided diagnosis software use Machine Learning (ML) techniques with hand-engineered features for decision-making [2, 3]. Many of previous works have used complex workflows for image processing and classification [4–6]. However, the process demands expertise in analyzing the variability in a predetermined set of measurements (e.g., intensity, shape, texture, size, background, angle, position of the region of interest) on the images.

To overcome the challenges of devising hand-engineered features that capture variations in the underlying data, Deep Learning (DL) is used with the purpose of reducing the overall task of developing an efficient feature extraction pipeline. Abu Seman et al. [7] used a multilayer perceptron network to classify P.falciparum, P.vivax and P.malariae to achieve 89.80% accuracy. Khot and Prasad [8] employed an artificial neural network to automate the assessment of parasite infection using morphological features and obtained an accuracy of 73.57%. Memeu [9] introduced an artificial neural network to detect and differentiate plasmodium parasite stages and species from the infected parasite and attained 79.7% accuracy.

Recently, Convolutional Neural Network (CNN) has been frequently used to analyze the behavior of microorganisms from its biomedical images. The pioneering is Liang et al. [10], who use a convolutional neural network to discriminate between infected and uninfected cells in thin blood smears, after applying a conventional level-set cell segmentation approach. They reported an accuracy of 97%. Jane Hung and Anne Carpenter [11] used a CNN-based model for detection and AlexNet for classification. They acquired an accuracy of 72% for finding malaria infected cells. Pan et al. [12] used LeNet-5 for classification of malaria infected cells and obtained the accuracy of 90%.

In summary, existing DL studies have reported quite satisfactory results. However, many of the articles just present performance numbers in terms of sensitivity and specificity for classification, representing only 1 operating point among many on a receiver operating characteristic. Furthermore, the data used for evaluation have very often been simply too small to allow a convincing statement about a system's performance. Besides that, none of the studies have reported the performance of the predictive models at the patient level. This technique can provide a more realistic performance evaluation of the predictive models as the images in the independent test set represent truly unseen images for the training process, with no information about staining variations or other artifacts leaking into the training data. Although the reported outcomes are promising, existing approaches need to substantiate their robustness on a larger set of images and perform cross-validation at the patient level.

3 Experiments

3.1 Data Collection

This study utilized four datasets. The first set [13] includes 1364 images of P.vivax infected blood smears with approximately 80,000 cells, consisting of two classes of uninfected cells and four classes of infected cells. The second set [14] includes 193 Giemsa-stained thin blood smear microscopy images with manual cell annotations for P.falciparum from 193 malaria patients, with 5 images per patient and approximately 162,700 cells. The other two [15, 16] manually annotated datasets include blood smear images from 300 malaria (P.falciparum or P.vivax) infected and 50 uninfected patients, with about 6000 images.

3.2 Exploratory Data Analysis

Labels

These sample datasets consist of blood smear samples that have been analyzed under a microscope.

In terms of labels, there are two classes of uninfected cells (Red Blood Cells - RBCs and leukocytes) and four classes of infected cells (gametocytes, rings, trophozoites, and schizonts). Annotators were allowed to mark some cells as difficult if they were not clearly in one of the cell classes.

In short, the above labels can be simplified into two classes:

- Healthy cells
- Infected cells.

The cells marked as "difficult" were attached to the "infected cells" label, as the system should mark any suspicious cells regardless of its certainty and give the information to the specialist who will later inspect and classify the cell manually for double cross-validation (Fig. 1).

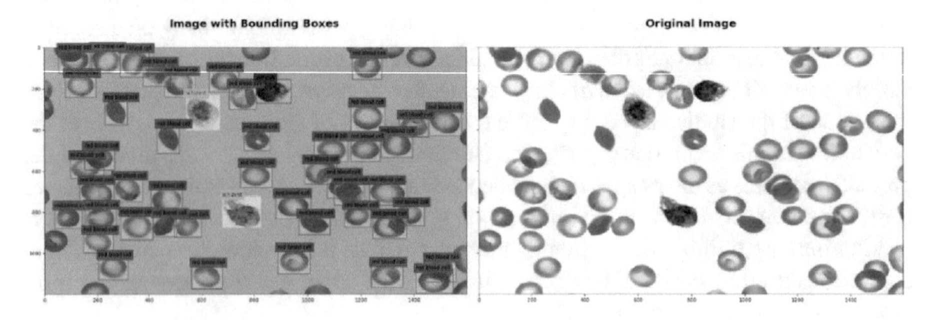

Fig. 1. Sample images with and without bounding boxes

The uninfected RBCs are visible as monochromatic oval shapes with a light center and smooth surface. The RBCs that are infected with malaria are larger, shape-distorted and have a pinkish-purple color. The parasites can take forms of single rings or multiple eosinophilic dots, which can be seen inside the RBCs. The blood smear samples may also contain singular White Blood Cells (WBCs), part of the body's immune system.

Data Preprocessing

As all images were in eight bits, the normalization after float conversion has been made by dividing all values by 255.0.

Data Augmentation

Image augmentation can artificially enlarge the training dataset and thus improve the training process and final accuracy. For simplicity, the basic augmentation has been used, such as:

- Horizontal flip.
- Vertical flip.
- Random rotation.
- Random zoom and crop.
- Random hue shift.
- Random contrast adjustment.

3.3 Two-Stage Approach

General

Take a blood smear as input, the Faster R-CNN detector will localize and classify if that cell is "healthy RBCs" or non- "healthy RBCs". Then, for all the non- "healthy RBCs", we will crop all the non- "healthy RBCs" and pass to our multiclass classifier to predict the exact categories (Fig. 2).

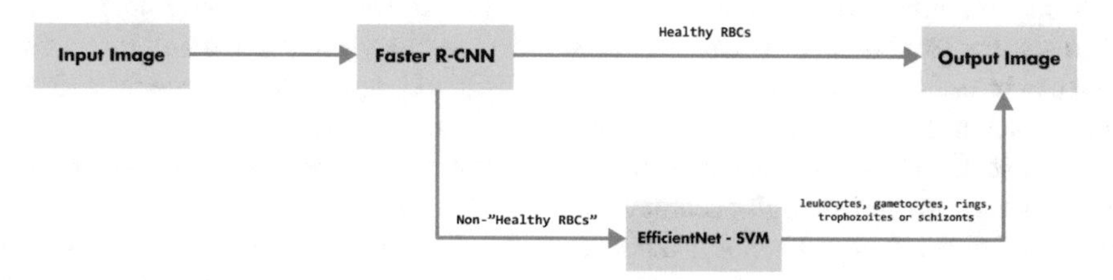

Fig. 2. Proposed framework

First Step: Binary Classification

Faster R-CNN (Faster Region-based Convolutional Neural Network) is a state-of-the-art object detection framework proposed by Shaoqing Ren, Kaiming He, Ross Girshick, and Jian Sun [17] in 2015.

Faster R-CNN is an extension of the R-CNN and Fast R-CNN frameworks, which were the first methods to achieve state-of-the-art performance on object detection benchmarks. However, Faster R-CNN improves upon these methods by introducing a region proposal network (RPN) that allows for much faster processing times.

The idea starts with an image, from which we want to obtain:

- A list of bounding boxes.
- A label assigned to each bounding box.
- A probability for each label and bounding box.

The input images are represented as *Height* × *Width* × *Depth* tensors (multidimensional arrays), which are passed through a pre-trained CNN, ending up with a convolutional feature map. We use this as the feature extractor for the next part.

Next, we use Region Proposal Network. to find up to a pre-defined number of regions (bounding boxes), which may contain objects. The problem of generating a variable-length list of bounding boxes is solved in the RPN using anchors or fixed-sized reference bounding boxes that are placed uniformly throughout the original image.

After having a list of possible relevant objects and their locations in the original image, we apply Region of Interest (RoI) Pooling and extract those features which would correspond to the relevant objects into a new tensor.

Finally, comes the R-CNN module, which uses that information to:

- Classify the content in the bounding box.
- Adjust the bounding box coordinates.

In this study, we use ResNet101 [18] as the base network for Faster R-CNN. Due to the large number of model parameters and limited device resources, we trained this Faster R-CNN model with $batch\_size = 1$ for 40,000 epochs. Initially, we use the model with pre-trained weights from ImageNet and froze the entire backbone for training for 20,000 epochs. After completion, the model was unfrozen for further fine-tuning for another 20,000 epochs.

Second Step: Multi-class Classification

Transfer learning is a technique of using a pre-trained model for a similar problem rather than starting from scratch. The pre-trained model has already learned important features that can be used to solve a new problem. The transfer of knowledge can occur at different levels such as feature extraction, fine-tuning, and multi-task learning.

EfficientNetV2-L [19] is a convolutional neural network architecture introduced in 2021 that is designed for high-accuracy image classification tasks. It is an extension of the original EfficientNet and includes a new compound scaling method that optimizes multiple network dimensions simultaneously. It also uses a Squeeze-and-Excitation module and a Swish activation function to improve accuracy.

The architecture can be divided into three components: the stem, which extracts low-level features from the input image using a combination of convolutional layers and grouped convolutions; the body, which consists of seven blocks that process the extracted features using depthwise convolutions, pointwise convolutions, and SE modules; and the head, which includes a global pooling layer and a fully connected layer for final classification probabilities. The network's design features, such as the compound scaling method, SE module, and Swish activation function, contribute to its high performance and computational efficiency.

In this study, we simply removed the head component of EfficientNetV2-L in order to use this model as a feature extractor.

When using the pre-trained EfficientNetV2-L as a feature extractor for multiclass classification, the resulting features are multi-dimensional, so a suitable classifier is required. Support Vector Machine (SVM) is a supervised learning algorithm that can handle such features. SVM constructs a hyperplane in a multi-dimensional space to separate classes by maximizing the margin, and it uses a kernel function to transform the lower dimensional data into a higher dimensional space. In this study, we use SVM with Radial Basis Function (RBF) kernel since it has a quality of kernel trick-based classification.

$$K(x_i, x_j) = exp\left(-\gamma \left|\left|x_i - x_j\right|\right|^2\right) \tag{1}$$

When training an SVM with the RBF kernel, two parameters must be considered are C and γ. For the hyperparameters, low C makes the decision surface smooth, while a high C aims at classifying all training examples correctly. γ defines how much influence a single training example has. The larger γ is, the closer other examples must be to be affected.

We use GridSearchCV to search over specified parameter values and choose the best estimator based on macro F1-score. Finally, we ended up with the best hyper-parameters $C = 10.0$ and $\gamma = 1e - 05$.

Because of the imbalanced dataset, we have also adjusted the different C for different classes using class weights. We set the hyperparameter C for class i as $classweight[i] * C$. Class weights are calculated by the following formula.

$$classweight = \frac{\text{n}}{N \times np.bincount(y)} \tag{2}$$

Besides that, we want to gain a less biased result of the model's performance. So, we have also used k-fold cross-validation strategy with $k = 10$.

4 Results and Discussions

4.1 Evaluation Metrics

Average Precision (AP) and mean Average Precision (mAP) are the most popular metrics used to evaluate object detection models.

The IoU is used when calculating mAP, a number from 0 to 1. It is calculated as follows:

$$IoU = \frac{area\ of\ overlap}{area\ of\ union} = \frac{area(groundtruth \cap prediction)}{area(groundtruth \cup prediction)} \tag{3}$$

A precision-recall (PR) curve, which is a plot of precision and recall at varying confidence values, is then generated for each object class and the AP is computed. The general definition for the AP is finding the area under the precision-recall curve above.

$$AP = \sum(recall_{n+1} - r_n)precision_{interp}(recall_{n+1}) \tag{4}$$

where

$$precision_{interp}(recall_{n+1}) = \max_{\widetilde{recall} \geq recall_{n+1}}\left(\widetilde{recall}\right) \tag{5}$$

Finally, the mean of the AP of all object classes is the mAP.

$$mAP = \frac{1}{N}\sum_{i=1}^{N} AP_i \tag{6}$$

The diagram below (Fig. 3) demonstrates just how many steps it takes to calculate the final mAP.

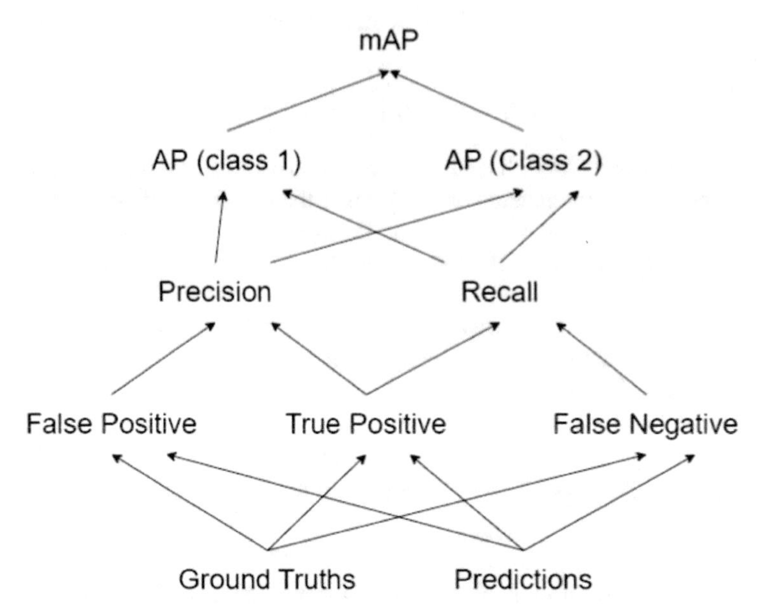

Fig. 3. mAP calculation.

On the side of the multiclass classification model, we used F1-score to evaluate its performance. The formula of F1-score is as follows.

$$F1 - score = \frac{2 \times precision \times recall}{precision + recall} \tag{7}$$

where

$$Precision = \frac{TP}{TP+FP} \tag{8}$$

$$Recall = \frac{TP}{TP+FN} \tag{9}$$

4.2 Results

In this study, we evaluated our Faster R-CNN model on the testing set using mAP@.50, mAP@.75, and mAP@[.5:.95] (the average AP over 10 IoU levels from 0.5 to 0.95 with a step size of 0.05). We have achieved the mAP@.50, mAP@.75, mAP@[.5:.95] of 0.91, 0.81, and 0.65, respectively. Figures 4, 5, and 6 show the performance results which tested on mAP@.50, mAP@.75, and mAP@[.5:.95], respectively.

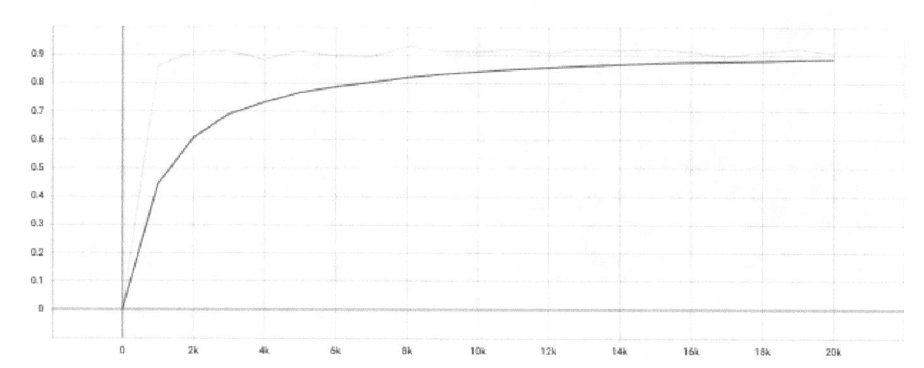

Fig. 4. mAP@.05 result.

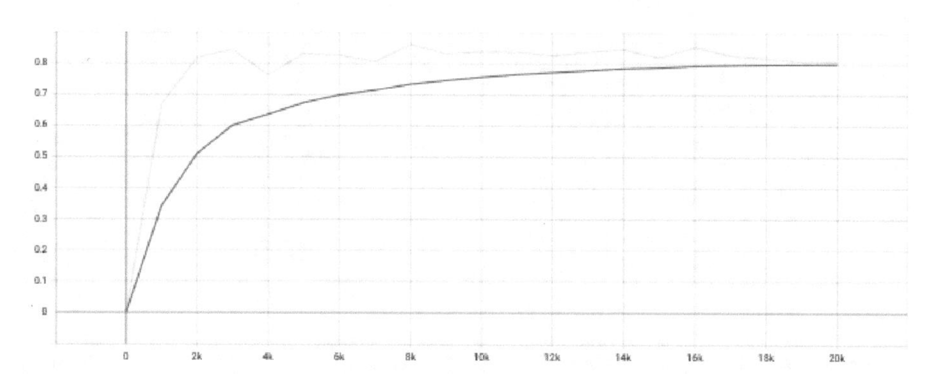

Fig. 5. mAP@.075 result.

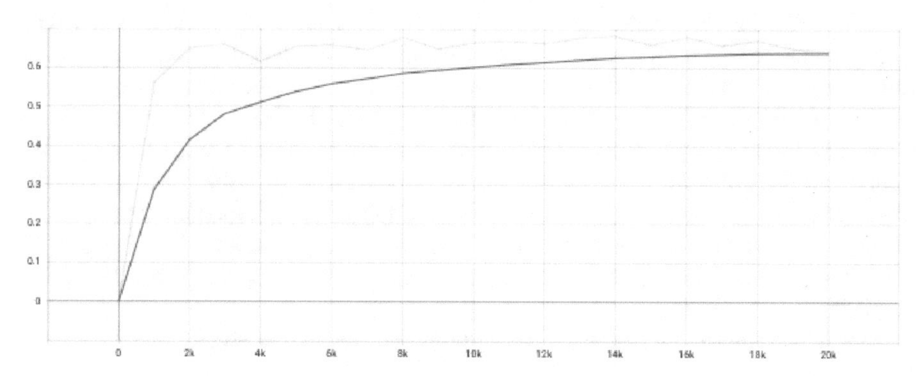

Fig. 6. mAP@[.5:.95] result.

Here is the visualization of a sample after going through the binary classification step using Faster R-CNN (Figs. 7 and 8).

Our multiclass classifier EfficientNetV2L-SVM also achieves quite high performance when the macro F1-score reaches 97% and the accuracy reaches 98%, which presented in Table 1.

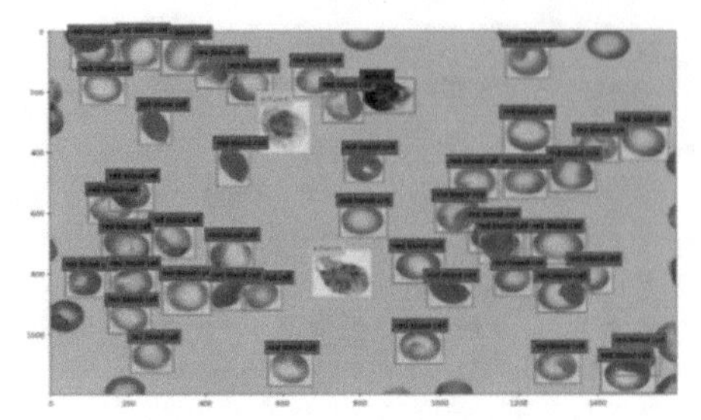

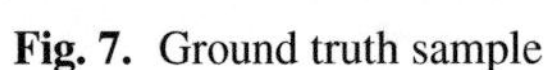

Fig. 7. Ground truth sample

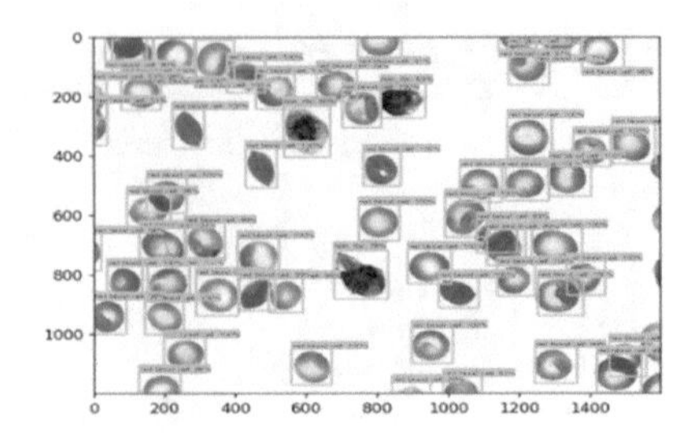

Fig. 8. Predicted sample

Table 1. EfficientNetV2L-SVM accuracy

Class	Precision	Recall	F1-score
Gametocyte	1.00	0.91	0.95
Leukocyte	1.00	0.93	0.96
Ring	1.00	0.96	0.98
Schizont	0.99	0.93	0.96
Trophozoite	0.97	1.00	0.98

We evaluated the performance of our approach using the mean average precision at IoU = 0.5 (mAP@0.5), IoU = 0.75 (mAP@.75), and mAP@[.5:.95] metrics. Our experimental results show that our proposed approach achieves the highest accuracy at mAP@.5 with the overall performance of 88.7%. This is a promising result with acceptable accuracy levels. It indicates the potential of deep learning-based approaches for accurate and efficient detection of malaria in peripheral blood smear images.

The obtained results show that our proposed approach has been validated on a larger dataset and achieves better accuracy than some previous related works. However, we found that the proposed framework does not work well on the minor class due to the data imbalance problem. The performance of our approach for the minor class was still much lower than that of the major class, as the recall of the gametocyte class is just 0.91, leukocyte and schizont are both achieved the recall of 0.93, compared with 0.96 of ring and 1.00 of trophozoite.

We have visualized a prediction of our proposed pipeline in Fig. 9.

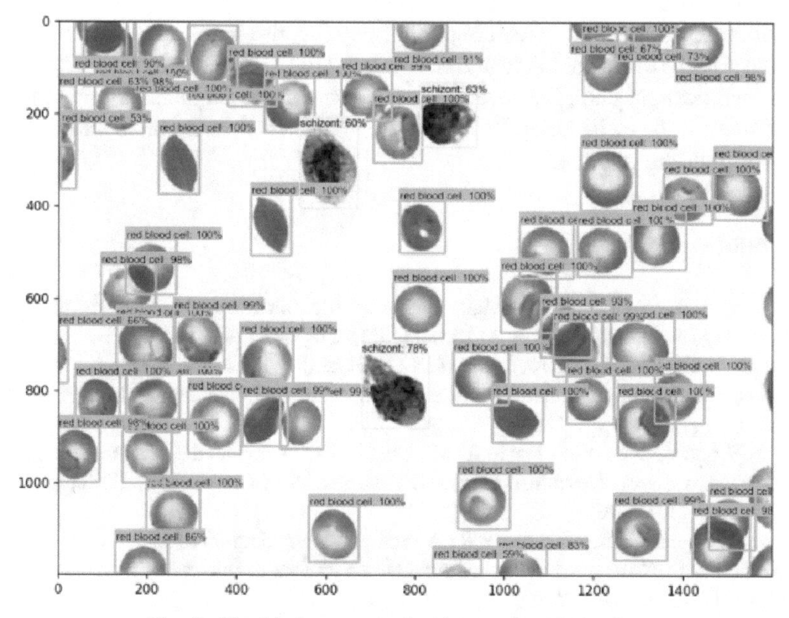

Fig. 9. Final inference result using proposed pipeline

5 Conclusion

This study developed a deep learning system to diagnose malaria from blood smear images, improving detection rate and performance. Moreover, our approach can be easily adapted to other related diseases, such as dengue and Zika, which also require the detection of parasites in blood samples. Our results also highlight the importance of addressing the data imbalance problem. The performance of our approach on the minor class was much lower than that of the major class, indicating the need for more data to improve the detection of the minor class. In addition to collecting more data, other approaches such as data augmentation and class weighting can be used to address the data imbalance problem.

The topic of an artificial intelligence model for automatically diagnosing malaria through peripheral blood smear images is a new and innovative study with great potential for widespread application in the field of healthcare. However, applying this model to real-life medical situations still poses many challenges.

One of the major challenges is ensuring the reliability and accuracy of the model, especially when used in various pathological cases or under different conditions. This research needs to be continued and validated on larger and more diverse datasets, along with the use of in-depth model evaluation methods to ensure reliability and effectiveness.

In addition, the deployment of the model into real healthcare environments needs to be carefully considered, especially regarding its ability to integrate with existing healthcare systems, as well as ensuring patient privacy and security. This requires close collaboration between medical experts and computer science experts, along with acceptance and support from regulatory and legal agencies.

In summary, this research is an important step forward in the application of artificial intelligence to the healthcare field and has the potential to bring many benefits to patients and the healthcare system. However, the development and implementation of this proposed model needs to be carefully and closely performed with collaboration between experts from different fields.

References

1. Chavan, S.N., Sutkar, A.M.: Malaria disease identification and analysis using image processing.Int. J. Lat. Trends Eng. Technol. **3**(3), 218–223 (2014)
2. Das, D.K., Ghosh, M., Pal, M., Maiti, A.K., Chakraborty, C.: Machine learning approach for automated screening of malaria parasite using light microscopic images. Micron **45**, 97–106 (2013)
3. Ross, N.E., Pritchard, C.J., Rubin, D.M., Duse, A.G.: Automated image processing method for the diagnosis and classification of malaria on thin blood smears. Med. Biol. Eng. Comput. **44**(5), 427–436 (2006)
4. Díaz, G., González, F.A., Romero, E.: A semi-automatic method for quantification and classification of erythrocytes infected with malaria parasites in microscopic images.J. Biomed. Inform. **42**(2), 296–307 (2009)
5. Linder, N., et al.: A malaria diagnostic tool based on computer vision screening and visualization of Plasmodium falciparum candidate areas in digitized blood smears. PLoS ONE **9**(8), e104855 (2014)
6. Tek, F.B., Dempster, A.G., Kale, I.: Parasite detection and identification for automated thin blood film malaria diagnosis. Comput. Vis. Image Underst. **14**(1), 21–32 (2010)
7. Seman, N.A., Isa, N.A.M., Li, L.C., Mohamed, Z., Ngah, U.K., Zamli, K.Z.: Classification of malaria parasite species based on thin blood smears using multilayer perceptron network.Int. J. Comput. Internet Manag. **16**(1), 46–52 (2008)
8. Khot, S., Prasad, R.: Optimal computer based analysis for detecting malarial parasites," trong. In: Satapathy, S., Biswal, B., Udgata, S., Mandal, J. (eds) Proceedings of the 3rd International Conference on Frontiers of Intelligent Computing: Theory and Applications (FICTA) 2014. Advances in Intelligent Systems and Computing, vol. 327, pp. 69–80. Springer, Cham (2015). https://doi.org/10.1007/978-3-319-11933-5_9
9. Memeu, D.M.: A rapid malaria diagnostic method based on automatic detection and classification of plasmodium parasites in stained thin blood smear images. University of Nairobi (2014)
10. Liang, Z., et al.: CNN-based image analysis for malaria diagnosis. In: 2016 IEEE International Conference on Bioinformatics and Biomedicine (BIBM). IEEE, pp. 493–496 (2016)
11. Hung, J., Carpenter, A.: Applying faster R-CNN for object detection on malaria images. In: Proceedings of the IEEE Conference on Computer Vision and Pattern Recognition Workshops, pp. 56–61 (2017)
12. Pan, W.D., Dong, Y., Wu, D.: Classification of malaria-infected cells using deep convolutional neural networks. Machine Learn.: Adv. Tech. Emerg. Appl. **159** (2018)
13. Ljosa, V., Sokolnicki, K.L., Carpenter, A.E.: Annotated high-throughput microscopy image sets for validation. Nat. Methods **9**(7), 637 (2012)
14. Kassim, Y.M., et al.: Clustering-based dual deep learning architecture for detecting red blood cells in malaria diagnostic smears. IEEE J. Biomed. Health Inform. **25**(5), 1735–1746 (2020)
15. Yang, F., et al.: Deep learning for smartphone-based malaria parasite detection in thick blood smears. IEEE J. Biomed. Health Inform. **24**(5), 1427–1438 (2014)

16. Kassim, Y.M., Yang, F., Yu, H., Maude, R.J., Jaeger, S.: Diagnosing malaria patients with plasmodium falciparum and vivax using deep learning for thick smear images. Diagnostics **11**(11), 1994 (2021)
17. Ren, S., He, K., Girshick, R., Sun, J.: Faster R-CNN: towards real-time object detection with region proposal networks. Adv. Neural Inf. Process. Syst. **28** (2015)
18. He, K., Zhang, X., Ren, S., Sun, J.: Deep residual learning for image recognition. In: Proceedings of the IEEE Conference on Computer Vision and Pattern Recognition (2016)
19. Tan, M., Le, Q.: Efficientnetv2: smaller models and faster training. In: International Conference on Machine Learning (2021)

An Improved Transfer Learning-Based Approach for Detecting Abnormal Cervical Cells from Pap Smear Microscopic Images

Nga Le-Thi-Thu(✉) and Vu-Thu-Nguyet Pham

The University of Danang – Vietnam-Korea University of Information and Communication Technology, Danang, Vietnam
{lttnga,pvtnguyet.19it1}@vku.udn.vn

Abstract. Cervical cancer is one of the most common and dangerous diseases for women's health. Automated abnormal cell screening will help detect early and improve the accuracy of cervical cancer diagnosis, from which women can plan treatment, preventing the development of cervical cancer and improving survival rates. This paper proposes an improved method to detect abnormal cervical cells from images of cells that are stained and examined under a microscope based on a transfer learning approach. Achieved results show that the proposed EfficientNetv2L-SVM model has an accuracy of 92% and an F1-score of 88%, higher than the ResNet50v2 deep learning model with 90% and 83%, respectively. The measurement of Precision and Recall also gives the same result. Experimental data were collected from the Cytopathology Laboratory of Binh Dinh province over ten years, from 2014 to 2023 with five different types of cervical cells based on the diagnosis of expert pathologists (*The dataset in this research is provided and allowed to use by the Cytopathology Laboratory, located at Quy Nhon City, Binh Dinh Province, VietNam.*).

Keywords: deep learning · transfer learning · screening · abnormalities detection · cervical cells

1 Introduction

Cervical cancer is one of the most common and dangerous diseases for women's health. Automated abnormal cell screening will help detect cancer at an early stage and improve the accuracy of cervical cancer diagnosis, from which women can plan treatment, preventing the development of cervical cancer and improving survival rates.

Abnormal cervical cell detectionis also an issue of concern to researchers in recent years. Research on artificial intelligence for screening and diagnosis of cervical cancer also has given positive results. Various machine learning models have been applied to detect abnormal cervical cells [1–7]. Commonly machine learning models include logistic regression, k-nearest neighbor, decision tree, support vector machine, and random forest. Deep learning models are also interested in research such as artificial neural networks, multilayer perceptrons, convolutional neural networks, deep neural networks,

N. T. Nguyen et al. (Eds.): CITA 2023, LNNS 734, pp. 126–137, 2023.
https://doi.org/10.1007/978-3-031-36886-8_11

and ResNet [8–13]. In general, the accuracy of the models in predicting cervical cancer varied from 67% to 92%. The effectiveness of predicting cervical cancer by each algorithm is markedly different.

In 2022, A. Leila and et al. review searches were performed on three databases: Medline, Web of Science Core Collection, and Scopus to find papers published until July 2022. Their systematic review highlights the acceptable performance of artificial intelligence models in the prediction, screening, or detection of cervical cancer and pre-cancerous lesions [14].

This paper proposes an improved method to detect abnormal cervical cells from images of cells that are stained and examined under a microscope based on a transfer learning approach. Achieved results show that the proposed EfficientNetv2L-SVM model has an accuracy of 92% and an F1-score of 88%, higher than the ResNet50v2 deep learning model with 90% and 83%, respectively. The measurement of Precision and Recall also gives the same result.

The rest of this paper is organized as follows: Sect. 2 presents abnormal cervical cells and the dataset used for training models. The proposed approach based on transfer learning (EfficientNetv2L-SVM), deep learning model used for comparing (ResNet50v2), and abnormal cervical cell detection system is described in Sect. 3. Experimental results are provided in Sect. 4. Finally, several concluding remarks are discussed, and future works are drawn in Sect. 5.

2 Abnormal Cervical Cells and Preprocessing Data

2.1 Abnormal Cervical Cells

Normal cervical cells appear flat, smooth, and thin when viewed under a microscope. The ratio between the nucleus and cytoplasm can vary depending on factors such as age, hormonal status, and health status. In general, normal cervical cells have a higher ratio of nucleus to protoplasm than other cell types.

Normal cervical cells are a sign of a healthy cervix and are typically seen in women who are not experiencing any symptoms or abnormalities in their cervical screening tests (Fig. 1).

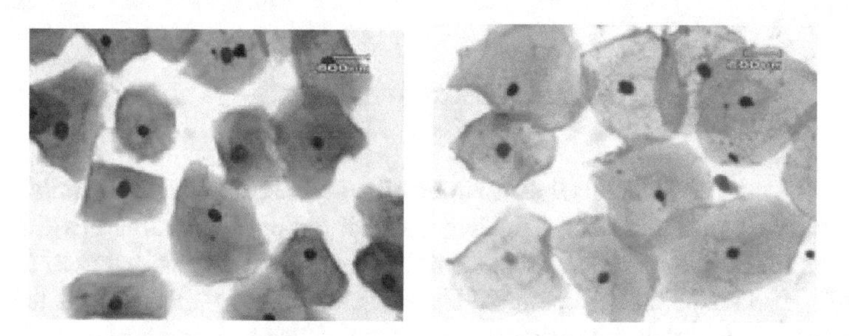

Fig. 1. Normal cervical cells

Abnormal cervical cells are cells in the surface lining of the cervix that appear abnormal under a microscope. These abnormal cells are not cancerous, but they do require testing, early detection, follow-up, and treatment to prevent the development of cervical cancer. Many types of abnormal cervical cells can be detected through a pap smear or cervical screening tests. This study focused on four types of abnormal cervical cells: cells infected with Actinomyces bacteria, Atrophy cells, cells infected with Candida albicans, and cells infected with Clue bacteria.

Actinomyces is a type of Gram-positive bacteria that alters the appearance and structure of cervical cells. Under a microscope, cells infected with Actinomyces may appear enlarged and irregularly shaped with a distorted nucleus, as well as the presence of sulfur granules or other bacterial colonies. These cells tend to inflammation and degeneration.

Candida albicans is a type of fungus that infects the cervix, they may be visible under a microscope as budding yeast cells or chains of elongated yeast cells (Fig. 2).

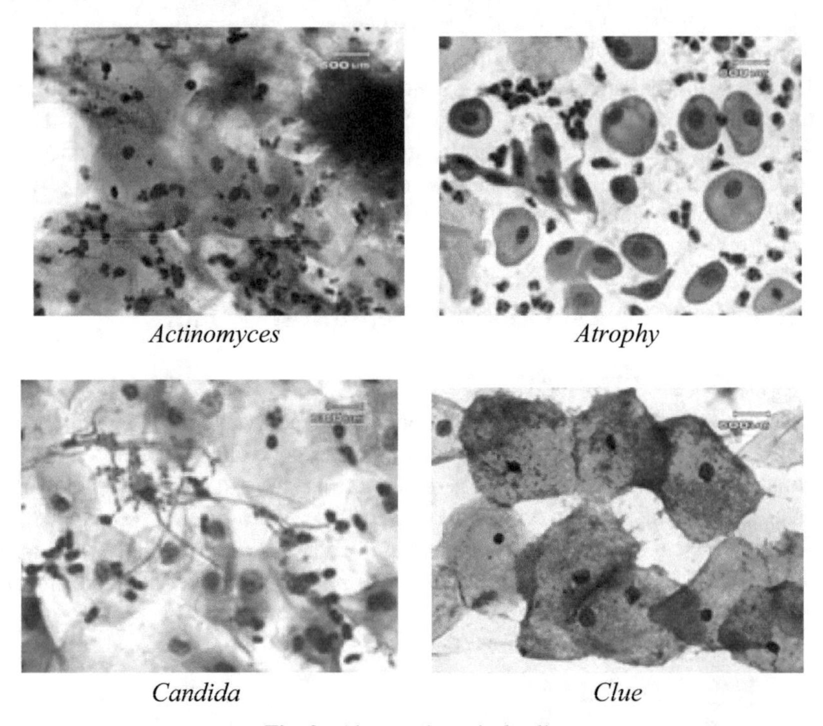

Fig. 2. Abnormal cervical cells

Clue cells are cervical cells that are infected with the bacteria Gardnerella vaginalis. Under a microscope, these cells are large, and irregular with numerous tiny dots or granules on the surface. The dots represent the bacteria that are adhering to the surface of the cell. The shape and structure of Clue cells can vary depending on the severity of the infection and other factors. In general, Clue cells are larger and more irregularly shaped compared to normal cells.

Cervical cells can be atrophied, which is a gradual loss of cell volume and size. As cervical cells atrophy, they tend to become smaller and deformed compared to normal cervical cells. The protoplasm may be shrunk or retracted, the nucleus appears smaller and denser, and the chromatin inside the nucleus may be denser or clumped together, giving the nucleus a darker color than in normal cervical cells. The most common cause of cervical cell atrophy is hormonal changes, medical conditions, or treatments.

2.2 Exploratory Data Analysis and Preprocessing

The dataset used in this study was collected from the Cytopathology Laboratory of Binh Dinh province, VietNam. The dataset consists of cervical cell images obtained over ten years, from 2014 to 2023. The samples were categorized into five different classes: actinomyces, atrophy, candida, clue, and normal, based on expert pathologist diagnosis. The details of the dataset are mentioned in Table 1.

The dataset comprises a total of several thousand images of cervical cells, with varying image resolutions and quality. The images were captured using different microscopy devices, including light microscopy and digital microscopy. In order to ensure consistency in the image quality, we preprocessed the images by applying filters and standardizing their sizes.

Table 1. Dataset of Cervical Cells

Class	Train	Test	No. Samples
Actinomyces	227	97	324
Atrophy	284	122	405
Candida	1354	580	1934
Clue	785	337	1122
Normal	2030	870	2900

3 Transfer Learning and Proposed Approach

3.1 ResNet50v2

ResNet50v2 is a convolutional neural network architecture that builds upon the ResNet model and is an improved variant of ResNet50. It comprises 49 convolutional layers, with one additional Batch Normalization layer inserted before each convolutional layer, one MaxPool, and one Average Pool layer. The model achieves a total of 4.1 x 10^9 floating-point operations (FLOPs) [15].

The main advantage of ResNet50v2 over ResNet50 is the inclusion of Batch Normalization layers, which help to stabilize the training process by reducing the effect of covariate shift. This is achieved by normalizing the input data to each layer to have zero mean and unit variance. The inclusion of Batch Normalization layers also enables the use of higher learning rates during training, resulting in faster convergence and better model performance.

Furthermore, ResNet50v2 employs an improved residual block design, known as the bottleneck residual block, which reduces the computational cost and memory usage of the model, making it more efficient. This design utilizes 1x1 convolutional layers to reduce the dimensionality of the input before the expensive 3x3 convolutional layer is applied, thereby reducing the number of computations required [15]. ResNet50v2 has demonstrated superior performance over ResNet50 and other state-of-the-art CNN models in various computer vision tasks, including object recognition, image classification, and object detection [15–17]. It has been widely adopted in research and industry applications, owing to its superior performance and efficiency.

3.2 EfficientNetv2L

EfficientNetv2L is a highly efficient convolutional neural network architecture that has been developed to achieve high accuracy with low computational cost [18]. This architecture is based on a compound scaling method that involves scaling the network depth, width, and resolution in a balanced manner. The EfficientNetv2L model achieves state-of-the-art performance on various image classification benchmarks, such as ImageNet and CIFAR-10, while being significantly smaller and faster than existing models. The architecture has significant implications for real-world applications where high accuracy and low computational cost are essential.

The EfficientNetv2L architecture employs several design choices that contribute to its efficiency and accuracy. One of these design choices is the use of squeeze-and-excitation blocks, which are able to adaptively recalibrate the feature maps of the network based on their importance. Another key design choice is the use of MBConv blocks, which are efficient building blocks that combine depthwise convolutions and pointwise convolutions to reduce computation without sacrificing accuracy. Additionally, EfficientNetv2L uses SE-Swish activation functions, which are a variant of the Swish activation function that is computationally efficient and achieves higher accuracy than the widely used ReLU activation function.

EfficientNetv2L is a significant advancement in the field of computer vision, as it enables the development of highly efficient models that can achieve state-of-the-art performance on image classification tasks. The architecture has several practical applications, such as in the development of mobile and embedded systems where computational resources are limited, or in real-time applications such as autonomous driving where low latency is critical. Overall, EfficientNetv2L represents a major contribution to the field of deep learning and has the potential to drive innovation in a wide range of applications.

3.3 Proposed Approach Based on Transfer Learning

The transfer learning model is a new model that uses part or all of a pre-trained model. Transfer learning can be defined as follows: Given a source domain D^{source} and a learning task T^{source}. We denote D^{target} and T^{target} are a target domain and a target learning task, respectively. Transfer learning technique aims to help improve the learning of the f prediction model for T^{target} utilizing knowledge in D^{source} and T^{source}, where $D^{source} \neq D^{target}$ or $T^{source} \neq T^{target}$. By re-weighting, the observations in D^{source}, the effects of the various samples are reduced. Conversely, similar instances will contribute more to

T^{target} and may lead to a more accurate prediction. The use of pre-trained models is a big step forward to inheriting the results of previous models, taking advantage of existing pre-trained models to create new models for more specific, applicable target tasks, more practical.

In this study, we propose a transfer learning-based approach that employs EfficientNetv2L as a feature extractor and Support Vector Machine (SVM) as its head for the classification of images. Figure 3 shows a detailed model of the Abnormal cervical cells detection system. We have also trained the ResNet50v2 model to compare its performance with our proposed approach.

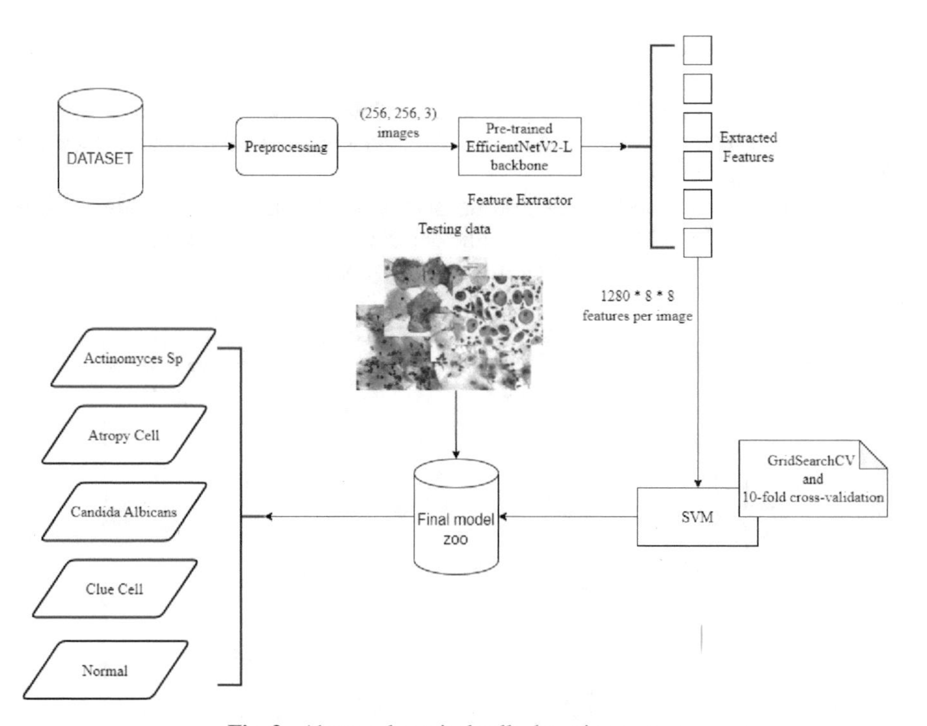

Fig. 3. Abnormal cervical cells detection system

EfficientNetv2L is an advanced neural network architecture that has been specifically designed for efficient image recognition. The architecture combines efficient building blocks and neural architecture search to achieve high accuracy with minimal computational cost. SVM is a widely-used machine learning algorithm known for its effectiveness in image classification. SVM works by identifying a hyperplane in a high-dimensional space that optimally separates the classes.

To use EfficientNetv2L as a feature extractor, the network is first pre-trained on a large dataset such as ImageNet to enable it to learn features that are beneficial for image recognition tasks. The last layer of the network, which typically maps the learned features to class probabilities, is removed. The output of the penultimate layer of EfficientNetv2L, which contains the learned features that are most pertinent for image recognition tasks, is then used as input to the SVM. The SVM is trained on the target dataset to learn the decision boundary that separates the classes.

During the testing phase, each image is fed through EfficientNetv2L to extract its features, which are then used as input to the SVM to determine the predicted class label.

This approach provides several benefits. Firstly, EfficientNetv2L is highly efficient and can extract features from images very quickly, making it suitable for real-time applications. Secondly, SVM is a well-established algorithm that has been shown to work well for image classification tasks. Finally, by leveraging transfer learning through EfficientNetv2L, the classifier's performance can be improved even with small datasets.

4 Experimental Results

All our experiments used NVIDIA GeForce GTX 1050Ti Laptop GPU 4GB VRAM with a system of 16 GB RAM and 8 CPU cores. We used Keras with Tensorflow backend as the deep learning framework.

Below Figs. 4 and 5 present the loss and accuracy of training and validating process of the ResNet50v2 model. The yellow line describes validating process, and the blue line illustrates training process of the ResNet50v2 model.

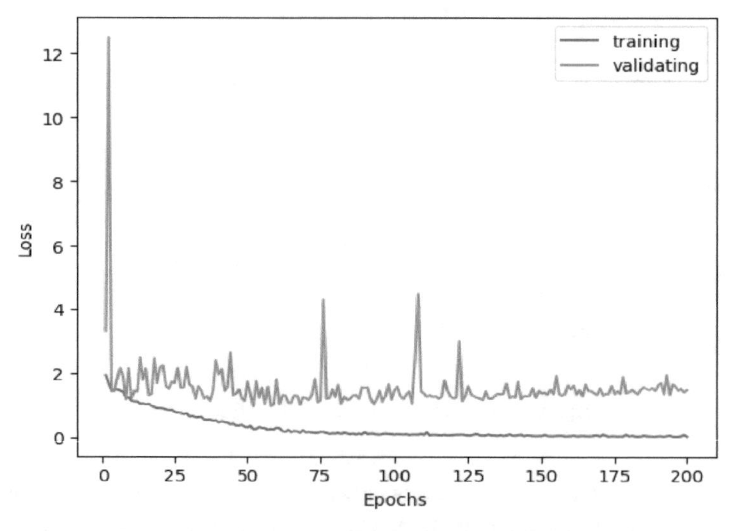

Fig. 4. Loss history of training and validating process of the ResNet50v2 model

On the side of the EfficientNetv2L-SVM model, we use k-fold cross-validation strategy with $k = 10$. The results are shown in Fig. 6. We can see that the proposed model gives a much more stable test result than the pre-trained ResNet50v2.

The detailed results in Tables 2 and 3 show that our proposed approach does better even in the minor classes, result in higher accuracy and F1-score. Looking at the first table, we can see that the model achieved high precision (above 0.75) for all classes, indicating that when it predicted a positive instance, it was usually correct. The highest

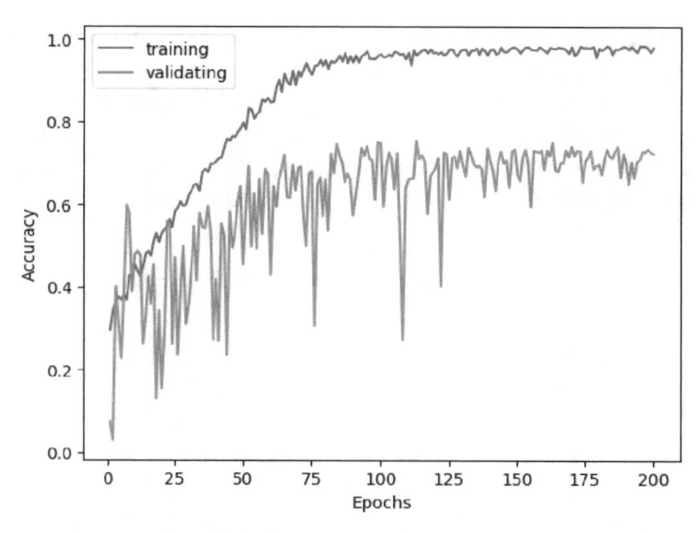

Fig. 5. Accuracy history of training and validating process of the ResNet50v2 model

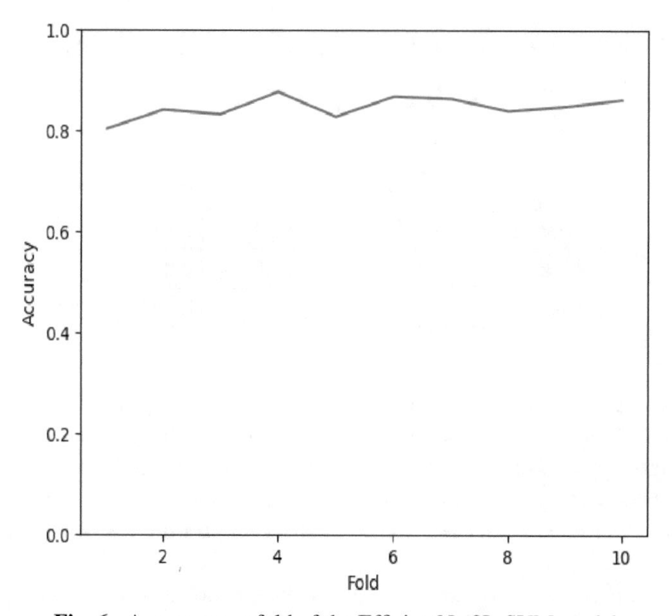

Fig. 6. Accuracy per fold of the EfficientNet2L-SVM model

precision was achieved for the "Normal" class, which means that the model was very accurate in identifying normal instances. The lowest precision was achieved for the "Actinomyces" class, which means that the model had more false positives for this class. The recall metric in the first table shows that the model was generally good at identifying positive instances, with recall above 0.8 for all classes except "Atrophy". The highest recall was achieved for the "Normal" class, which means that the model was very good at identifying normal instances. The lowest recall was achieved for the "Atrophy" class, which means that the model missed many positive instances of this class. The F1-score metric in the first table shows that the model had the best overall performance for the "Normal" class, with an F1-score of 0.95, indicating high accuracy and completeness. The lowest F1-score was achieved for the "Atrophy" class, indicating that the model had difficulty identifying positive instances of this class.

Table 2. Detail scores of classes using ResNet50v2

Class	Precision	Recall	F1-score
Actinomyces	0.76	0.88	0.82
Atrophy	0.78	0.69	0.73
Candida	0.84	0.83	0.83
Clue	0.79	0.85	0.82
Normal	0.95	0.95	0.95

Table 3. Detail scores of classes using proposed EfficientNetv2L-SVM

Class	Precision	Recall	F1-score
Actinomyces	0.96	0.83	0.89
Atrophy	0.90	0.83	0.86
Candida	0.86	0.84	0.85
Clue	0.81	0.89	0.84
Normal	0.95	0.96	0.96

Turning to the second table, we see that the precision and recall metrics for each class have changed from the first table. The precision for the "Actinomyces" class has increased significantly from 0.76 to 0.96, indicating that the model is now much more accurate in predicting positive instances of this class. The precision for the "Atrophy" class has also increased from 0.78 to 0.90, indicating an improvement in the accuracy of positive predictions.

The recall metric in the second table shows that the model is now better at identifying positive instances of the "Atrophy" class, with an increase from 0.69 to 0.83. The recall for the "Actinomyces" class has decreased from 0.88 to 0.83, indicating that the model is now missing more positive instances of this class.

The F1-score metric in the second table shows that the model has improved its performance for the "Atrophy" class, with an increase from 0.73 to 0.86, indicating a better balance between precision and recall. The F1-score for the "Actinomyces" class has also improved, from 0.82 to 0.89.

Turning to the second table, we see that the precision and recall metrics for each class have changed from the first table. The precision for the "Actinomyces" class has

Table 4. Inference results

Model	Accuracy	Precision	Recall	F1-score
ResNet50v2	0.90	0.82	0.84	0.83
Proposed model	**0.92**	**0.90**	**0.87**	**0.88**

increased significantly from 0.76 to 0.96, indicating that the model is now much more accurate in predicting positive instances of this class. The precision for the "Atrophy" class has also increased from 0.78 to 0.90, indicating an improvement in the accuracy of positive predictions.

The recall metric in the second table shows that the model is now better at identifying positive instances of the "Atrophy" class, with an increase from 0.69 to 0.83. The recall for the "Actinomyces" class has decreased from 0.88 to 0.83, indicating that the model is now missing more positive instances of this class.

The F1-score metric in the second table shows that the model has improved its performance for the "Atrophy" class, with an increase from 0.73 to 0.86, indicating a better balance between precision and recall. The F1-score for the "Actinomyces" class has also improved, from 0.82 to 0.89.

Table 4 shows the inference results of the two models. Here, we use accuracy, precision, recall, and F1-score metrics to evaluate the performance of the models. Achieved results show that the proposed EfficientNetv2L-SVM model has an accuracy of 92% and an F1-score of 88%, higher than the ResNet50v2 deep learning model with 90% and 83%, respectively.

Looking at the table, we can see that the proposed model achieved a higher accuracy than ResNet50v2, with an accuracy of 0.92 compared to 0.90. This means that the proposed model made more correct predictions than ResNet50v2 on the given dataset.

The precision of the proposed model is also higher than ResNet50v2, with a precision of 0.90 compared to 0.82. This indicates that the proposed model had fewer false positives than ResNet50v2.

The recall of the proposed model is slightly higher than ResNet50v2, with a recall of 0.87 compared to 0.84. This means that the proposed model correctly identified more positive instances compared to ResNet50v2.

Based on the evaluation metrics provided in the table, it can be concluded that the proposed model outperformed ResNet50v2 in terms of accuracy, precision, and F1-score, while having a slightly higher recall value. This means that the proposed model made fewer false positive predictions, had a better balance between precision and recall, and identified more actual positive instances compared to ResNet50v2. However, it should be noted that the relative importance of these metrics may vary depending on the specific application or requirements of the task. Therefore, further analysis may be needed to determine which model is more appropriate for specific use cases. Overall, the results indicate that the proposed model is a promising candidate for classification tasks that require high accuracy and precision, while maintaining a good balance between recall and F1-score.

5 Conclusion

In conclusion, the proposed approach of utilizing Efficientnetv2L as a feature extractor and SVM as the head to classify cervical cell abnormalities is a promising technique for improving the accuracy of detection. The results of our experiments demonstrate that the proposed approach achieves a higher accuracy rate of 92% as compared to the state-of-the-art ResNet50v2 model. This indicates that the proposed model has the potential to improve the efficiency and accuracy of cervical abnormalities diagnosis using deep learning techniques.

The study highlights the importance of selecting an appropriate feature extractor and classification head for the given task, as it significantly affects the performance of the model. Efficientnetv2L, being a relatively recent architecture, demonstrates its effectiveness as a feature extractor for this task. Additionally, the use of SVM as the head classifier results in an improved performance, which can be attributed to its ability to handle high-dimensional data and reduce overfitting.

Overall, the proposed approach provides a promising solution to the problem of abnormalities detection in cervical cell images using deep learning techniques. Further research could focus on investigating the effectiveness of the proposed approach on larger datasets and exploring the possibility of utilizing other deep learning architectures as feature extractors. The findings of this study have the potential to contribute to the development of more accurate and efficient diagnostic tools for early cervical cancer detection.

Acknowledgments. *Thank the Cytopathology Laboratory, located at Quy Nhon City, Binh Dinh Province, VietNam, for providing and allowing us to use the dataset in this research.*

References

1. Ali, M., Ahmed, K., Bui, F.M., Paul, B.K., et al.: Machine learning-based statistical analysis for early stage detection of cervical cancer. Comput. Biol. Med. **139**, 104985 (2021)
2. Sarwar, A., Suri, J., Ali, M., Sharma, V.: Novel benchmark database of digitized and calibrated cervical cells for artificial intelligence based screening of cervical cancer. J. Ambient Intell. Human. Comput. (2016)
3. Nithya, B., Ilango, V.: Evaluation of machine learning based optimized feature selection approaches and classification methods for cervical cancer prediction. SN Appl. Sci. (2019)
4. Shanthi, P., Hareesha, K., Kudva, R.: Automated detection and classification of cervical cancer using pap smear microscopic images: a comprehensive review and future perspectives. Eng. Sci. (2022)
5. Wesabi, Y., Choudhury, A., Won, D.: Classification of cervical cancer dataset. arXiv (2018)
6. Asadi, F., Salehnasab, C., Ajori, L.: Supervised algorithms of machine learning for the prediction of cervical Cancer. J. Biomed. Phys. Eng. (2020)
7. Bao, H., Sun, H., Zhang, Y., et al.: The artificial intelligence-assisted cytology diagnostic system in large-scale cervical cancer screening: a population-based cohort study of 0.7 million women. Cancer Med. **9**, 6896–6906 (2020)
8. Nieminen, P., Hakama, M., Viikki, M., Tarkkanen, J., Anttila, A.: Prospective and randomised public-health trial on neural network-assisted screening for cervical cancer in Finland: Results of the first year. Int. J. Cancer **103**,422–426 (2002)

9. Hu, L., Bell, D., Antani, S., et al.: An observational study of deep learning and automated evaluation of cervical images for cancer screening. Natl. Cancer Inst. (2019)
10. Sanyal, P., Ganguli, P., Barui, S.: Performance characteristics of an artificial intelligence based on convolutional neural network for screening conventional Papanicolaou-stained cervical smears. Med. J. Armed Forces India (2019)
11. Xiang, Y., Sun, W., Pan, C., Yan, M., Yin, Z., Liang, Y.: A novel automation-assisted cervical cancer reading method based on convolutional neural network. Biocybern. Biomed. Eng. (2020)
12. Cheng, S., Liu, S., Yu, J., et al.: Robust whole slide image analysis for cervical cancer screening using deep learning. Nat. Commun. (2021)
13. Tang, H., Cai, D., Kong, Y., et al.: Cervical cytology screening facilitated by an artificial intelligence microscope: A preliminary study. Cancer Cytopathol. **129**, 693–700 (2021)
14. Leila, A., et al.: Diagnosis of Cervical Cancer and Pre-Cancerous Lesions by Artificial Intelligence: A Systematic Review. Diagnostics (2022)
15. He, K., Zhang, X., Ren, S., Sun, J.: Deep residual learning for image recognition. In: Proceedings of the IEEE Conference on Computer Vision and Pattern Recognition (2016)
16. Tan, M., Le, Q.: Efficientnet: rethinking model scaling for convolutional neural networks. In: International Conference on Machine Learning (2019)
17. Chollet, F.: Xception: Deep learning with depthwise separable convolutions. In: Proceedings of the IEEE Conference on Computer Vision and Pattern Recognition (2017)
18. Tan, M., Le, Q.: Efficientnetv2: smaller models and faster training. In: International Conference on Machine Learning (2021)

Digital Economy

Emotional Intelligence and Social Media Site Experiences' Effects on College Choice Behavior: The Mediating Role of Brand Attitude

Thi Thanh Minh Dang(✉), Phuoc Cuu Long Le, and Thi Thuy Trinh Tran

The University of Da Nang, Vietnam – Korea University of Information Technology and Communication, Danang, Vietnam
{dttminh,lpclong,ttttrinh}@vku.udn.vn

Abstract. This study examines how students use social networks to choose their universities, looking at the effects of online experience, Emotional intelligence, and brand attitude as a mediating variable. The postulated connections between emotional intelligence, social media site experiences, brand attitude, and decision-making have led to the construction of a model. Students who used to base their institution choice on information found on social networks comprised the 310 respondents to a study that was conducted. Utilizing a quantitative method, the findings show that the more pleasant a student's social media site experiences are, the more positive a student's opinion toward the social media site is, which affects the student's behavior when choosing a university. Also, it was revealed by the results that emotional intelligence influences students' university decisions in a good way.

Keywords: Social media site experiences · Emotional intelligence · choice behavior · brand attitude

1 Introduction

The past two decades have experienced a dramatic shift in Internet applications. E-Commerce is becoming increasingly common; Digital marketing tools gradually replace traditional tools. In particular, the explosion of social media channels has changed how many consumers interact with each other and businesses. This has led to a significant shift in the operations of organizations in all sectors as well as changes in consumer behavior.

According to Cho & Park, 2001 [5]; an online customer is not simply a shopper but also an information technology user. So, the online experience is a more complicated issue than the physical shopping experience. There are many surveys of adult internet users showed that the website experience had changed their opinion of brands, transforming brands during purchases. For example, according to Nua Internet Surveys, 2002 [6], 60% of respondents (adult Internet users) whose opinions changed switched brands at purchase due to negative online experience, whether they bought via the Net or at

N. T. Nguyen et al. (Eds.): CITA 2023, LNNS 734, pp. 141–153, 2023.
https://doi.org/10.1007/978-3-031-36886-8_12

a brick-and-mortar store. The survey result also underlines the synergic effects of the positive Web experience on the decision to use Website next to traditional channels. Moreover, academic studies have found a relationship between website experience and buying behavior [14, 16].

In the field of education in Vietnam, universities are constantly expanding their online presence, especially their presence on social networks. According to the Vietnam Digital Development Overview Report 2021 published by we are social and Kepios in early 2022 [23], Vietnam has 76.95 million social media users, and the number of Vietnamese social network users is mainly aged 18-34. The potential and existing customers of universities are mostly Generation Z, belonging to the group of people with a high rate of social media use. Therefore, using social media channels to reach learners is necessary, and the quality of online experience is a problem requiring special attention at universities. However, only some studies fully describe how web experiences impact learners' responses in this field. Whether poorly designed or dysfunctional social media Pages provides a negative experience and influence learners' decision to choose a school.

Besides the website experience, emotional intelligence (EI) is also a factor that is being emphasized in research, and the impact of EI on decision-making is noteworthy. A better understanding of EI can expand our knowledge of customer behavior, providing answers to questions such as People with high EI versus low EI, who make easier decisions? How might consumer EI impact relationships between crucial consumer aspects such as Attitude and purchase intention? Previous studies on EI identified the relationships between people's emotional processing abilities and the quality of decision-making and performance. Higgs, 2004 [11] and Rozell et al., 2006 [21] found that higher EI results in higher performance. Kidwell et al, 2008 [13] identified the EI to be related to shopping behavior; Consumers with higher EI make better choices than those with lower EI. Therefore, there is a difference in the consumer's choice decision based on their level of EI.

Until now, there exist many studies on consumers' psychological status by taking an emotive approach in the e-commerce context. The studies also highlight the importance of emotional intelligence in consumer behavior. However, there needs to be a framework to reflect the impact of these influences, especially in the field of education. From the gaps indicated above, we constructed a model based on the proposed relationships of emotional intelligence, website experience, brand attitude, and choice behavior. This work aims to explain the effects of website experience, emotional intelligence, and brand attitude as mediating variables on the university choice behavior of students through social media channels. A survey was carried out with a sample size of 310 students who used to select a university based on information provided on social media channels.

2 Theoretical Background and Hypotheses

2.1 Website Experiences and Emotional intelligence

Website Experiences

Both academics and practitioners have identified the "online experience" or "virtual experience" as a significant issue in the digital marketing field. The Web experience is defined as the consumer's overall impression of the online company, resulting from exposure to a combination of virtual marketing tools under the marketer's direct control, likely to influence the buying behavior of the online consumer (Lorenzo, Carlota, et al., 2007) [16]. The Web experience embraces elements like searching, browsing, finding, selecting, comparing, evaluating information, interacting, and transacting with the online firm. The virtual customer's total impression and actions are influenced by design, events, emotions, atmosphere, and other elements experienced during interaction with a given Web site; features aim to create customer goodwill and affect the outcome of the online exchange. Researchers and practitioners have used several terms, such as Online Web Experience, Site Experience and Online Experience, to describe Consumers' experiences with websites. Generally, these terms have the same meaning. They are the overall perception of consumers of the site.

Emotional Intelligence

In recent years, emotional intelligence (EI) has created much interest and concern within the scientific community. EI is also known as emotional quotient (EQ). It refers to positively understanding, utilizing, and managing emotions to decrease stress, communicate effectively, empathize with others, overcome challenges, and defuse conflict. Similarly to this point of view, Goleman, 1995 [9] defined Emotional intelligence as the capacity to distinguish, be aware of, comprehend, and regulate our own emotions and sentiment as well as other people's sentiment to inspire ourselves and manage feelings very well concerning ourselves and our interactions.

According to Mayer et al., 1997 [18] the consumer's emotional Intelligence (EI) is "a person's ability to use emotional information to achieve a desired consumer outcome, comprised as a set of first-order emotional abilities that allow individuals to recognize the meanings of emotional patterns that underlie consumer decision making and to reason and solve problems based on them."

2.2 The Attitude Toward Website Model – AWS Model

The Attitude toward the advertising model (Aad) is a popular framework used to measure the effectiveness of traditional advertising on consumers' Attitudes and behavior. This model is developed into four different forms based on four alternative hypotheses: ATH (affect transfer hypothesis), DMH (Dual mediation hypothesis), RMH (reciprocal mediation hypothesis), and IIH (independent influences hypothesis). All four models describe the relationship between brand cognitions, attitudes toward the brand, and behavior intention differently. Mackenzie et al., 1986 [17] and Homers, 1990 [12] performed comparative studies between these four models based on experiment data, the results of which indicated that the DMH model best matched the data. This again confirms that attitudes towards advertising have both direct and indirect influence through brand cognitions on attitudes towards the brand.

In the past few years, the intense and rapid development of information technology has shifted the way companies do business; traditional advertising has changed by online advertising, and digital Marketing tools play an essential role in marketing activities. In the web environment, Attitude toward the website (Aws), a new variable is added to evaluate the effectiveness of advertising. Similar to Aad, Aws is defined as "predispositions to respond favorably or unfavorably to web content in natural exposure situations" of the web user (Chen & Wells, 1999) [3]. In the same view, Bruner & Kumar, 2020 [2] proposed that customer's reactions to the context where an advertisement is displayed also affect how consumer react to the advertising. They tested their new model of web advertising effectiveness. Their findings showed that one's web experience plays a vital role in Attitude toward the web along with webpage complexity and interestingness. More specifically, the relationship between Attitude toward the website, Attitude toward the advertising, brand attitude, and purchase intention was found. Thus, this work provided evidence that the Aad model can be extended into research concerning the Web and marketing communication. Integrating the three aspects of Aws (entertainment, informativeness, and organization) proposed by Chen & Wells, 1999 [3] and the research results of Bruner & Kumar 2020 [2], Poh & Adam, 2002 [22] developed the Attitude toward the website model. This model can explain the relationship between Attitude toward the website, Attitude toward the advertising, Brand attitude, and purchase intention in the Internet context (Fig. 1).

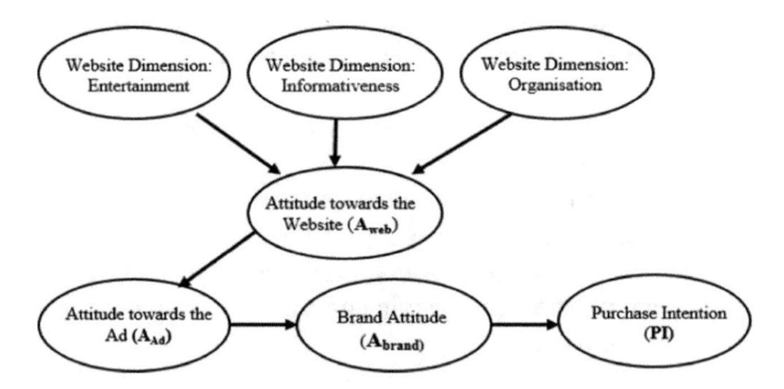

Fig. 1. The Attitude toward website Model

2.3 The Research Model and Hypotheses

As analyzed above, the authors suggested that the Aws model can be applied as a theoretical framework to explain the influence of website experiences on university-choosing behavior. This study will focus on testing the influence of website experience on attitudes toward advertising and brand attitude, which influences university choice behavior. Social networks are interactive applications based on Web 2.0 (Obar, Jonathan A., Wildman, Steve, 2015) [20]. Therefore, Social network is considered a type of website. In this study, the concept of Website experiences is actual social-media-site experiences, and Attitude toward the website is Attitude toward the social media site.

Besides, the study added emotional intelligence, which is believed to be an essential factor in making a difference in consumer behavior. Accordingly, authors can expect that

EI has both direct and indirect effects through Attitude toward brand on the university choosing behavior of the student. Specifically, five hypotheses were proposed:

+ **H1:** The more positive student's social media site experiences, the more positive student's Attitude toward the social media site
+ **H2:** The more positive student's Attitude toward the social media site, the more positive student's Attitude toward the university brand
+ **H3:** The more positive a student's Attitude toward a university brand, the higher student's university-choosing behavior
+ **H4:** Emotional intelligence has a positive influence on student's Attitudes toward the university brand
+ **H5:** Emotional intelligence has a positive influence on student's university choice behaviour

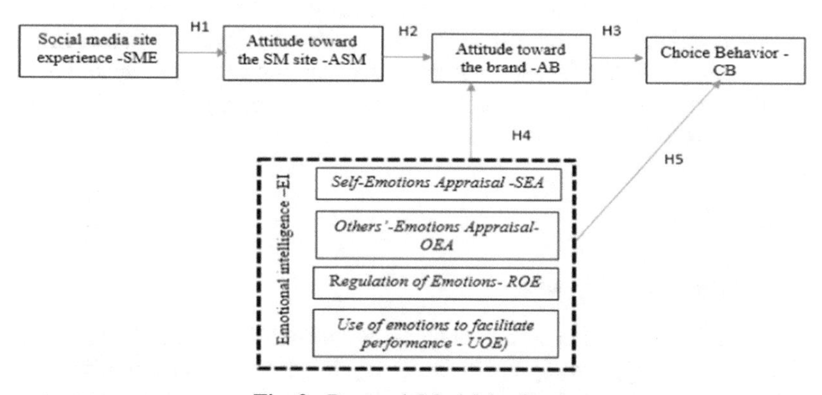

Fig. 2. Research Model (authors)

A path model was constructed (Fig. 2) to examine these hypotheses, and the scale of Variables was also specifically proposed as follows:

Social media site experience (SME) factor and factor-related Attitude (ASM, AB) were measured by 5-point semantic differential scales anchored by not at all (1) and very much (5). The social media site experience scale has 14 items developed by Vaughan, Edwards, and Peres, 2009 [19]; the Aad scale has eight items (Choi, Y. K., 2000) [7]; and the Attitude toward brand scale has six items (Chiang & Jang, 2006) [4].

Emotional intelligence and choosing behavior were measured by a 5-point Likert scale. Emotional intelligence was measured based on the Wong and Law (2002) [24] developed scale. This scale is based on the four dimensions, including SEA, OEA, ROE, and UOE, with 16 items. Choosing behavior was developed by Chiang and Jang (2006) [4] with three items.

3 Research Method

Google Forms conducted the online survey in January 2023. The authors implemented cluster sampling. Using this strategy, the authors divide the students in Da Nang into two groups, equivalent to 310 students belonging to 2 areas: Public universities and Private universities. Then, the research utilized Random Sampling technique with the Forms link was shared on the social media page of colleges and institutions. A total of 310 participants completed the survey within a month of data collection.

The questionnaires consisted of two sections. The first part included several demographic questions regarding gender, the usage extent, and the purpose of social media. The second section consists of questions regarding Social media site experience, Attitude toward the social media site, Attitude toward the brand, emotional intelligence, and choice behavior with 43 observed variables. Which, 24 items belonging to Social media site experience and factors related to Attitude (ASM, AB) were measured by 5-point semantic differential scales anchored by not at all (1) and very much (5). All items belonging to emotional intelligence and choice behavior were measured by 5-point Likert scales anchored by strongly disagree (1) and strongly agree (5).

After screening and coding, data were entered into SPSS and smartPLS software for measurement and analysis, including (1) Analysis of Cronbach's Alpha reliability coefficient. The scale has acceptable reliability when Cronbach's Alpha coefficient is more significant than 0.6; (2) Exploratory factor analysis EFA to evaluate the scale's validity; (3) Partial Least Squares Structural Equation Modeling (PLS-SEM) technique was utilized to test the research model.

4 Research Result

4.1 Descriptive Data Statistics

With 310 samples included in the data analysis, 86 males (27.7%), 215 females (42.5%), and 09 unknown (2.9%) participated in the survey. Regarding major, the authors divided into five fields: 126 Social Sciences (40.6%), 104 Engineerings and technology (33.5%), 24 Medical sciences, pharmacy (7.7%), 09 Agricultural Sciences (2.9%), 47 Humanities (15.2%). The results also showed that 100% of students participating in the survey used social networks. The rate of social media use from 1 hour to less than 3 hours/day is the highest (accounting for 69%); over 3 hours accounts for 25%, the level of social media use under 1-hour accounts for the lowest rate (6%). This result is similar to the results of other studies on the level of social media use among Vietnamese students. Most students utilize social channels to contact relatives and friends (94.8%), followed by Entertainment (reading news, watching videos/movies, games...) and updating information about ongoing social activities and events with 93.9% and 84.2%, respectively.

4.2 Reliability Test Results

According to the Cronbach's Alpha reliability test findings, all observable variables of the Attitude toward social media (ASM), Attitude toward brand (AB), Emotional intelligence, and Choice behavior have a Corrected item-total Correlation higher than 0.30. All scales have a Cronbach's Alpha coefficient better than 0.6. Therefore, these scales ensure reliability.

The reliability test findings also showed Cronbach's Alpha for the Social media site experience factor is 0.95. According to Hair et al., 2006 [10], Cronbach's Alpha

>= 0.95 may not be good. This suggests that may have "overlapping" phenomena between observed variables. Therefore, the authors recheck the scale and remove the duplicate observation variables (remove EMS7, EMS10). After deleting the SMS7 and SMS10 variables from the model. The Corrected Item – Total Correlation for all observed variables is more significant than 0.3. The alpha coefficient of the scale is 0.946, more than 0.6, and less than 0.95. Thus, the scale guarantees dependability. Therefore, all remaining 41 observed variables will be used for exploratory factor analysis (EFA) (Table 1).

Table 1. Cronbach's Alpha of the scales

Observed variables	Scale Mean if Item Deleted	Scale Variance if Item Deleted	Corrected Item-Total Correlation	Cronbach's Alpha if Item Deleted
Social media site experience (SME) = **0.946** (N=12)				
SME1	42.49	68.613	.704	.945
SME2	42.50	67.435	.809	.943
SME3	42.60	68.984	.752	.945
SME4	42.55	68.526	.784	.944
SME5	42.68	68.200	.716	.946
SME6	42.71	69.074	.718	.946
SME8	42.69	68.242	.736	.946
SME9	42.79	69.624	.671	.946
SME11	42.48	67.849	.838	.942
SME12	42.75	68.436	.757	.945
SME13	42.60	67.276	.840	.942
SME14	42.62	67.886	.792	.944
Attitude toward Social media site (ASM) = 0.915 (N=4)				
ASM1	10.64	7.681	.832	.880
ASM2	10.61	7.358	.837	.878
ASM3	10.74	7.373	.828	.881
ASM4	10.51	8.251	.727	.914
Attitude toward brand (AB) = 0.934 (N=6)				
AB1	19.00	16.689	.756	.928
AB2	19.06	16.284	.845	.917
AB3	19.08	16.851	.773	.926
AB4	18.94	16.384	.822	.920

(*continued*)

Table 1. (*continued*)

Observed variables	Scale Mean if Item Deleted	Scale Variance if Item Deleted	Corrected Item-Total Correlation	Cronbach's Alpha if Item Deleted
AB5	19.05	16.363	.819	.920
AB6	19.08	16.579	.815	.921
Emotion intelligence (EI)				
Self-Emotions Appraisal –SEA = 0.887 (N=4)				
SEA1	11.26	6.839	.746	.857
SEA2	11.25	6.328	.810	.831
SEA3	11.12	6.688	.792	.839
SEA4	11.15	7.199	.664	.886
Others'- Emotions Appraisal – OEA = 0.831 (N=4)				
OEA1	11.25	5.925	.637	.796
OEA2	11.21	5.435	.735	.752
OEA3	11.00	5.647	.613	.808
OEA4	11.42	5.512	.656	.788
Regulation of Emotions – ROE = 0.907 (N=4)				
ROE1	10.55	7.290	.780	.883
ROE2	10.58	7.216	.810	.873
ROE3	10.56	7.011	.774	.887
ROE4	10.55	7.213	.799	.877
Use of Emotions to facilitate performance – UOE = 0.872 (N=4)				
UOE1	11.52	6.037	.703	.844
UOE2	11.61	5.558	.729	.836
UOE3	11.37	5.948	.739	.830
UOE4	11.26	6.124	.737	.832
University choice behavior- CB = 0.917 (N=3)				
CB1	7.41	3.672	.793	.912
CB2	7.36	3.564	.851	.865
CB3	7.38	3.381	.854	.862

4.3 Exploratory Factor Analysis (EFA) Results

EFA was performed by the principal axis factoring method with Promax rotation to examine the dimensionality of independent factors. As a result of the first EFA analysis, there were four observable variables SME6, SME12, SME9, and ASM4, with a factor loading < 0.5. Therefore, the authors removed those variables from the model.

After deleting, the factor analysis findings indicated that KMO = 0.944, Sig. = 0.000, demonstrating that the data is consistent with the EFA analysis; 34 observed variables were extracted into seven factors at Eigenvalues = 1.060; the total variance extracted reached 69.14% > 50%; the observable variables all have a Factor loading factor >0.5, and each observable variable is guaranteed to be uploaded for a factor.

The findings indicated that Bartlett's test had Sig 0.05 for dependent variables. KMO coefficient = 0.749 > 0.5. This demonstrated that factor analysis is compatible with research data. The EFA analysis yielded one extracted component. The total variance extracted reached 78.87% > 50%, Eigenvalues of all factors greater than one, and factor loading coefficients greater than 0.5. According to the above findings of the factor analysis, all study model variables had convergent and discriminant values. Therefore, the authors decided to stop the EFA analysis moving on to the next analysis steps.

4.4 Research Model Testing

The findings identified the composite Reliability (CR) of all concepts as more than 0.7; the average variance extracted as more than 0.5. This proves scales of the variables in the model ensure reliability and convergence values. The structural model after analysis for the VIF coefficient of the variables is at the level of <5, so it can be concluded that there is no multicollinearity phenomenon (Table 2).

Table 2. Results of analysis of concepts in the model

Variables	CR	AVE	VIF	R^2	R^2adj
SME	0.829	0.658	2.401		
ASM	0.828	0.592	2.351	0.674	0.670
AB	0.832	0.607	1.945	0.626	0.621
EI	0.838	0.658	1.134		
CB	0.790	0.608		0.470	0.467

R^2adj coefficient of ASM is 0.670, so the SME independent variable explains 67% of the variation of the ASM variable. R2adj coefficient of AB is 0.621, which shows that the two variables ASM and EI explain 62.1% of the variation of AB. Finally, the R2adj coefficient of CB is 0.467, and the variables AB and EI explain 46.7% of the variation of the choice behavior.

After analyzing Bootstrap for relationships in the study model, the results are shown in Fig. 3 and Table 3. The relationships are accepted and statistically significant. The social media site experience positively affects Attitude toward the social media site with a normalized coefficient (original sample) equal to 0.821. In other words, the more positive student's social media site experiences, the more positive student's Attitude toward the social media site. The two factors, the Attitude toward social media and Emotional intelligence had a positive effect on the AB variable with impact coefficients is 0.557 and 0.265, respectively. In turn, AB affects the students' behavior when they select a

Table 3. Results of analysis of hypothetical relationships

Hypo-theses	Relationship	Original sample	Sample mean	Standard deviation	T statistics	P values	Result
H1	SME -> ASM	0.821	0.822	0.019	43.610	0.000	Supported
H2	ASM -> AB	0.557	0.555	0.059	9.378	0.000	Supported
H3	AB -> CB	0.395	0.394	0.066	5.989	0.000	Supported
H4	EI -> AB	0.265	0.266	0.058	4.560	0.000	Supported
H5	EI -> CB	0.386	0.389	0.068	5.672	0.000	Supported

university (0.395). This means that the more positive the student's Attitude towards the university brand, the higher the student's university choice behavior. In addition, EI also has a direct positive impact on CB with an Original sample coefficient = 0.386.

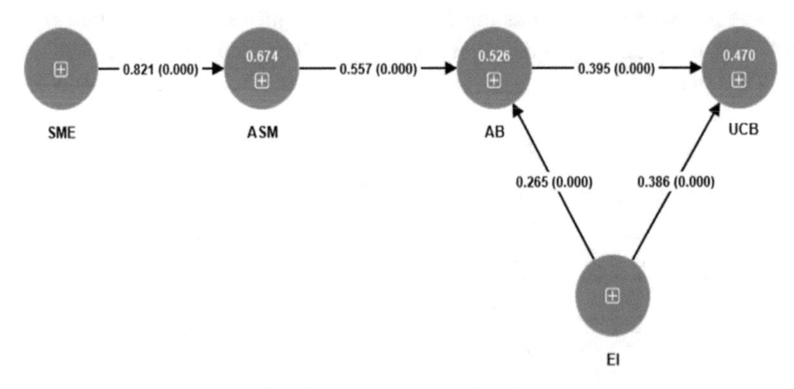

Fig. 3. Model testing Results

5 Discussion

In this paper, the authors provided a complete research methodology to demonstrate the relationships between social media site experience, emotional intelligence, and choice behavior via mediating role of factors related to Attitude. An online survey was conducted by Google Forms to test the hypothesized model. The data analysis findings illustrated the validity and dependability of the model's scales. All proposed hypothesis relationships are accepted and statistically significant at 5%.

The construction of the proposed research model is based on the Attitude toward the website Model of Poh & Adam, 2002 [22] and documents explaining the relationship between emotional intelligence and consumer decision-making (Se Hun Lim & Dan J. Kim, 2020; Anastasiadis, Lazaros, 2020) [1, 15]. This research contributed to the current literature on the influence of social media marketing efforts on behavior in Vietnam's education sector. Colleges and institutions are constantly expanding their online presence, especially their presence on social networks. Utilizing social media channels to reach learners is necessary, and determining the quality of the social media experience and the emotion' effects is crucial. This is the foundation upon which managers plan communication strategies, create conditions for the organization to quickly reach target customers-high school students, easily collect feedback, and provide timely helpful information to increase experiences and positive emotion. This contributes to forming a positive attitude towards the university brand and promoting student behavior when choosing a university.

A prior study by Leung et al., 2015 [14] explored the social media marketing effectiveness of two different social media platforms (Facebook and Twitter) in the hotel industry. The results demonstrated that the customer's social media experiences affect their attitudes toward social media site, which impacts attitudes toward hotel brand. Thereby toward hotel brand has a positive effect on booking intention. Although, the authors conduct research in other areas (Education) along with other data analysis techniques - PLSSEM compared with studies of Xi Y. Leung. However, the findings once again confirm that the more positive student's social media site experiences, the more positive student's Attitudes toward the social media site. This finding is also consistent with Bruner & Kumar, 2000 [2]. Their study suggested that an individual's web experiences positively affect their Attitude toward website.

The support of the hypothesis regarding the relationship between attitudes toward the social media site and attitudes toward the brand illustrates that if a student has a favorable attitude toward the university's social channel, the student's Attitude toward the university brand is more positive. This result suits all studies related to the Attitude toward the website model, which examines by studies by Poh & Adam, 2002; Bruner& Kumar, 2000 [2, 22]. It defines the significance of social media marketing in education organizations. Especially it is essential to leverage social media pages to impress students and change their attitudes to become more positive.

Besides that, the acceptance of the hypothesis "the more positive student's Attitude toward the University brand, the higher choice behavior" proposed that if a student has a positive attitude toward a University brand; they are more likely to decide to choose this institution when they select the university. Both Aad and Aws model results suggested that attitudes toward brand positively influence purchase intention. According to the Theory of Reasoned Action – TRA (Fishbein & Ajzen, 1975) [8], behavior intention is a powerful predictor and explanation of actual behavior, and intention is a part of behavior-oriented attitudes. However, in this study, the author ignores behavior intention and attempts to explain the influence of attitudes on behavior. Empirical results showed that attitudes toward brand directly affect choice behavior.

Furthermore, there was one new finding from SEM analysis. This study expands the examination of emotional intelligence's influences on students' university choice behavior. According to the data research findings, emotional intelligence has both direct and indirect influence through Attitude toward brand on college choice behavior. This result contributes to the theoretical foundations of current research on emotion. The practical results again demonstrate the role of the emotion aspect on consumer behavior, as mentioned in several previous studies (Se Hun Lim & Dan J. Kim, 2020; Rozell et al., 2006; Kidwell et al., 2008) [13, 15, 21].

6 Conclusion

The study provides practical evidence to support utilizing the Aws model to explain the relationship between emotional intelligence, social media site experiences, brand attitude, and decision-making. The findings found that the model's application is entirely consistent with the survey data in Vietnam. The findings identified that the social media site experience has a positive effect on Attitude toward the social media site with a normalized coefficient (original sample) equal to 0.821. The two factors, attitudes toward social media site and Emotional intelligence, positively affect the attitudes toward brand, with impact coefficients of 0.557 and 0.265, respectively. In turn, attitudes toward brand affect the student's behavior when they choose a university (0.395). This means that the more positive the student's Attitude towards the university brand, the higher the student's university choice behavior. Additionally, emotional intelligence also has a direct positive impact on choice behavior, with an original sample coefficient = 0.386.

From a theoretical perspective, research on the experiences and emotional intelligence's effects in the education area is minimal. Moreover, the paper has valuable contributions to both academic and practical researchers. Our study is one of the efforts to apply attitudes toward the website model in social media context to explore student choice behavior in the education sector.

Practically, the universities' managers can utilize these findings to improve their social media communication strategies. Organizations should make their social media page more attractive and appealing, provide timely helpful information, and user-centric aims to increase interaction and promote members' engagement. The university should then concentrate on Social media page design to enhance user experiences. This facilitates the organization to quickly reach target customers (high school students) and create and maintain positive emotions and experiences for users. This will contribute to form a positive attitude towards the university brand, and promoting student's behavior when choosing a university.

Similar to another study, there are several limitations are noted in this examination. The major limitation of our study is that data just collected in Da Nang as a sample. Because this is an exploratory study in the education area – University education, further work is going to expand collecting the research data to universities in another region of Viet Nam as Ha Noi, Ho Chi Minh. Besides, the study tests the relationship between emotional intelligence, social media site experiences, brand attitude, and decision-making with data from social media platform – Facebook. Currently, the organizations have been integrated many platforms to reach the target customer. So, the examination of the relationships with data from different social media channels is necessary in future.

References

1. Anastasiadis, L.: Emotional intelligence influences on consumers consumer behavior (2020)

2. Bruner, G.C., Anand, K.: Web commercials and advertising hierarchy-of-effects. J. Advert. Res. **40**(1-2), 35-42 (2000)
3. Chen, Q., Wells, W.D.: Attitude toward the site. J. Advert. Res. **39**(5), 27–38 (1999)
4. Chiang, C., Jang, S.: The effects of perceived price and brand image on value and purchase intention: Leisure travelers' attitudes toward online hotel booking. J. Hosp. Leis. Mark. **15**(3), 49–69 (2006)
5. Cho, N., Park, S.: Development of electronic commerce user-consumer satisfaction index (ECUSI) for Internet shopping. Ind. Manage. Data Syst. **101**(8), 400–406 (2001)
6. Constantinides, E.: Influencing the online consumer's behavior: the Web experience. Internet Res. **14**(2) (2004)
7. Choi, Y.K.: Effects of presence on the effectiveness of web site advertising. Michigan State University (2000)
8. Ajzen, I., Fishbein, M.: A Bayesian analysis of attribution processes. Psychol. Bull. **82**(2), 261 (1975)
9. Goleman, D.: Emotional Intelligence, New York, NY, England (1995).
10. Hair, E., et al.: Children's school readiness in the ECLS-K: Predictions to academic, health, and social outcomes in first grade. Early Childhood Res. Q. **21**(4), 431-454 (2006)
11. Higgs, M.: A study of the relationship between emotional intelligence and performance in UK call centres. J. Manage. Psychol. **19**(4) (2004)
12. Homer, P.M.: The mediating role of attitude toward the ad: some additional evidence. J. Mark. Res. **27**(1), 78–86 (1990)
13. Kidwell, B., Hardesty, D.M., Childers, T.L.: Emotional calibration effects on consumer choice. J. Consum. Res. **35**(4), 611–621 (2008)
14. Leung, X.Y., Bai, B., Stahura, K.A.: The marketing effectiveness of social media in the hotel industry: a comparison of Facebook and Twitter. J. Hosp. Tour. Res. **39**(2), 147–169 (2015)
15. Lim, S.H., Kim, D.J.: Does emotional intelligence of online shoppers affect their shopping behavior? From a cognitive-affective-conative framework perspective. Int. J. Hum.–Comput. Interact. **36**(14) (2020)
16. Psaila, G., Wagner, R. (eds.): EC-Web 2007. LNCS, vol. 4655. Springer, Heidelberg (2007). https://doi.org/10.1007/978-3-540-74563-1
17. MacKenzie, S.B., Lutz, R.J., Belch, G.E.: The role of attitude toward the ad as a mediator of advertising effectiveness: a test of competing explanations. J. Mark. Res. **23**(2), 130–143 (1986)
18. Mayer, J.D., et al.: Emotional intelligence as a standard intelligence (2001)
19. Moital, M., Vaughan, R., Edwards, J., Peres, R.: Determinants of intention to purchase over the Internet. Anatolia **20**(2), 345–358 (2009)
20. Obar, J.A., Wildman, S.: Social media definition and the governance challenge: an introduction to the special issue (2015)
21. Pettijohn, C.E., Rozell, A.N.: The relationship between emotional intelligence and customer orientation for pharmaceutical salespeople: a UK perspective. Int. J. Pharm. Healthc. Mark. (2010)
22. Poh, D., Stewart, A.: An exploratory investigation of attitude toward the website and the advertising hierarchy of effects (2002)
23. We are social & Kepios: Vietnam Digital Development Overview Report 2021 (2022). https://vtc.edu.vn/wp-content/uploads/2022/03/digital-in-vietnam-2022.pdf
24. Wong, C.-S., Law, K.S.: Wong and law emotional intelligence scale. Leadersh. Q. (2002)

Image and Natural Language Processing

A Multi Context Decoder-based Network with Applications for Polyp Segmentation in Colonoscopy Images

Ngoc-Du Tran, Dinh-Quoc-Dai Nguyen, Ngoc-Linh-Chi Nguyen, Van-Truong Pham, and Thi-Thao Tran(✉)

Department of Automation Engineering, School of Electrical and Electronic Engineering, Hanoi University of Science and Technology, Hanoi, Vietnam
thao.tranthi@hust.edu.vn

Abstract. Polyp Segmentation is important in helping doctors diagnose and provide an accurate treatment plan. With the emerging of deep learning technology in the last decade, deep learning models especially Unet and its evolved versions, for medical segmentation task have achieved superior results compared to previous traditional methods. To preserve location information, Unet-based models use connections between feature maps of the same resolution of encoder and decoder. However, using the same resolution connections has two problems: 1) High-resolution feature maps on the encoder side contain low-level information. In contrast, high-resolution feature maps on the decoder side contain high-level information that leads to an imbalance in terms of semantic information when connecting. 2) In medical images, objects such as tumours and cells often have diverse sizes, so to be able to segment objects correctly, the use of context information on a scale of the feature map encoder during the decoding process is not enough, so it is necessary to use context information on full-scale. In this paper, we propose a model called CTDCFormer that uses the PvitV2_B3 model as the backbone encoder to extract global information about the object. In order to exploit the full-scale context information of the encoder, we propose the GCF module using the lightweight attention mechanism between the decoder's feature map and the encoder's four feature maps. Our model CTDCFormer achieves superior results compared to other state of the arts, with the Dice scores up to 94.1% on the Kvasir-SEG set, and 94.7% on the CVC-ClinicDB set.

Keywords: Polyp Segmentation · Multi Context Decoder · Attention

1 Introduction

Medical segmentation plays a vital role in the medical field. It provides useful information to help doctors diagnose the disease and make an accurate treatment plan. Segmentation models using convolution neural networks have

N. T. Nguyen et al. (Eds.): CITA 2023, LNNS 734, pp. 157–168, 2023.
https://doi.org/10.1007/978-3-031-36886-8_13

yielded impressive results in recent years. For example, the U-net [1] model using encoder-decoder architecture and skip-connections to preserve spatial information has yielded impressive results in medical segmentation tasks. However, using skip connections causes an information imbalance when connecting the low-level features in the encoder with the high-level features in the decoder. To solve this problem, improved variants of the Unet model were born, such as ResUnet [2], ResUnet++ [3], MultiResUnet [4] using ResPath, or Unet ++[5] using nested connections and dense connections to balance the semantic information between the encoder and decoder feature maps. Unet3++ model [6] improved the Unet++ model [5] by redesigning the connections between the encoder-decoder and the connections between the decoder-decoder to capture information across the entire scale. Although the models based on the CNN method have provided amazing performance, they have limited ability to learn long-range dependencies because the convolution operations are only good at learning local information compared to global information.

The success of the Transformer [7] models in the NLP domain, due to the use of self-attention mechanisms for parallel computation and better learning of long-range dependencies between tokens, promotes studies applying them to the computer vision domain. Some well-known models such as Vision transformer (VIT) [8], Swin Transformer [9], PvitV2 [10], etc achieved SOTA results on image classification tasks on the ImageNet dataset. In medical segmentation, models using transformer-backbone, such as TransUnet [11], SwinUnet [9], SWTR-Unet [12], etc., have shown superior results compared to the methods using CNN-backbone. However, these models have the limitation of only making connections between the feature map of the encoder and the decoder of the same resolution. Therefore, they cannot extract context information across the full scale of the feature map encoder. To solve this problem, we propose a model called CTDCFormer. CTDCFomer uses the PvitV2_B3 model [10] as backbone encoder and CNN blocks as decoder so that CTDCFormer is capable of capturing both global and local information. In order to selectively extract information on the full scale of the feature map encoder, we propose the GCF module to create feature map contexts that store the extracted context information on the full scale. These feature map contexts will be used for the decoding process to generate highly accurate prediction masks. A summary of our contributions is as follows:

- **Propose module GCF:** We propose the GCF module use light-weight attention mechanism as a bridge between encoder and decoder to replace connections with the same resolution between feature map of encoder and decoder in Ushape models, which often suffer from imbalance problem with semantic information and lack of context information on full-scale.
- **Propose model CTDCFormer:** We introduce a new model called CTDCFormer that uses PvitV2_B3 [10] model as encoder to take advantage of transformer model type in learning long-range dependencies and use CNN blocks in decoding process to learn detailed information. The GCF module is used to extract full-scale information.

- **Improve the performance:** We perform experiments on Polyp datasets including Kvasir [13], CVC-ClinicDB [14], CVC-ColonDB [15], CVC-T [16], ETIS -LaribPolypDB [17]. The results show that our model CTDCFormer achieves SOTA results on Kvasir and CVC-ClinicDB sets.

2 Related Work

2.1 Vision Transformer

For a long time, CNN has been considered the standard for Vision tasks due to its advantage in learning local information. However, recent studies have shown that models using transformer architecture have superior results compared to the CNN mechanism in computer vision domains. Dosovitskiy et al. first introduced the Vision Tranformer (VIT) [8] model for the image classification task. The Vision transformer breaks the image down into non-overlapping patches and then passes them through linear projections to create patches embedding. These embedding patches along with positional embedding and token cls will be passed through the transformer model to create a classifier model based on the cls token. However, VIT [8] is computationally expensive due to performing self-attention on all input patches. Moreover, using fixed resolution across stages makes it unsuitable for dense prediction tasks. To address that problem a series of improvement models were born such as Swin Transformer [9], Pvit [18], SegFormer [19], etc. Wang et al. presented the Pyramid Vision Transformer (Pvit) model [18] for the first time using Pyramid architecture for dense prediction tasks. Pvit used a progressive shrinking strategy to create the Pyramid architecture with patch embedding. To reduce the computational cost, Pvit introduces the spatial-reduction attention (SRA) module to replace the traditional multi-head attention (MSA) module. To improve the performance of Pvit [18], Wang et al. introduced PvitV2 [10] with three additional improvements, namely linear complexity attention layer, overlapping patch embedding and convolutional feed-forward network. In polyp segmentation, models built on transformers backbone are increasingly popular and yield impressive results such as PVT-Cascade [20], SSFormer [21], MSMA-Net [22], etc.

2.2 Fusion On Multi Resolution Feature Map

Because the objects in medical images are often varied in size, using information on a multi resolution-scale is critical. Huimin Huang et al. suggest Unet3++ [6] uses connections from the encoder's shallow layers and the decoder's deep layers to create full-scale connections that can capture raw and detailed information on the full scale. In MSRF-Net [23], Abhishek Srivastava et al. introduce Dual-Scale Dense Fusion (DSDF) block capable of exchanging information on two different scaled features and MSRF subnetwork consisting of multiple DSDF blocks stacked together sequentially to merge feature multi-scale. In UCTransNet [24], Haonan Wang introduces the CTrans module, which includes two sub-modules,

Channel Cross fusion with Transformer (CCT) and Channelwise Cross-Attention (CCA), to replace skip-connection in Unet to solve the semantic gaps in medical image segmentation.

3 Methodology

3.1 Overview

As illustrated in Fig. 1, our CTDCFormer model has a Ushape structure in which the Pretrained model PvitV2_B3 [10] is the encoder and Decoder Block (DB) 4b blocks are the decoder. The GCF module is prposed for extracting information on the multi-scale feature map of the encoder to add context information to the decoding process reasonably. Our GCF module uses only standard convolution and fully-connected layers and thus has a lower computational cost than UCTransNet [24]'s CCT module and connections in Unet3++. The detail is presented in the following sections.

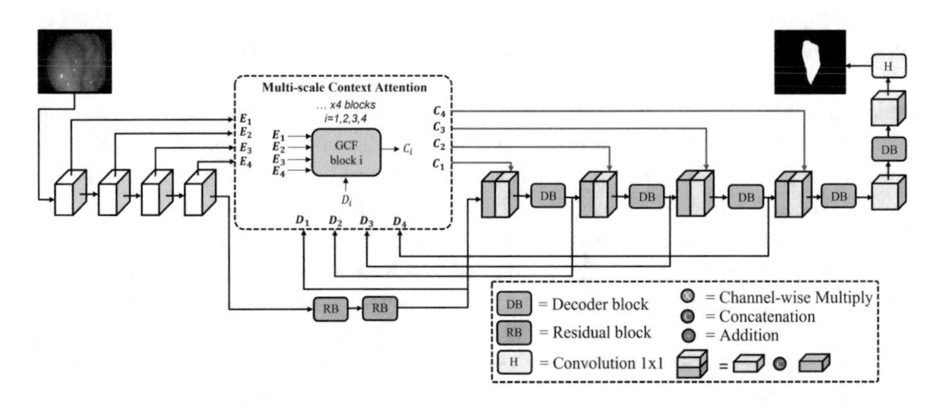

Fig. 1. Overview Architecture of the proposed CTDCFomer model. E_1, E_2, E_3, E_4 are features on 4 stages of model PvitV2_B3 [10]. D_i is the feature map generated from the ith Decoder Block (DB) block. C_i is feature map context generated from the ith GCF module.

3.2 Generate Context Feature Map (GCF)

Most of the previous full-scale information extraction models [6, 24, 25] only used features in the encoder without using features on the decoder side, therefore, the information on the entire scale lacks efficiency. To solve that problem, during the decoding process at each feature map $D_i, i = \overline{1,4}$ generated by Decoder Block (DB) 4b, the GCF module will use 4 feature maps E_1, E_2, E_3, E_4 of encoder and feature map D_i to create feature map context C_i containing detailed information and global information extracted on full-scale feature map of encoders. Feature context C_i will then be concatenated with feature map D_i as input for the next Decoder Block (DB) during decoding. The operation of the GCF module is illustrated in Fig. 2 and will be described in detail in the Multi-Scale Context Attention section.

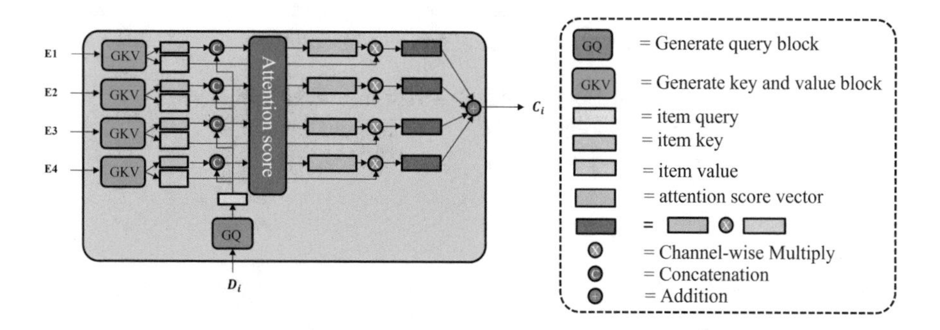

Fig. 2. Structure of the proposed GCF module.

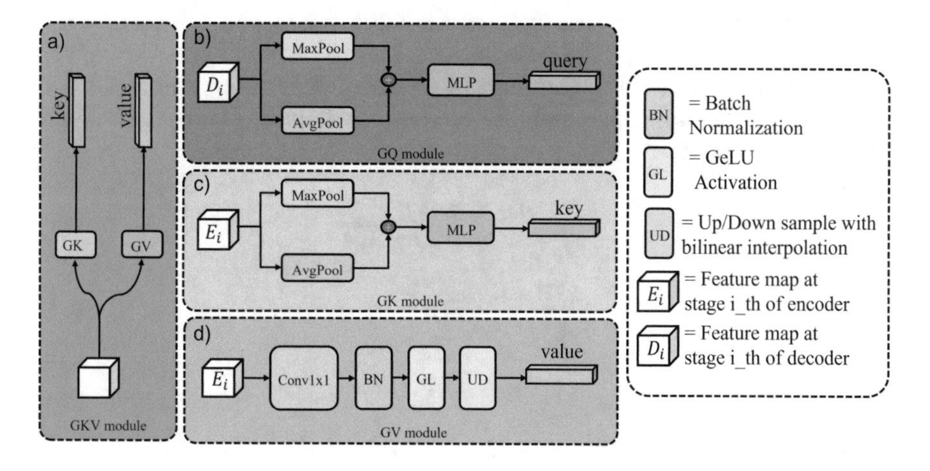

Fig. 3. Structure of a) Generate key and value module (GKV), b) Generate query module (GQ), c) Generate Key module (GK), d) Generate value module (GV)

Generate Query Key and Value. Inspired by the transformer model, our GCF 2 module uses three items: query, key and value. Fig. 3a illustrates the process of generating two items key and value from the GK and GV modules respectively. Module GQ and GK as illustrated in Fig. 3b and Fig. 3c respectively includes max pooling, average pooling and fully connected layer. The GQ module

3b takes the feature map of the decoder as input to generate the item query. The GV module 3c takes the feature map of the encoder as input to generate the item key. Item value is generated by the GV module as illustrated in Fig. 3d, where feature map of encoder is passed a pointwise convolution block followed by batch normlization and a Gelu activation function to change the channel number and an up/down sample block to make the feature maps have the same resolution. The formula of query, key and value is described as follows:

$$\begin{aligned} q_i &= FC(MaxPooling(D_i) + AveragePooling(D_i)) \\ k_i &= FC(MaxPooling(E_i) + AveragePooling(E_i)) \\ v_i &= UpDown(GeLu(BN(C1(E_i)))) \end{aligned}$$

with $E_i \in \mathbf{R}^{N_e^i \times H_e^i \times W_e^i}, D_i \in \mathbf{R}^{N_d^i \times H_d^i \times W_d^i}, q_i, k_i \in \mathbf{R}^{N_d^i}, v_i \in \mathbf{R}^{N_d^i \times H_d^i \times W_d^i}, i \in [1, 4]$

where BN is Batch normalization, $UpDown$ is up-sample or down-sample by bilinear interpolation, $C1$ is convolution with kernel size 1×1, E_i, D_i is the ith feature map of encoder, decoder respectively.

Multi-Scale Context Attention. To create a feature map context for the ith decoder layer we use the GCF module as shown in Fig. 2. First, the encoder's four feature maps E_j will pass through the GKV module 3a to generate item key k_j and item value v_j. The decoder's D_i feature map will pass through the GQ 3b module to generate item query q_i . Finally, the feature map context C_i is created by calculating the attention between the items q_i, k_j, v_j according to the following formulas:

$$\begin{aligned} m_j &= [q_i; k_j] \\ h_j &= FC(m_j), j = \overline{1,4} \\ a_{kj} &= \frac{e^{h_{jk}}}{\sum_{n=1}^{4} h_{nk}} \\ C_{ik} &= \sum_{j=1}^{4} v_{jk} * a_{kj} \\ M_i &= [D_i; C_i] \end{aligned} \tag{1}$$

where $m_j \in \mathbf{R}^{2N_d^i}$, $h_j \in \mathbf{R}^{N_d^i}$, $a_k \in \mathbf{R}^4$, $C_i \in \mathbf{R}^{N_d^i \times H_d^i \times W_d^i}$, $M_i \in \mathbf{R}^{2N_d^i \times H_d^i \times W_d^i}$, $j \in [1, 4]$, [;] denotes the concatenation operator, and FC(.) denotes for fully conected layer to reduce from $2N_i$ dimension vector to N_i dimension vector. a_{kj} and C_{ik} denote the jth value and kth channel of the vector a_k and feature map context C_i respectively . The feature map M_i obtained from concatenating feature D_i and C_i will be input to the next Decoder Block (DB) 4b block in the decoding process. From the 1 equation, we can see that our GCF Module is more general than the skip connections of the Unet model by instead of hard concatenating feature maps of the same resolution from the encoder, the feature context of we are more selective by paying attention between the information from the decoder and the entire feature map of encoder.

3.3 Decoder Block

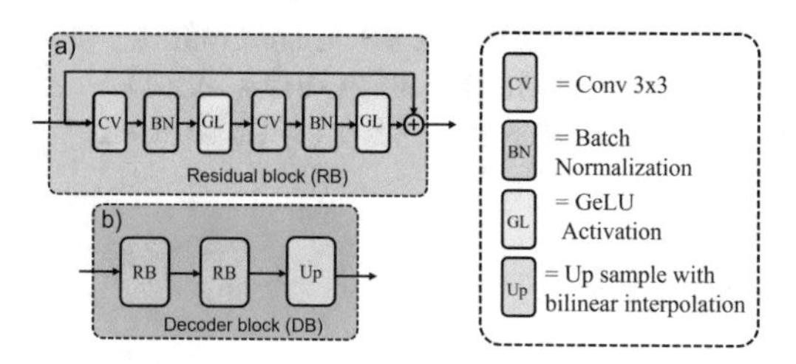

Fig. 4. Structure of a) Residual block (RB), and b) Decoder Block (DB)

To overcome the weakness of transformer models in learning local information. In the decoding process, we use Decoder Block (DB) blocks as illustrated in Fig. 4b consisting of two blocks Residual Block 4a and Upsample block uses bilinear interpolation. The operation of the RB block is described as follows:

$$x' = x + GeLu(BN(C3(GeLu(BN(C3(x)))))) \tag{2}$$

where x is input, $BN, C3$ is Batch Normalization and convolution respectively. The input feature map will go through the convolution layer with kernel size of 3×3 and padding of 1 to extract local information. Followed by batch normalization layer and the GeLU activation function. To avoid vanishing gradient, skip connection is used.

4 Experiments

4.1 Dataset

To evaluate the performance of the proposed model in medical image segmentation, we perform experiments on five data sets: Kvasir-SEG [13], CVC-ClinicDB [14], CVC-T [16], CVC-ColonDB [15], and ETIS-LaribPolypDB [17]. The Kvasir-SEG data set is an open data set containing 1000 medical polyp images with resolutions ranging from 332×487 to 1920×1072 pixels with corresponding masks manually labelled by an experienced gastroenterologist. The CVC-ClinicDB data set includes 612 images extracted from colonoscopy videos. CVC-T, CVC-ColonDB, and ETIS include 60, 380, and 196 images collected from different medical sources. In this paper, we set up two different types of experiments:

Learning Ability Experiment: In this experiment, we will perform experiments on Kvasir-SEG, and CVC-ClinicDB sets separately. We randomly divide the data set at the rate of 80%, 10%, 10% for train, validate, and test respectively. Models will be evaluated on the test set after reaching maximum Dice coefficient on val set.

Generalizability Experiment: To evaluate the model's generalization ability on unseen datasets, in this experiment, we will randomly divide the dataset as suggested by PraNet [25], in which the training set includes 1450 images (90% Kvasir- SEG + 90%CVC-ClinicDB), the val set includes the remaining 10% of images.. Models will be tested on CVC-T, CVC-ColonDB, ETIS-LaribPolypDB set after reaching Dice coefficient max on set validate set.

4.2 Evaluation Metrics

In the experiments, we use two metrics, Dice coefficient and IoU, to evaluate the performance of the model. These metrics measure the similarity between ground truths and predicted maps. They will be closer to 1 if ground truths and predicted maps are the same and closer to 0 if the ground truth is different from the predicted map.

$$Dice = \frac{2TP}{2TP + FP + FN + \epsilon} \tag{3}$$

$$IoU = \frac{TP}{TP + FP + FN + \epsilon} \tag{4}$$

where TP is true positive, FP is false positive, and FN is false negative, ϵ is the constant added to the denominator to avoid potential Divide-by-zero.

4.3 Implementation Details

We implement the CTDCFormer model based on the Pytorch framework version 1.10.0. The model was trained for 80 epochs using AdamW optimizer with initial learning rate of 0.0001 and Learning rate scheduler StepLR with step size 13 for training on Kvasir-SEG set and step size 15 for training on CVC-ClinicDB set. All photos are resized to 352×352. All models were trained on Kaggle free computational resource with Nvidia-Tesla P100 16G Ram GPU. We use a combination of two loss functions, binary cross entropy (BCE) and dice loss:

$$L_{BCE} = -\sum_{i=1}^{N} \frac{1}{N} [y_i \times \log(\hat{y}_i) + (1 - y_i) \times \log(1 - \hat{y}_i)] \tag{5}$$

$$L_{Dice} = 1 - \frac{2 \times \sum_{i=1}^{N} y_i \times \hat{y}_i}{\sum_{i=1}^{N} y_i + \hat{y}_i} \tag{6}$$

$$L_{total} = L_{BCE} + L_{Dice} \tag{7}$$

where N is the number of pixel in the image, $y_i \in \{0; 1\}$ is ground truth, $\hat{y}_i \in (0, 1)$ is the predicted mask.

4.4 Representative Results

In this section, we will visualize the prediction of the CTDCFormer model on 5 Polyp test data sets to verify the segmentation ability of the model. To test the GCF module's ability to focus attention on object containers, we visualized heat maps in four-layer decoders using the GradCam [26] algorithm.

From Fig. 5a it is shown that the predicted mask of the model is close to the ground truth. The heat map from Fig. 5b shows that in the 1st and 2nd decoder layers the GCF module focuses attention on the neighborhoods of the object to extract context information, while in the 3rd and 4th decoder layers the GCF module successfully focuses attention on the object.

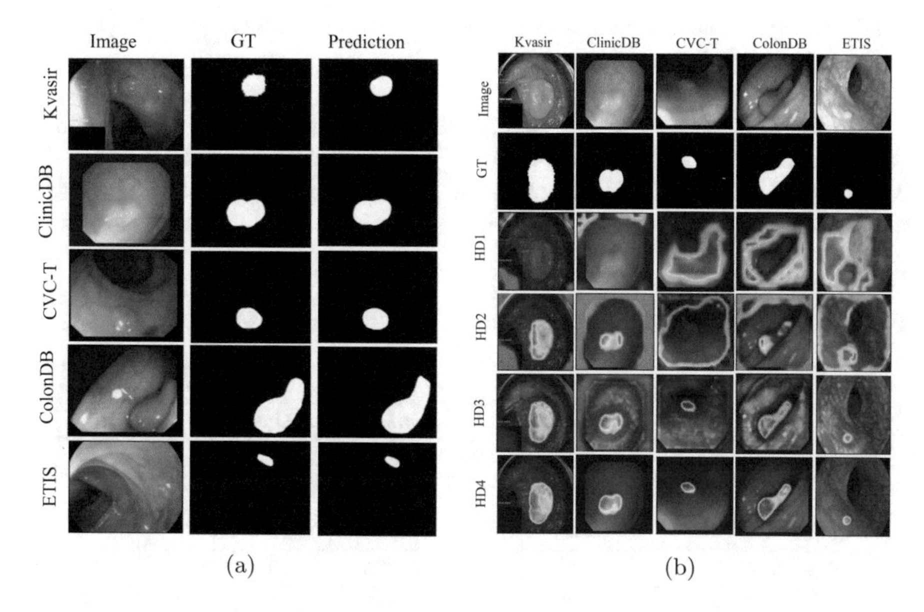

Fig. 5. Qualitative segmentation results of the proposed CTDCFormer model. a) The predicted results on 5 datasets. b) Attention heatmaps of four decoder layers connected to GCF module module (HD_i is the heatmap of the ith decoder layer)

4.5 Comparative Results

We compare our CTDCFormer model with SOTA models on the Polyp segmentation task. The mean scores of Dice coefficients, mDice, and mean values of Intersection Over Union, mIOU are used for validation. From Table 1, it is shown that our CTDCFormer model achieved SOTA results in the learning ability experiment with 94.1% Dice and 89.4% IOU on Kvasir-SEG, 94.7% Dice and 90.2% on CVC-ClinicDB data sets, which proves that the CTDCFormer model with the GCF module has strong learning ability.

Table 1. Experiment comparing learning ability of CTDCFormer with other models.

Models	Type	FLOPs (G)	Kvasir		CVC-ClinicDB	
			mDice↑	*mIOU*↑	*mDice*↑	*mIOU*↑
Unet++	CNN-based	55.9	0.821	0.743	0.794	0.729
Unet3++	CNN-based	151.7	0.831	0.740	0.855	0.769
FCMD-Net	CNN-based	–	0.898	0.835	0.936	0.888
MSRF-Net	CNN-based	84.2	0.922	0.891	0.942	0.904
UCTransNet	Transformer-based	32.9	0.918	0.860	0.933	0.860
PVT-CASCADE	Transformer-based	15.4	0.926	0.877	0.943	0.899
SSFormerPVT	Transformer-based	34.8	0.936	0.891	0.945	0.900
CTDCFormer	Transformer-based	26.9	**0.941**	**0.894**	**0.947**	**0.902**

The Table 2 and showed that our model has the highest generalization on all three datasets with 92.1% Dice and 86% IOU on CVC-T, 83.4% Dice and 75.5% IOU on CVC-ColonDB, 81.3% Dice and 73.1% IOU on ETIS-LaribPolypDB, which which proves that our model has good generalization ability on unseen datasets.

Table 2. Experiment the generalizability of the CTDCFormer model on unseen datasets.

Models	CVC-T		CVC-ColonDB		ETIS	
	mDice↑	*mIOU*↑	*mDice*↑	*mIOU*↑	*mDice*↑	*mIOU*↑
Unet++	0.707	0.624	0.624	0.410	0.401	0.344
Unet3++	0.802	0.728	0.720	0.633	0.596	0.508
CaraNet	0.903	0.838	0.773	0.689	0.747	0.672
UCTransNet	0.851	0.775	0.768	0.680	0.654	0.565
PVT-CASCADE	0.905	0.838	0.825	0.745	0.801	0.726
SSFormerPVT	0.895	0.823	0.793	0.706	0.783	0.701
CTDCFormer (Ours)	**0.921**	**0.860**	**0.834**	**0.755**	**0.813**	**0.731**

5 Conclusion

In this work, we have proposed the CTDCFormer model for the Polyp segmentation task. With the use of GCF module to replace traditional encoder-decoder connections, the model can effectively exploit context information on the full-scale of the encoder. Our model has achieved impressive results in both leaning ability and generalizability experiments on Polyp data sets. In the future, we will improve the GCF module to be able to extract context information more efficiently thereby increasing the accuracy of the model even more.

Acknowledgment. This research is funded by Vietnam National Foundation for Science and Technology Development (NAFOSTED) under grant number 102.05-2021.34.

References

1. Ronneberger, O., Fischer, P., Brox, T.: U-Net: convolutional networks for biomedical image segmentation. In: Navab, N., Hornegger, J., Wells, W., Frangi, A. (eds.) Medical Image Computing and Computer-Assisted Intervention– MICCAI 2015. MICCAI 2015. Lecture Notes in Computer Science, vol. 9351, pp. 234–241. Springer, Cham (2015). https://doi.org/10.1007/978-3-319-24574-4_28
2. Jha, D., et al.: Kvasir-seg: a segmented polyp dataset. In: International Conference on Multimedia Modeling. Springer, pp. 451–462 (2020)
3. Jha, D. , et al.: Resunet++: an advanced architecture for medical image segmentation. In: 2019 IEEE International Symposium on Multimedia (ISM), pp. 225–2255. IEEE (2019)
4. Ibtehaz, N., Rahman, M.S.: Multiresunet: rethinking the u-net architecture for multimodal biomedical image segmentation. Neural Netw. **121**, 74–87 (2020)
5. Zhou, Z., Rahman Siddiquee, M.M., Tajbakhsh, N., Liang, J.: Unet++: a nested u-net architecture for medical image segmentation. In: Deep learning in medical image analysis and multimodal learning for clinical decision support, pp. 3–11. Springer (2018)
6. Huang, H.: Unet 3+: a full-scale connected unet for medical image segmentation. In: ICASSP 2020–2020 IEEE International Conference on Acoustics, Speech and Signal Processing (ICASSP), pp. 1055–1059. IEEE (2020)
7. Vaswani, A.: Attention is all you need. Adv. Neural Inf. Process. Syst **30** (2017)
8. Dosovitskiy, A., et al.: An image is worth 16×16 words: Transformers for image recognition at scale, arXiv preprint arXiv:2010.11929 (2020)
9. Cao, H.: Swin-unet: Unet-like pure transformer for medical image segmentation. arXiv preprint arXiv:2105.05537 (2021)
10. Wang, W., et al.: Pvt v2: improved baselines with pyramid vision transformer. Comput. Vis. Media **8**(3), 415–424 (2022)
11. Chen, J.: Transunet: Transformers make strong encoders for medical image segmentation. arXiv preprint arXiv:2102.04306 (2021)
12. Hille, G., Agrawal, S., Wybranski, C., Pech, M., Surov, A., Saalfeld, S.: Joint liver and hepatic lesion segmentation using a hybrid CNN with transformer layers. arXiv preprint arXiv:2201.10981 (2022)
13. Jha, D.: Kvasir-seg: a segmented polyp dataset. In: International Conference on Multimedia Modeling. Springer, pp. 451–462 (2020)
14. Bernal, J., Sánchez, F.J., Fernández-Esparrach, G., Gil, D., Rodríguez, C., Vilariño, F.: WM-dova maps for accurate polyp highlighting in colonoscopy: validation vs. saliency maps from physicians. Comput. Med. Imaging Graph. **43**, 99–111 (2015)
15. Tajbakhsh, N., Gurudu, S.R., Liang, J.: Automated polyp detection in colonoscopy videos using shape and context information. IEEE Trans. Med. Imaging **35**(2), 630–644 (2015)
16. Vázquez, D.: A benchmark for endoluminal scene segmentation of colonoscopy images. J. Healthcare Eng. (2017)
17. Silva, J., Histace, A., Romain, O., Dray, X., Granado, B.: Toward embedded detection of polyps in WCE images for early diagnosis of colorectal cancer. Int. J. Comput. Assist. Radiol. Surg. **9**(2), 283–293 (2014)

18. Wang, W.: Pyramid vision transformer: a versatile backbone for dense prediction without convolutions. In: Proceedings of the IEEE/CVF International Conference on Computer Vision, pp. 568–578 (2021)
19. Xie, E., Wang, W., Yu, Z., Anandkumar, A., Alvarez, J.M., Luo, P.: Segformer: simple and efficient design for semantic segmentation with transformers. Adv. Neural Inf. Process. Syst. **34**, 12077–12090 (2021)
20. Rahman, M.M., Marculescu, R.: Medical image segmentation via cascaded attention decoding. In: Proceedings of the IEEE/CVF Winter Conference on Applications of Computer Vision, pp. 6222–6231 (2023)
21. Wang, J., Huang, Q., Tang, F., Meng, J., Su, J., Song, S.: Stepwise feature fusion: Local guides global. arXiv preprint arXiv:2203.03635 (2022)
22. Le, T.-K., Tran, T.-T., Pham, V.-T., et al.: Msma-net: a multi-scale multidirectional adaptation network for polyp segmentation. In: 2022 RIVF International Conference on Computing and Communication Technologies (RIVF). IEEE, pp. 629–634 (2022)
23. Srivastava, A., et al.: MSRF-net: a multi-scale residual fusion network for biomedical image segmentation. IEEE J. Biomed Health Inform. **26**(5), 2252–2263 (2021)
24. Wang, H., Cao, P., Wang, J., Zaiane, O.R.: Uctransnet: rethinking the skip connections in u-net from a channel-wise perspective with transformer. Proc. AAAI Conf. Artif. Intell. **36**(3), 2441–2449 (2022)
25. Fan, D.-P., et al.: Pranet: parallel reverse attention network for polyp segmentation. In: Medical Image Computing and Computer Assisted Intervention-MICCAI 2020: 23rd International Conference, Lima, Peru, October 4–8, 2020, Proceedings, Part VI 23, pp. 263-273. Springer (2020)
26. Selvaraju, R.R., Cogswell, M., Das, A., Vedantam, R., Parikh, D., Batra, D.: Grad-cam: visual explanations from deep networks via gradient-based localization. In: Proceedings of the IEEE International Conference on Computer Vision, pp. 618–626 (2017)

AMG-Mixer: A Multi-Axis Attention MLP-Mixer Architecture for Biomedical Image Segmentation

Hoang-Minh-Quang Le, Trung-Kien Le, Van-Truong Pham, and Thi-Thao Tran(✉)

Department of Automation Engineering, School of Electrical and Electronic Engineering, Hanoi University of Science and Technology, Hanoi, Vietnam
thao.tranthi@hust.edu.vn

Abstract. Previously, Multi-Layer Perceptrons (MLPs) were primarily used in image classification tasks. The emergence of the MLP-Mixer architecture has demonstrated the continued efficacy of MLPs in other visual tasks. To obtain superior results, it is imperative to have pre-trained weights from large datasets, and the Cross-Location (Token Mix) operation must be adaptively modified to suit the specific task at hand. Inspired by this, we proposed AMG-Mixer, an MLP-based architecture for image segmentation. In particular, recognizing the importance of positional information, we proposed AxialMBconv Token Mix utilizing Axial Attention. Additionally, to reduce Axial Attention's receptive field constraints, we proposed Multi-scale Multi-axis MLP Gated (MS-MAMG) block which employs Multi-Axis MLP. The proposed AMG-Mixer architecture outperformed State-of-the-Art (SOTA) methods on benchmark datasets including GLaS, Data Science Bowl 2018, and Skin Lesion Segmentation ISIC 2018, even without pre-training. The proposed AMG-Mixer architecture has been confirmed effective and high performing in our study. The code is available at https://github.com/quanglets1fvr/amg_mixer

Keywords: MLP-Mixer · image segmentation · Axial Attention · Multi-axis MLP

1 Introduction

In the past decade, convolutional neural networks have made remarkable advancements in the field of computer vision, particularly in the domain of image segmentation. Image segmentation represents a significant and challenging problem in computer vision, with the objective of simplifying the complexity of an image to facilitate its further processing and analysis. The seminal work of AlexNet, introduced in 2012 [1], served as a catalyst for the advancements in convolutional networks. Typical architectures in image segmentation usually

N. T. Nguyen et al. (Eds.): CITA 2023, LNNS 734, pp. 169–180, 2023.
https://doi.org/10.1007/978-3-031-36886-8_14

consist of encoder, decoder, and skip connection for helping to retain rich information such as Unet [2] and its variant UNet++ [3], DoubleUNet [4],...Although convolution is very good at extracting features, it performs poorly at building long range-dependendies. Many research works have shown how to overcome the shortcoming of convolution, in which the most prominent is the application of Transformer [5] in visual tasks. The Vision Transformer [6] architecture achieves efficiency in various visual tasks by partitioning the input image into multiple non-overlapping patches, flattening them, and applying a self-attention scheme. This architecture has demonstrated superior performance in numerous computer vision tasks due to its ability to build long-range dependencies and dynamically calculate weights to emphasize regions of interest. However, the large computation requirements of this architecture present a significant challenge. To address this limitation, various modifications of self-attention have been proposed, including Axial Attention [7], the Swin Transformer [8], and the Multi-Axis Attention [9]. Recently, the MLP-Mixer architecture [10], which utilizes multi-layer perceptrons (MLPs) exclusively, has garnered significant attention in the computer vision community. The MLP-Mixer architecture achieves competitive results by dividing the image into patches and performing both per-location and cross-location operations. However, it requires pre-trained weights to attain optimal performance.

Nowadays, various deep learning models for image segmentation have been developed, combining convolutional, transformer and MLPs studies, such as TransUNet [11]: deep features of convolution fed into transformer layers, UNext [12]: pass through three Convolution block and then patch fed into two MLPs Block, AxialAttMLPMixer [13]: uses Axial Attention for token-mix block and also has convolution branch. While AxialAttMLPMixer has achieved remarkable results without needing any pre-training, its limitations include the limited receptive field of axial attention, the extraction of redundant information by convolution, and its focus on learning at a single scale which is incompatible with image segmentation problems. This combination of MLPs, Convolution, and Transformer brings many benefits, however, it should be noted that they are also known to inhibit one another.

Based on the motivation discussed above, this study introduces AMG-Mixer, a hybrid MLP-based architecture that combines the strengths of Transformer and Convolution in an efficient manner. The main contributions of this work are presented below:

- We proposed a new token mixer block - AxialMBConvTokenMix, which leverages and preserves the position encoding of Axial Attention. This was achieved with the intention of enhancing generalization and attention across multiple dimensions.
- We proposed Multi-scale Multi-axis MLP Gated (MS-MAMG) block on skip connection, which helps improve Axial Attention receptive fields.
- Our proposed method, AMG-Mixer, has proven to be effective with a small number of parameters through evaluations on the GLaS, Data Science Bowl 2018, and Skin Lesion ISIC 2018 datasets.

2 Related Work

Axial Attention MLP Mixer: It was introduced in [13], an MLP-based method that utilizes Axial Attention to construct Axial Token Mix to overcome limitations of MLP-Mixer as well as Transformer. Despite not being pre-trained on a large dataset, Axial Attention MLP Mixer still demonstrates remarkable performance. This is due to its capability to learn global and local information effectively and its ability to establish long-range dependencies through its per-location and cross-location operating mechanisms.

Multi-Axis MLP Gated: It was introduced in [14], consisting of 2 processes(Local Branch and Global Branch) using only MLP. In the Global Branch, input $x^{C,H,W}$ is firstly divided into patches $g \times g$ and swap axes so x will become $x^{(g^2, \frac{HW}{g^2}, C)}$, then pass through Fully Connected Layer, and finally reshaped to original shape. In the Local Branch, input $x^{C,H,W}$ is firstly divided into patches $b \times b$ so x will become $x^{(\frac{HW}{b^2}, b^2, C)}$, and the next steps same to the Global Branch. It is effective in gathering information in both local and global contexts.

Progressive Atrous Pyramid Pooling (PASPP): PASPP [15] is a pooling technique used in computer vision tasks, especially in semantic segmentation problems, by using dilated convolution with different dilation rates. Therefore, the model can flexibly extract features and gather more global information without increasing computational complexity. PASPP is usually located at the junction between Encoder and Decoder

MBConv with Squeeze and Excitation (SE) [16]**:** This is a structure using the sequence of depthwise convolution, pointwise convolution, SE and residual connection. It has been applied in MnasNet [17], which optimizes both accuracy and latency. Besides, the ability to generalize and train is also improved.

3 Proposed Network Architecture

This study proposes a novel approach to image segmentation, namely AMG-Mixer. The architecture of AMG-Mixer, depicted in Fig. 1, comprises both encoding and decoding sections, similar in design to the UNet architecture. In the encoder, the image $x^{H \times W \times C}$ is first divided into many non-overlapped patches $p \times p$, then passes through mixer building blocks (containing AxialMBconv token mix and channel mix) and patch merging layers - a technique that was introduced in SwinUNet [18]. The resolution of the image is reduced in order to create pyramid features. The depth of the mixer building blocks is $(2, 2, 6, 2)$. Before being put into the decoder, we apply PASPP [15] for improving the model's receptive fields. The decoder consists of convolution layers and a bilinear upsampling layer, which upscales and aggregates information from the encoder. Additionally, on the skip connection, MS-MAMG improves the information from the encoder before it is put into the decoder stream. We will detail this further in the following sections.

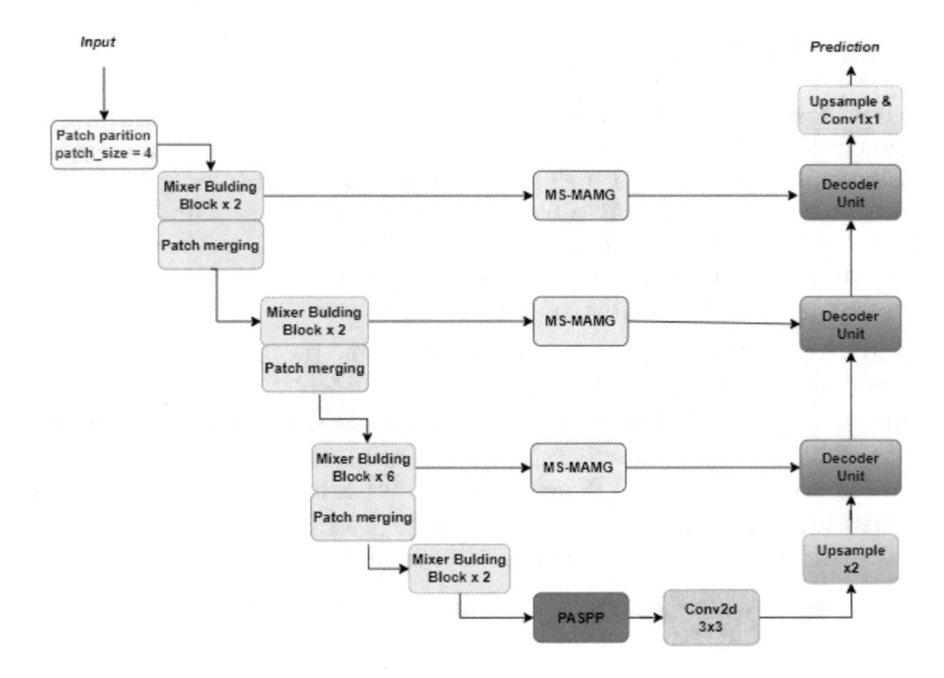

Fig. 1. The proposed AMG-Mixer architecture

3.1 AxialMBConv Token Mix and Mixer Building Block

Recently, token-mix block [10] substitution has been of great interest to researchers [12,13].... In this study, we proposed Axial MBConv Token Mix, an advancement from Axial Token Mix [13], as demonstrated in Fig. 2.

Axial Attention is a variant of Self-Attention that reduces computational complexity by enabling the implementation of Self-Attention mechanisms on rows or columns separately while being able to learn positional information from input images. In Axial Token Mix, the Squeeze and Excitation (SE) [16] block follows Height-Axis Attention and Width-Axis Attention and helps improve channel interdependence. However, methods for exploiting the positional information of previous Axial Attention layers are also of great importance, and so we have utilized the MBConv to address this issue by replacing the SE block of the Axial Token Mix with an MBConv. The MBConv, comprising of the PointWise Conv, SE and Depthwise Conv, when placed after the Axial Attention acts as a Conditional Position Encoding (CPE) [19] before being fed into the Channel Mix. As a result, the AxialMBConv Token Mix allows the model to be able to identify objects in varying positions, creating interdependencies on three dimensions (channels-height-width) of features. The formulation of the Mixer Building Block is presented below here.

$$Y = AxialMBConvTokenMix(LayerNorm(X)) + X \tag{1}$$

$$Z = ChannelMix(LayerNorm(Y)) + Y \tag{2}$$

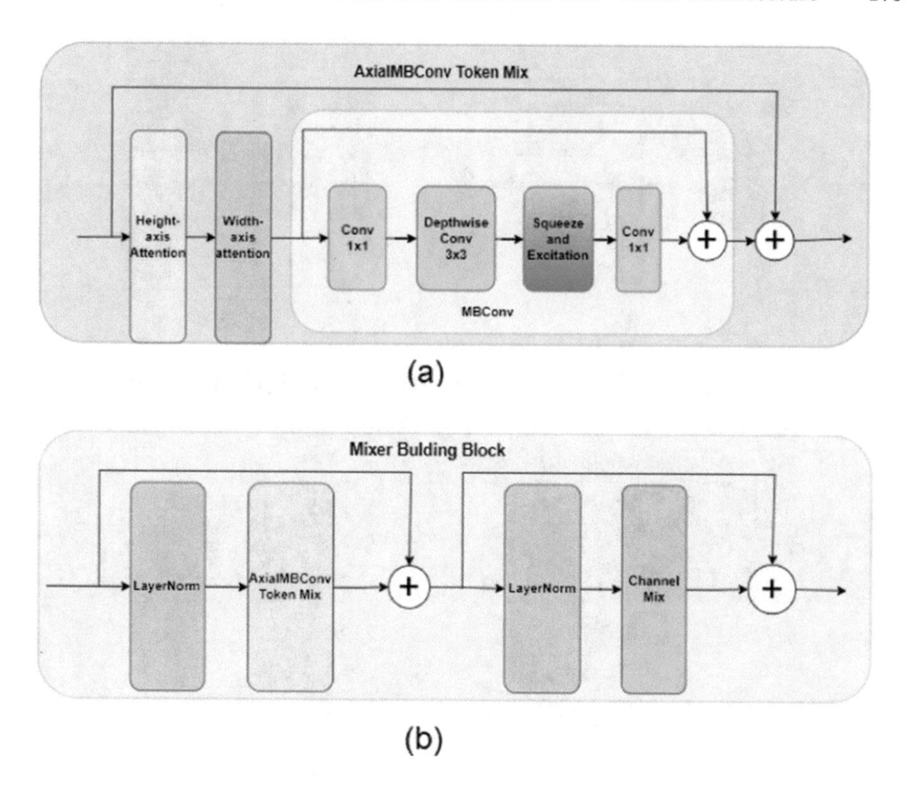

Fig. 2. (a) AxialMBConv Token Mix Block, and (b) Mixer Building Block

3.2 Multi-scale Multi-axis MLP Gated (MS-MAMG)

In Axial Attention, a pixel limited in a region can receive information from diagonal or other regions. Figure 3 illustrates that Axial Attention comprises two stages, the first of which is column-wise and the second is row-wise. Similarly, Multi-Axis MLP Gated has two stages, beginning with the Global-Branch and followed by the Local-Branch. The figure indicates that Multi-Axis MLP has a sparser and wider receptive field than Axial Attention.

Therefore, we proposed MS-MAMG, has shown in Fig. 4 by using Multi-axis Gated MLP for improving Axial Attention's receptive field.

Inspired by Maxim [14], we design MAG Block with Global Branch and Local Branch in series. The main structure of MS-MAMG Block contains three MAG Blocks, each block has *grid* and *block* depending on each other ($grid \times block = 16$), and has inspiration from CM-MLP [20]. About the effect of 3 MAG Blocks, the size *grid* in the global branch decrease in order 8, 4, 2 that makes the information in the spatial domain more and more sparse, and the size *block* in the local branch increase in order 2, 4, 8 makes the perceptual fields expand. Along with using depthwise convolution and pointwise convolution, MS-MAMG allows model to gain more enriched information and aggregate multi-scale local information without increasing computational complexity.

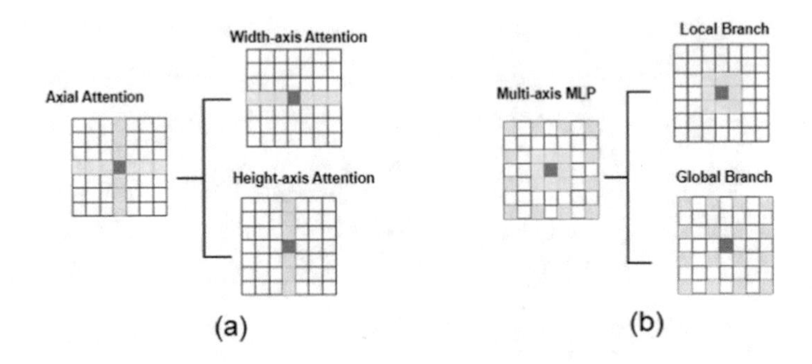

Fig. 3. Comparison between (a) Axial Attention, and (b) Multi-axis MLP

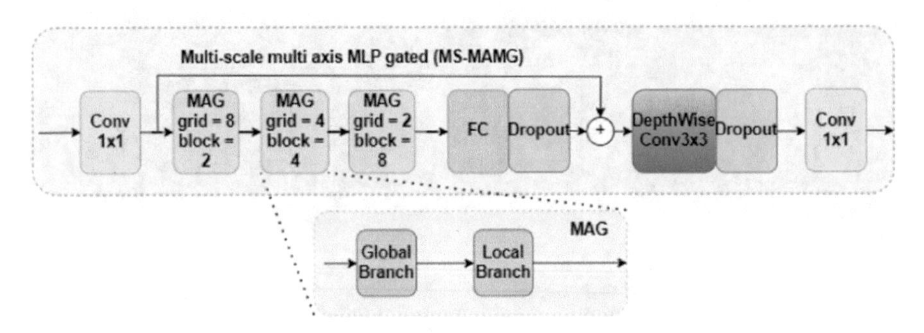

Fig. 4. Multi-scale - Multi-axis MLP Gated

3.3 Decoder Unit

Unlike almost all methods that use either addition or concatenation for skip connections, we apply both addition and concatenation in the Decoder Unit shown in Fig. 5. This approach helps to retain rich information from MS-MAMG Blocks while synthesizing it with information from the decoder stream.

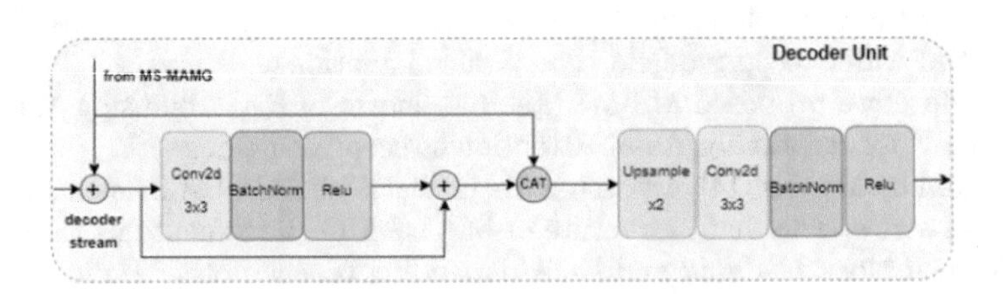

Fig. 5. The proposed Decoder Unit

4 Experiments

4.1 Datasets

In order to evaluate the effectiveness of AMG-Mixer, we use three datasets . The Gland Segmentation [28] dataset comprises 165 microscopic images and accompanying masks, with 85 images designated for training and 80 images reserved for testing. The Data Science Bowl 2018 [27] dataset comprises 671 nuclei images and annotations for each image's target masks, but the absence of test masks limits evaluation to the training set alone. In this study, the Bowl data was divided into 534 images for training and 137 images for testing; The ISIC 2018 Challenge [25] consists of 2594 images and accompanying masks, with 80% of the data designated for training, 10% for validation, and 10% for testing. It is noted that the proposed method and all compared methods mentioned below use the same training set and testing set for all databases.

4.2 Training Strategy

The proposed AMG-Mixer architecture was implemented using Pytorch and trained on a 16 GB NVIDIA Tesla T4 GPU. Data augmentation techniques including random rotation, horizontal flip, and vertical flip were utilized. The AMG-Mixer model contains very deep features with a size of $\frac{H}{16} \times \frac{W}{16}$. When passing through the MS-MAMG block, it will undergo a process of partitioning into blocks with sizes of 2, 4, and 8. Therefore, to ensure the resolution and efficiency for the process in the deep layers of the MS-MAMG block, the resolution of the input image must be sufficiently large and a power of 2, we choose the input size 256×256 for all experiments.

The Adam optimization algorithm was employed with a learning rate of $1e^{-3}$ and a step size of $1e^{-4}$. Additionally, the Stochastic Weight Average (SWA) [26] method was applied with a step size of 0.0001 after 100 epochs. To prevent overfitting on smaller datasets, L2 regularization with a weight decay of $1e^{-2}$ was employed. The loss function was a combination of Dice Loss and BCE Loss, the BCE Loss measures the pixel-wise error between the prediction and ground truth, while the Dice Loss evaluates the overlap between the prediction and ground truth. By combining both loss functions, the model can balance precision and recall and achieve better results. The formula is presented below here:

$$\ell = 0.5 \times \ell_{BCE} + 0.5 \times \ell_{Dice} \tag{3}$$

$$\ell_{BCE} = \frac{1}{N} \sum_{i=1}^{N} \left((y_i \log \hat{y}_i + (1 - y_i) \log(1 - \hat{y}_i) \right) \tag{4}$$

$$\ell_{Dice} = 1 - 2 \frac{\sum_{i=1}^{N} y_i . \hat{y}_i}{\sum_{i=1}^{N} y_i + \hat{y}_i} \tag{5}$$

where: N is the number of pixels in the image , y_i is the ground truth label of the i-th pixel, $\hat{y}_i$ is the predicted label of the i-th pixel.

4.3 Evaluation Metrics

In order to assess the efficiency of our model, we use Dice Coefficient (Dice) and Intersection over Union (IoU), two of the most common metrics to evaluate the performance of a segmentation model, particularly in medical image segmentation. They are sensitive to both false positive and false negative errors, making them useful for detecting areas where the model needs improvement. The formula of Dice and IoU is below here:

$$Dice(A, B) = \frac{2|A \cap B|}{|A| + |B|} \tag{6}$$

$$IoU(A, B) = \frac{|A \cap B|}{|A \cup B|} \tag{7}$$

where A and B are sets of pixels representing the prediction and ground truth respectively.

4.4 Representative Results

Fig. 6 presents the results obtained by the proposed AMG-Mixer method. It can be observed that the predictions made by AMG-Mixer are highly consistent with the ground truth. In comparison to Axial Attention MLP Mixer, improvements in both True Negative and True Positive can be seen, demonstrating the significance of the proposed method in addressing the problem.

4.5 Comparative Results

To quantitatively assess the performance of the proposed approach, we provided the average values of Dice and IoU scores, denoted as mDice and mIoU in Tables 1 and 2. From these tables, we can see that the proposed AMG-Mixer achieved competitive results on Bowl 2018, GlaS and ISIC 2018 without pre-training, with 92.47% Dice, 86.04% IoU on Bowl; 87.24%, 77.61% on GlaS; 91.12%Dice, 83.91% IoU on ISIC 2018, has superior results compared to CNN-based, Transformer - based and get a better result than AxialAttMLPMixer. Our method has demonstrated a better mix of Axial Attention with MLP, as evidenced by its Dice results of 0.75% and 6.22% higher than MedT (a transformer-based model using Axial Attention) on GlaS and Data Science Bowls 2018, as shown in Table 1. Furthermore, our method is equipped with a reasonable number of parameters and yields higher results on the ISIC 2018 set compared to several traditional CNN networks as shown in Table 2, indicating that the encoder's feature extraction capacity is outstanding. It proved that:

- Using Multi-axis MLP has been effective in enhancing information for Axial Attention.
- The interaction between the two operations: $per-action$ and $cross-location$ is more effective thanks to MBConv.

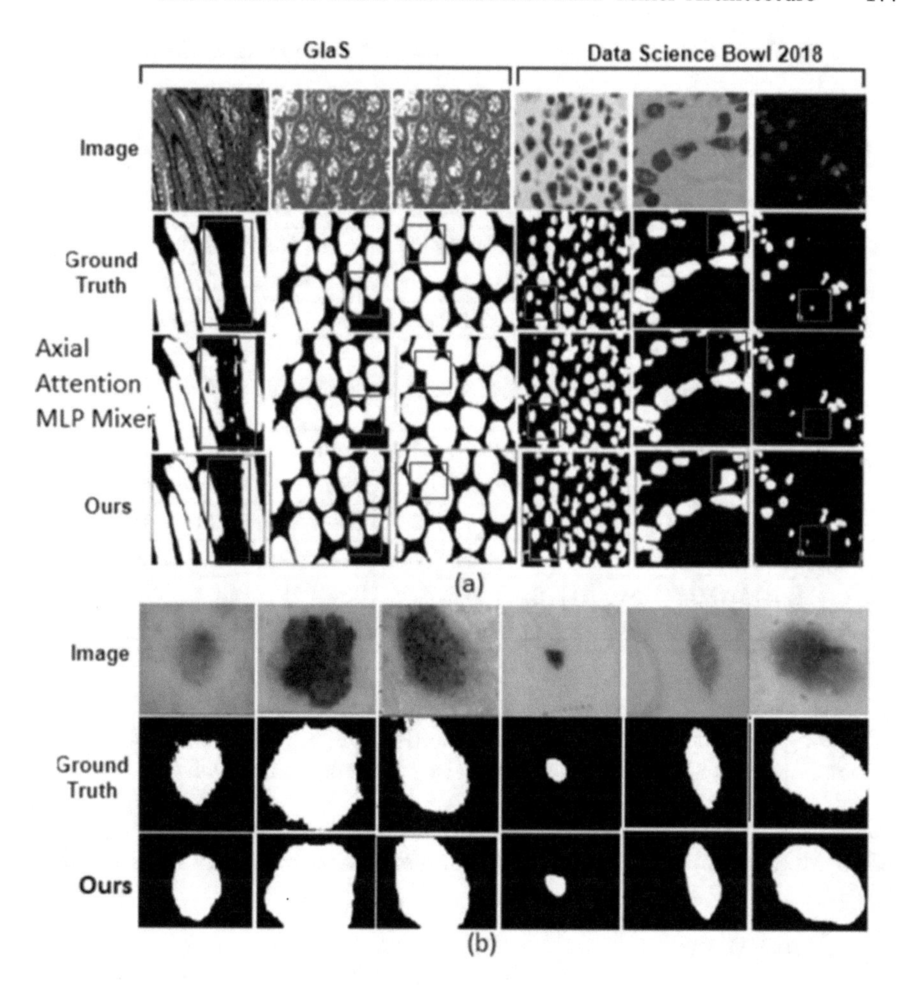

Fig. 6. Representative results by the proposed method (a), Data Science Bowl 2018 and GlaS (b), Skin Lesion Segmentation ISIC 2018

Table 1. Quantitative Evaluation of Diverse Models on Bowl and Glas Datasets

Methods	Types	Bowl		GlaS	
		mDice	***mIOU***	***mDice***	***mIOU***
UNET [2]	CNN-based	75.73	91.03	77.78	65.34
UNet++ [3]	CNN-based	89.74	**92.63**	78.03	65.55
MedT [21]	Transformer-based	91.72	84.75	81.02	69.61
AxialAttMLPMixer [13]	MLP-based	92.35	85.83	84.99	73.97
Ours (AMG-Mixer)	MLP-based	**92.47**	86.04	**87.24**	**77.61**

Table 2. Performances Comparison of Various Models of Skin Lesion ISIC 2018 Dataset

Methods	Type	Params	mDice	mIOU
Unet [2]	CNN-based	7.76 M	85.54	78.47
UNet++ [3]	CNN-based	9.04 M	80.94	72.88
ResUnet++ [24]	CNN-based	4.07 M	85.57	81.35
DoubleUnet [4]	CNN-based	29.3 M	89.62	82.12
Ours (AMG-Mixer)	MLP-based	22.3 M	**91.12**	**83.91**

Table 3. Comparisons with Variants of MLP-MIXER on Glas Dataset

Methods	Depth Layers	Params (M)	Input Size	GlaS	
				mDice	*mIOU*
MLP-Mixer [10]	24	100.1	224×224	82.83	70.81
Permutator [22]	36	88	224×224	84.21	72.80
AxialAttMLPMixer [13]	24	29.2	256×256	84.99	73.97
Ours(AMG-Mixer)	12	22.3	256×256	87.24	77.61

- Having pyramid features, reducing the dependency of convolutional features has improved the results better for the segmentation model

Table 3 shows that with far fewer parameters, training time as well as depth layer, AMG-Mixer still has superior results compared to other MLP-based methods.

5 Conclusion

In this work, we have proposed a new MLP-based architecture AMG-Mixer for image segmentation. It is a hybrid architecture of Convolution, MLP and Transformer. Specifically, the new token-mix block using Axial Attention design has also made the interaction functions between per-location and cross-location more efficient. Besides, we offer a method to use Multi-axis MLP gated to reduce the limitation of Axial Attention. In that way, without CNN features and pre-training, AMG-Mixer still achieves outstanding results compared to SOTA methods on GlaS, Bowl 2018 and ISIC 2018. In the future, we want to develop a pure MLP model based on AMG-Mixer with fewer parameters but still have the same or even better with AMG-Mixer.

Acknowledgements. This research is funded by Vietnam National Foundation for Science and Technology Development (NAFOSTED) under grant number 102.05-2021.34.

References

1. Krizhevsky, A., Sutskever, I., Hinton, G.: ImageNet classification with deep convolutional neural networks. In: NIPS (2012)
2. Ronneberger, O., Fischer, P., Brox, T.: U-net: convolutional networks for biomedical image segmentation. In: Proceedings Medical Image Computing Computer-Assisted Intervention, pp. 234–241 (2015)
3. Zongwei, Z., Md, M.R.S., Nima, T., Jianming, L.: UNet++: a nested u-net architecture for medical image segmentation. In: Deep Learning in Medical Image Analysis and Multimodal Learning for Clinical Decision Support (2018)
4. Jha, D., Riegler, M., Johansen, D., Halvorsen, P., Johansen, H.: Doubleu-net: a deep convolutional neural network for medical image segmentation. In: 2020 IEEE 33rd (CBMS), pp. 558–564 (2020)
5. Vaswani, A., et al.: Attention is all you need. Adv. Neural Inf. Process. Syst. **30** (2017)
6. Dosovitskiy, A., et al.: Image is worth 16×16 words: transformers for image recognition at scale. In: Proceedings of the 9th International Conference on Learning Representations (2021)
7. Wang, H., Zhu, Y., Green, B., Adam, H., Yuille, A., Chen, L.: Axial-deeplab: stand-alone axial-attention for panoptic segmentation. In: ECCV, pp. 108–126 (2020)
8. Liu, Z., et al.: Swin transformer: hierarchical vision transformer using shifted windows. In: Proceedings of the IEEE/CVF International Conference on Computer Vision (ICCV), pp. 10012–10022 (2021)
9. Tu, Z.: Maxvit: Multi-axis vision transformer In: ECCV 2022 (2022)
10. Tolstikhin, I., et al.: MLP-Mixer: an all-MLP architecture for vision. Adv. Neural Inf. Process. Syst. **34**, 24261–24272 (2021)
11. Jieneng, C., et al.: Transunet: Transformers make strong encoders for medical image segmentation. arXiv preprint arXiv:2102.04306 (2021)
12. Jeya, M.J.V., Vishal, M.P.: Unext: Mlp-based rapid medical image segmentation network. In: Medical Image Computing and Computer Assisted Intervention - MICCAI 2022 (2022)
13. Lai, H.P., Tran, T.T., Pham, V.T.: Axial attention MLP-mixer: a new architecture for image segmentation. In: ICCE (2022)
14. Tu, Z.: Maxim: Multi-axis mlp for image processing. In: Proceedings of the IEEE/CVF Conference on Computer Vision and Pattern Recognition (CVPR) (2022)
15. Yan, Q., et al.: COVID-19 chest CT image segmentation-a deep convolutional neural network solution, Jin, arXiv preprint arXiv:2004.10987 (2020)
16. Hu, J., Shen, L., Sun, G.: Squeeze-and-excitation networks. In: Proceedings of the IEEE Conference on Computer Vision and Pattern Recognition, pp. 7132–7141 (2018)
17. Tan, M., et al.: MnasNet: platform-aware neural architecture search for mobile. In: CVPR, pp. 2820–2828 (2019)
18. Cao, H.: Swin-unet: unet-like pure transformer for medical image segmentation. In: Computer Vision - ECCV (2022)
19. Chu, X., et al.: Conditional positional encodings for vision transformers. In: ICLR (2023)
20. Jinkai, L., et al.: CM-MLP: cscade multi-scale MLP with axial context relation encoder for edge segmentation of medical image. In: 2022 IEEE International Conference on Bioinformatics and Biomedicine (BIBM), pp. 1100–1107 (2022)

21. Valanarasu, J., Oza, P., Hacihaliloglu, I., Patel, V.: Medical transformer: gated axial-attention for medical image segmentation. In: International Conference on Medical Image Computing and Computer Assisted Intervention, pp. 36–46 (2021)
22. Hou, Q., Jiang, Z., Yuan, L., Cheng, M., Yan, S., Feng, J.: Vision permutator: a permutable MLP-like architecture for visual recognition. IEEE Tran. Pattern Analy. Mach. Intell. **45**(1), 1328–1334 (2022)
23. Badrinarayanan, V., Kendall, A., Cipolla, R.: Segnet: a deep convolutional encoder-decoder architecture for image segmentation. In: PAMI (2017)
24. Jha, D., et al.: ResUNet++: an advanced architecture for medical image segmentation. In: Proceedings of International Symposium Multimedia, pp. 225–230 (2019)
25. Codella, N.C., et al.: Skin lesion analysis toward melanoma detection: a challenge at the 2017 international symposium on biomedical imaging (ISBI), hosted by the international skin imaging collaboration (ISIC). In: Proceedings International Symposium Biomedical Imaging, pp. 168–172 (2018)
26. Izmailov, P., Podoprikhin, D., Garipov, T., Vetrov, D., Wilson, A.: Averaging weights leads to wider optima and better generalization. ArXiv Preprint ArXiv:1803.05407 (2018)
27. Rashno, A., et al.: Fully automated segmentation of fluid/cyst regions in optical coherence tomography images with diabetic macular edema using neutrosophic sets and graph algorithms. IEEE Trans. Biomed. Eng. **65**, 989–1001 (2017)
28. Malık, P., Kristofık, S., Knapov a, K.: Instance segmentation model ' created from three semantic segmentations of mask, boundary and centroid pixels verified on GlaS dataset. In: 2020 15th Conference On Computer Science And Information Systems (FedCSIS), pp. 569–576 (2020)

Investigating YOLO Models for Rice Seed Classification

Thi-Thu-Hong Phan(✉), Huu-Tuong Ho, and Thao-Nhien Hoang

FPT University, Da Nang, Vietnam
hongptt11@fe.edu.vn, tuonghhde170471@fpt.edu.vn

Abstract. Rice is an important staple food over the world. The purity of rice seed is one of the main factors affecting rice quality and yield. Traditional methods of assessing the purity of rice varieties depend on the decision of human technicians/experts. This work requires a considerable amount of time and cost as well as can lead to unreliable results. To overcome these problems, this study investigates YOLO models for the automated classification of rice varieties. Experiments on an image dataset of six popular rice varieties in Vietnam demonstrate that the YOLOv5 model outperforms the other YOLO variants in both accuracy and time of training model.

Keywords: Rice seed classification · Deep learning · YOLOv5 · YOLOv6 · YOLOv7

1 Introduction

Rice is one of the primary and most essential sources of food consumed by over half of people around the world [5]. Controlling rice quality is a very important task but the mixing of varieties occurs frequently in each stage of rice cultivation, production, and circulation. This affects the rice quality and yield. Therefore, identifying the purity of rice seeds is an essential task. However, this task is primarily performed manually based on the decision of human technicians in distinguishing the appearance characteristics of paddies [27]. This can lead to inaccurate results because of subjective factors like experience and tiredness. For that reason, it is crucial to develop a fast, accurate automatic method for classifying rice seed purity.

In the past decades, computer vision technology has made some breakthroughs and has been widely used in agriculture, such as detecting diseases in plants [4], evaluating food products, and food grain quality [14,22], classification of rice varieties [18]. This approach provides an alternative solution for automatically classifying rice seeds that is a cost-effective, rapid, accurate, and non-destructive technique.

Along with the rapid development of computer vision, numerous machine learning (ML) and deep learning (DL) techniques have been proposed to improve accuracy as well as save time and human effort in yield prediction, classifying

N. T. Nguyen et al. (Eds.): CITA 2023, LNNS 734, pp. 181–192, 2023.
https://doi.org/10.1007/978-3-031-36886-8_15

diseases, and seed identification. Nguyen et al. applied image processing algorithms, convolutional neural networks (CNN), and machine learning methods to recognize and classify whole rice and broken rice. The authors pointed out that CNN achieved an accuracy of 93.85% [16].

Koklo et al. used Artificial Neural Network (ANN), Deep Neural Network (DNN), and CNN algorithms to classify five different varieties of rice in Turkey. The classification results were achieved as 99.87% for ANN, 99.95% for DNN, and 100% for CNN. In [21], the authors developed CNNs with 34-layers and obtained an average accuracy of 99.92% for recognizing five rice seeds classes. Qadria et al. [18] proposed to use of five ML methods to classify six common Asian rice varieties and a maximum overall accuracy (MOA) was reached 97.4% by the LMT-Tree model.

YOLO (You only look once) network is a one-stage deep learning algorithm based on CNN initially proposed for object detection, also used for classification tasks. In [10], YOLO model was applied to detect disease on the leaves at an earlier stage before it affected the whole tree. This method achieved good efficiency in accuracy and speed. The classification of healthy and faulty maize seeds in [11] produced outstanding results, with accuracy, recall, and mAP of 99%, 99%, and 98%, respectively.

The success of YOLO models for the classification task in exceptional accuracy and speed criteria leads us to apply this model to identifying rice seed purity. In this study, we focus on the feasibility of YOLOv5, YOLOv6, and YOLOv7 for classifying six popular rice seed varieties in Vietnam in order to find out which YOLO model is suitable for the rice classification. These models are selected because they are state-of-the-art real-time deep learning algorithms used for object detection.

The rest of the paper is organized as follows. In Sect. 2, we describe shortly the theoretical overview and architecture of YOLO algorithm. Experiment setting and results are discussed in Sect. 3. Finally, some concluding remarks and perspectives have been given in Sect. 4.

2 Methodology

Recently, object detection has been one of the scorching topics in deep learning due to its high applicability and outstanding results. Object detection algorithms can be divided into two main categories i) one-shot object detection and ii) two-shot object detection. The last one has two steps which are region proposal and detection. Algorithms of this class, such as Faster-RCNN [20], usually provide high accuracy and performance, but they require much more executable time and large labeled dataset [1,9].

YOLO belongs to the single-shot object detection category which is capable of detecting objects in an image or video in a single pass. It uses a CNN to directly predict bounding boxes and class probabilities in one evolution [19] (Fig. 1).

YOLO takes an image as input, divides it into a fixed-size grid of cells, and extracts features from each cell using a CNN. It predicts a set of bounding

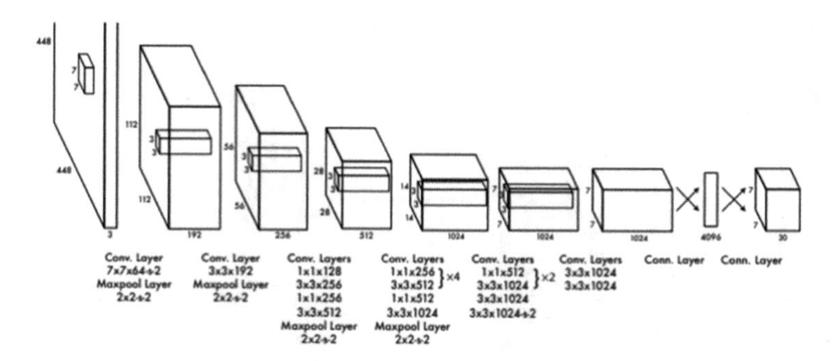

Fig. 1. The Architecture of YOLO model. The detection network has 24 convolutional layers followed by two fully connected layers. 1×1 convolutional layers help to reduce the number of features from preceding layers. The convolutional layers are pre-trained on the ImageNet with 224×224 input image size and then double the resolution for detection [19]

boxes and class probabilities of the bounding boxes for each cell in the grid. To remove overlapping detections and choose the most likely bounding box for each object, YOLO used a post-processing algorithm, called non-max suppression. The final result is a set of selected bounding boxes and their corresponding class probabilities for each object in the image.

Several versions of YOLO have been developed such as YOLOv1, YOLOv2, YOLOv3, YOLOv4, YOLOv5, YOLOv6, and YOLOv7. A later version usually improves and modifies the previous version with enhanced performance such as improved accuracy, faster processing, and better handling of small objects. In this paper, we focus on the most recent state-of-the-art object detectors which are YOLOv5, YOLOv6, and YOLOv7 due to their high accuracy, fast detection speed, adequate training time, and real-time applicability [8,15,25,28]. In the next section, these versions are briefly described.

2.1 YOLOv5

YOLOv5, an extended version from YOLOv4, was introduced by Jocher et al. from Ultralytics LLC (Los Angeles, CA - USA) [26] published another version of YOLO called YOLOv5 in a GitHub repository.

Figure 2 illustrates the overview architecture of YOLOv5. In this version, the main difference between YOLOv5 architecture and previous versions is that YOLOv5 uses "Focus structure with CSPdarknet53" as a backbone. This backbone reduces the repetitious gradient information in large backbones and integrates gradient change into a feature map that reduces the inference speed, increases accuracy, and reduces the parameters therefore smaller model size. Thanks to the Focus structure, the number of layers and parameters is reduced with increments of forward and backward speed while minimally affecting mAP [6].

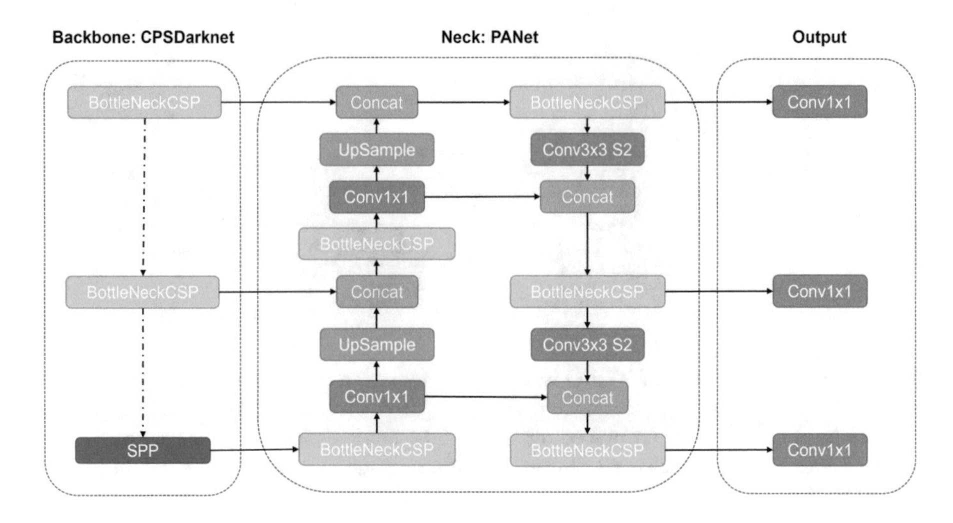

Fig. 2. Overview of YOLOv5 structure

2.2 YOLOv6

YOLOv6 is a new version of YOLO that has been developed to propagate its use for industrial applications. This model is a suite of deployment-ready networks at various scales to accommodate diversified use cases. YOLOv6 demands good performance on various hardware and real-world scenarios. The YOLOv6 has three components: the Backbone, the Neck, and the Head of the network. Firstly, YOLOv6 uses an efficient re-parameterizable backbone denoted as EfficientRep.

For small models, the main component of the backbone is RepBlock [3] during the training phase, and in the reference phase, this changes to simple 3×3 convolutional (RepConv) blocks and Relu activation.

For the Medium and large models, YOLOv6 architecture uses reparameterized versions of the CSP backbone [24], CSPStackRep Block, because the computation cost and the number of parameters in the single-path plain network grow exponentially when expanding model capacity.

For the neck architecture, similar to Yolov5 [7], YOLOv6 still uses the modified PAN topology [13] as the base of the detection neck. The CSPBlock used in YOLOv5 will change to the RepBlock (for small models)or CSPStackRep Block (for large models).

It is denoted as Rep-PAN, unlike YOLOv5, the YOLOv6 architecture used the Efficient Decoupled Head. This modification reduces computations and provides higher accuracy as well. In addition, instead of using anchor-base methods for object detection, YOLOv6 opts for the anchor-free method because of its better generalization ability and simplicity in decoding prediction results. The time cost of its post-processing is substantially reduced. The overall architecture of YOLOv6 is shown in Fig. 3.

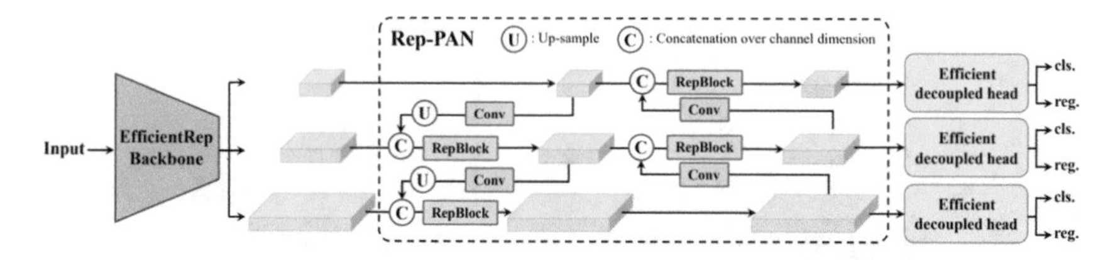

Fig. 3. The YOLOv6 framework [12]

2.3 YOLOv7

YOLOv7 is the latest version in the YOLO family up to now. Based on the architectures from the previous versions as the basis and undergoing testing to develop and improve performance, YOLOv7 is proposed to surpass the preceding object detection models and previous YOLO versions both in speed and accuracy. Unlike some previous YOLO versions, YOLOv7 backbones do not use ImageNet pre-trained backbones. For the architecture, YOLOv7 utilizes Extended-ELAN (E-ELAN) based on ELAN 4 to expand, shuffle, and merge cardinality in order to achieve the ability and continuously enhance the learning ability without destroying the original gradient path [23]. The author integrated the BoF methods and developed a new compound scaling method for the model, the corresponding compound model scaling method for a concatenation-based model5. The proposed method can maintain the properties that the model had at the initial design and maintains the optimal structure [23].

3 Experiments and Results

In this section, we describe the dataset, introduce metrics of evaluation and discuss experimental results.

3.1 Data Description

In order to evaluate and compare different variants of YOLO we use an image dataset of rice seed varieties that was introduced in our previous study [17]. This dataset contains six popular rice seed varieties in the North of Vietnam including BC15, Huong thom 1, Nep87, Q5, Thien uu 8, Xi 23. The rice seeds were collected by a rice seed company with standard conditions for the production of rice seeds in Vietnam. Figure 6 gives an example of rice seeds from the dataset. The number of rice seeds for each variety and our experimental setup in this study are shown in Table 1.

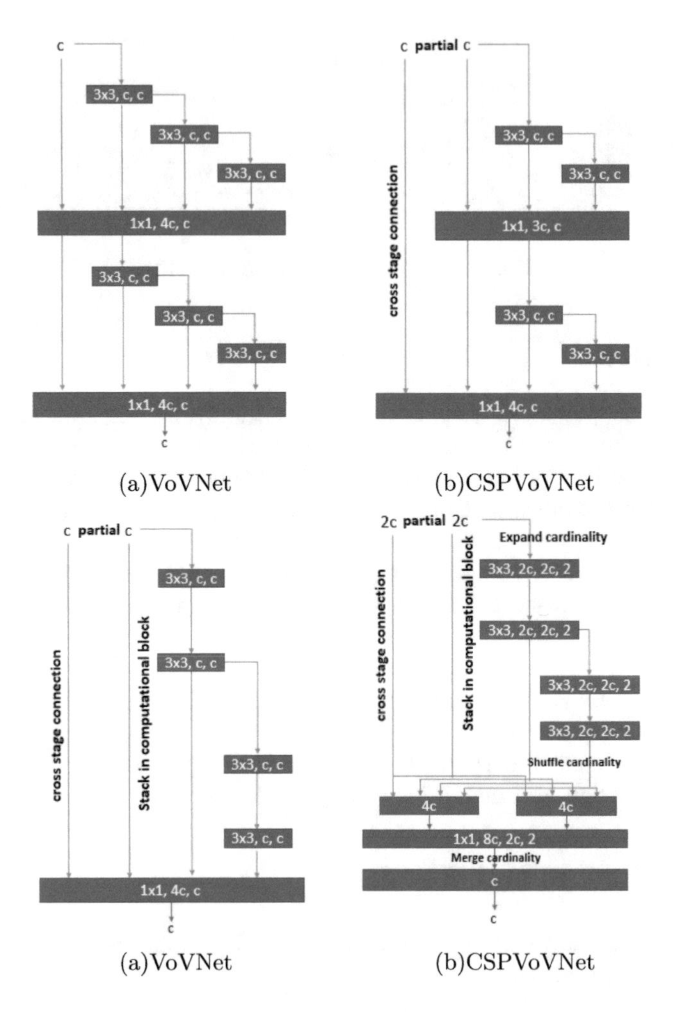

(a)VoVNet (b)CSPVoVNet

(a)VoVNet (b)CSPVoVNet

Fig. 4. Extended efficient layer aggregation networks [23]

3.2 Performance Indicator

In order to evaluate the performance of the three different YOLO models (YOLOv5, YOLOv6 and YOLOv7) we use five indices including Precision, Recall, F1 score, FPS, and Accuracy. The YOLO model is initially proposed for detecting objects but in this study, the purpose is object classification that is why the accuracy metric is added to assess the models. These indicators are briefly described as follows:

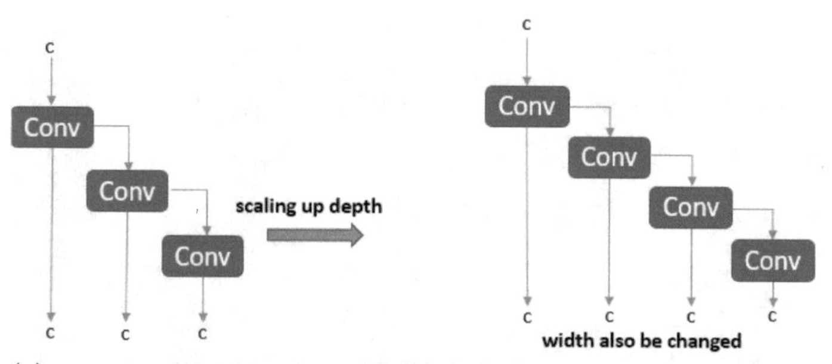

(a) concatenation-based model (b) scaled-up concatenation-based model

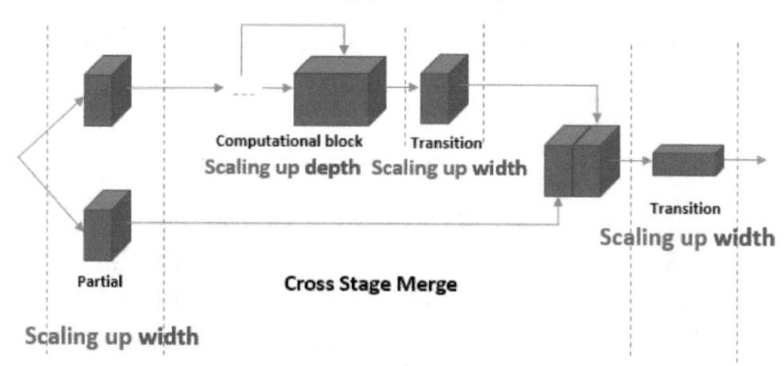

(c) compound scaling up depth and width concatenation-based model

Fig. 5. Model scaling for concatenation-based models [23]

1. Precision (P): indicates how many detected data points are truly relevant in the set of retrieved data as Eq. 1. A higher precision means that a model gives more relevant results than irrelevant ones.

$$Precision = \frac{TP}{TP + FP} \tag{1}$$

   where TP (True Positive) is the number of positive samples classified correctly, TN (True Negative) is the number of negative samples classified accurately; FP (False Positive) represents the number of actual negative samples classified as positive; and FN (False Negative) points out the number of actual positive examples classified as negative.
2. Recall (R) or Sensitivity: indicates the performance of the model to detect positive samples. It is computed as follows:

$$Recall = \frac{TP}{TP + FN} \tag{2}$$

3. F1: is calculated as the harmonic mean of precision and recall [2] as shown in Eq. 3

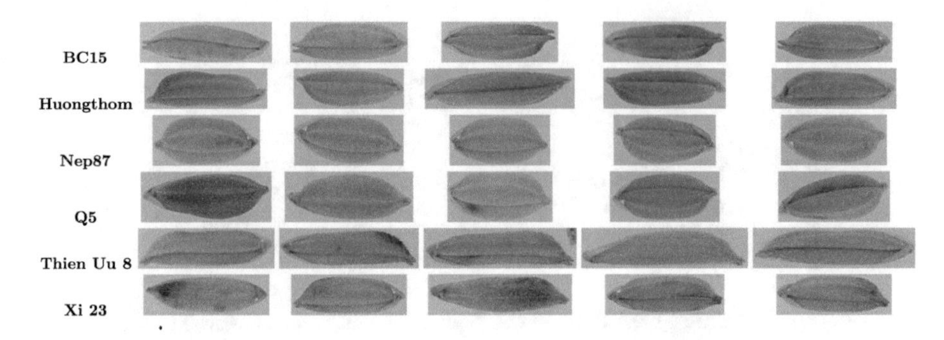

Fig. 6. An example of rice seed images from the dataset

$$F1 = \frac{2Precision * Recall}{Precision + Recall} \tag{3}$$

4. FPS (Frame per second): determines the speed at which a model processes the data and delivers the required result. This indicator reflects how quick the method is, hence the greater the FPS, the faster the model generates desired output. It is calculated by the following formula:

$$FPS = \frac{1000}{pre - process + inference + NMS} \tag{4}$$

where $pre - process$, $inference$, NMS are pre-processing, inference, and Non-Maximum Suppression time, respectively for each image.

5. Accuracy (Acc): is calculated by dividing the number of correct predictions by total the total prediction number as the following equation:

$$Accuracy = \frac{\text{TP} + \text{TN}}{\text{TP} + \text{TN} + \text{FP} + \text{FN}} \tag{5}$$

Accuracy is one of the most commonly used metrics while performing classification. Higher accuracy is a better model.

Table 1. Summary of the rice seed database

Name	Training set	Validation set	Test data set	Total
BC 15	1100	183	551	1834
Huong thom 1	1257	209	630	2096
Nep87	839	139	421	1399
Q5	909	151	455	1515
Thien uu 8	615	102	309	1026
Xi 23	1337	222	670	2229

3.3 Result and Discussion

For the experiment, we first split the data into the training set, validation set and testing set with the ratio shown in Table 1. That is, about 60% of the data is to train the machine learning models, 10% of the data is for the validation phase while the rest of 30% of the data is to test the performance of the trained models.

We then determine the input image size for the YOLO model. The default input size of the models is 640×640. However, the size of our images is smaller so some tests are conducted to find out which suitable image size to trade off between model training time and classification results. Table 2 demonstrates the efficacy of YOLOv5 with input sizes of square and rectangle. It clearly shows that the results of using a rectangle input size (96×224), non-square image size as suggested by the authors, are not good the results of using a square image size. This is because YOLO automatically disables mosaic when the argument "rect" is used during the training process, although other augmentation techniques, such as albumentations, and a normalization function are applied. The training phase often performs worse without mosaic.

Moreover, the training time when using the input size of 222×224, is not too much compared to the training time of the model when the input size of 224×96 (Table 2). Therefore, we use the input image with the size of 224×224 for the next experiments.

Table 2. Performance of YOLOv5 with input sizes (224×96) and (224×224)

Input size	P	R	F1	Acc	Training time
96×224	0.93	0.97	0.95	0.913	1.172h
224×224	0.996	0.996	0.996	0.992	1.842h

Table 3 presents the performance of all the YOLO models for identifying rice seeds. As shown in the table, the classification accuracy of models of all varieties is pretty good. YOLOv5 model achieves the best detection results with 99.2% for the average accuracy, the highest value of 99.8% for Huong thom 1 variety, and the lowest one of 99.3% for Q5. Considering the accuracy index, YOLOv6 is following YOLOv5 with an average value of 99%. The highest classification accuracy when classifying was 100% for Thien uu variety and the lowest value was 97.9% for Nep87 variety. Finally, YOLOv7 provides 97.2% classification results, lower than YOLOv5 and YOLOv6. Figure 7 shows an example of the classification result generated from YOLOv5. Regarding precision, recall, and F1 values, YOLOv5 once again proves its ability for the task of rice seed recognition. The YOLOv5 model yields better results than YOLOv6 and YOLOv7 for these four indices.

In addition, we also compare the training time and the speed of processing at the detection phase of the three YOLO versions. As shown in Table 4, YOLOv5 just only takes 1.84 h for the training model, lower 2.5 times than YOLOv7 with 4.674 h. YOLOv6 takes a little longer to train the model (approximately

6 minutes), and the classification performance is slightly lower than YOLOv5, but the detected results are still satisfactory, reaching 99%. And YOLOv7 has the slowest processing speed, with 166.7 fps at the detection phase, the longest model computing time, and its accuracy is lower than that of the YOLOv5 and YOLOv6 models.

Table 3. The performance of different YOLO models with input size 224 × 224 (best results are in bold)

Rice seed name	YOLOv5				YOLOv6				YOLOv7			
	P	R	F1	Acc	P	R	F1	Acc	P	R	F1	Acc
BC-15	0.997	0.987	0.991	0.991	0.996	0.983	0.989	0.991	0.998	0.967	0.982	0.964
Huong thom 1	0.997	0.999	0.997	0.998	1	0.997	0.998	0.998	0.985	1	0.992	0.997
Nep87	1	0.995	0.997	0.993	1	0.979	0.989	0.979	1	0.981	0.990	0.952
Q5	0.99	0.996	0.992	0.982	0.929	0.982	0.954	0.982	0.967	0.987	0.976	0.963
Thien uu 8	0.996	1	0.997	0.994	0.997	0.986	0.991	1	0.993	1	0.996	0.977
Xi 23	0.999	0.998	0.998	0.993	0.995	0.982	0.988	0.991	0.973	0.997	0.984	0.988
Average	**0.996**	**0.996**	**0.996**	**0.992**	0.986	0.985	0.985	0.99	0.986	0.989	0.987	0.976

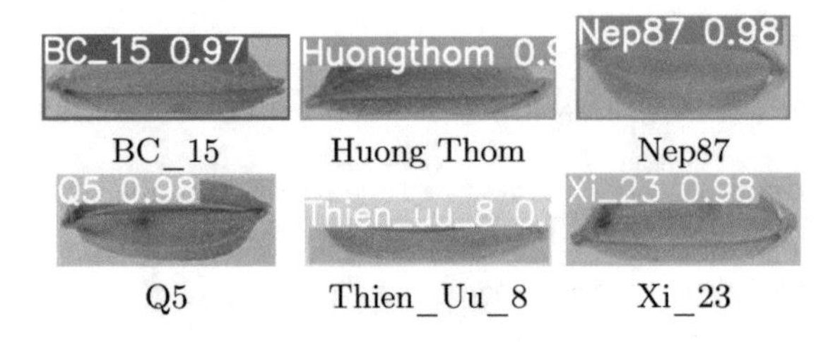

Fig. 7. Example of rice seeds classification using YOLOv5

Table 4. Comparison of the execution time and speed of processing of different models with input size 224 × 224

Model	FPS	Training Time (hours)
YOLOv5	303.03	1.842
YOLOv6	310.56	1.934
YOLOv7	166.67	4.674

4 Conclusion

In this paper, we investigate different algorithms, including YOLOv5, YOLOv6, and YOLOv7, for classifying varieties of rice seed purity based on a dataset of 10.099 images collected from a rice seed company in Vietnam. The experimental results using various evaluation measures clearly show that the YOLO is completely feasible for the classification problem, although the YOLO model was initially proposed for the task of object detection. Furthermore, among several latest models of the YOLO family, YOLOv5 achieves the best classification results with an average accuracy of 99.3% with the shortest execution time of 1.84 h. In addition, experimental results also demonstrate that, with a square image size, YOLO model provides better results than using a rectangle image size. In the future, we plan to combine the YOLO model with CNN applied to real-time classification problems such as leaf disease identification.

References

1. Daoud, E., Khalil, N., Gaedke, M.: Implementation of a one stage object detection solution to detect counterfeit products marked with a quality mark. **17**, 37–49 (2022)
2. Davis, J., Goadrich, M.: The relationship between precision-recall and roc curves. In: Proceedings of the 23rd International Conference on Machine Learning, pp. 233–240 (2006)
3. Ding, X., Zhang, X., Ma, N., Han, J., Ding, G., Sun, J.: REPVGG: making VGG-style convnets great again. In: Proceedings of the IEEE/CVF Conference on Computer Vision and Pattern Recognition, pp. 13733–13742 (2021)
4. Enkvetchakul, P., Surinta, O.: Effective data augmentation and training techniques for improving deep learning in plant leaf disease recognition. Appl. Sci. Eng. Progress (2021). https://doi.org/10.14416/j.asep.2021.01.003, http://ojs.kmutnb.ac.th/index.php/ijst/article/view/3810
5. FAOSTAT: Crop prospects and food situation. in Rome: Food and agriculture organization of the united nation. http://faostat.fao.org (2019). Accessed 7 Dec 2019
6. Jocher, G.: Yolov5 focus() layer #3181. https://github.com/ultralytics/yolov5/discussions/3181m1 (2021). Accessed 20 Oct 2021
7. Jocher, G.: Yolov5. code repository. https://www.github.com/ultralytics/yolov5
8. Jubayer, F., et al.: Detection of mold on the food surface using yolov5. Curr. Res. Food Sci.**4**, 724–728 (2021)
9. Koay, H.V., Chuah, J.H., Chow, C.O., Chang, Y.L., Yong, K.: Yolo-rtuav: towards real-time vehicle detection through aerial images with low-cost edge devices. Remote Sens. **13** (2021). https://doi.org/10.3390/rs13214196
10. Kodandaram, S.R., Honnappa, K.: Crop infection detection using yolo (2021). https://doi.org/10.13140/RG.2.2.27776.97281
11. Kundu, N., Rani, G., Dhaka, V.: Seeds classification and quality testing using deep learning and yolo v5. In: Proceedings of the International Conference on Data Science, Machine Learning and Artificial Intelligence, pp. 153–160 (2021). https://doi.org/10.1145/3484824.3484913
12. Li, C., et al.: Yolov6: a single-stage object detection framework for industrial applications. arXiv preprint arXiv:2209.02976 (2022)
13. Liu, S., Qi, L., Qin, H., Shi, J., Jia, J.: Path aggregation network for instance segmentation. In: Proceedings of the IEEE Conference on Computer Vision and Pattern Recognition, pp. 8759–8768 (2018)

14. Narendra, V.G., Hareesh, K.S.: Prospects of computer vsion automated grading and sorting systems in agricultural and food products for quality evaluation. Int. J. Comput. Appl. **1**(4), 1–12 (2010). https://doi.org/10.5120/111-226
15. Nepal, U., Eslamiat, H.: Comparing yolov3, yolov4 and yolov5 for autonomous landing spot detection in faulty UAVS. Sensors **22**(2), 464 (2022)
16. Nguyen, H.S., Nguyen, T.N.: Deep learning for rice quality classification. In: 2019 International Conference on Advanced Computing and Applications (ACOMP), pp. 92–96. IEEE (2019)
17. Phan, T.T.H., Hai, T.T.T., Hoang, V.T., Hai, V., Nguyen, T.T., et al.: Comparative study on vision based rice seed varieties identification. In: 2015 Seventh International Conference on Knowledge and Systems Engineering (KSE), pp. 377–382. IEEE (2015)
18. Qadri, S., et al.: Machine vision approach for classification of rice varieties using texture features. Int. J. Food Prop. **24**(1), 1615–1630 (2021). https://doi.org/10.1080/10942912.2021.1986523
19. Redmon, J., Divvala, S., Girshick, R., Farhadi, A.: You only look once: unified, real-time object detection. In: Proceedings of the IEEE Conference on Computer Vision and Pattern Recognition, pp. 779–788 (2016)
20. Ren, S., He, K., Girshick, R., Sun, J.: Faster r-cnn: Towards real-time object detection with region proposal networks (2015). https://doi.org/10.48550/ARXIV.1506.01497, https://arxiv.org/abs/1506.01497
21. Satoto, B.D., Anamisa, D.R., Yusuf, M., Sophan, M.K., Khairunnisa, S.O., Irmawati, B.: Rice seed classification using machine learning and deep learning. In: 2022 Seventh International Conference on Informatics and Computing (ICIC), pp. 1–7 (2022). https://doi.org/10.1109/ICIC56845.2022.10006960
22. Vithu, P., Moses, J.A.: Machine vision system for food grain quality evaluation: a review. Trends Food Sci. Technol. **56**, 13–20 (2016). https://doi.org/10.1016/j.tifs.2016.07.011
23. Wang, C.Y., Bochkovskiy, A., Liao, H.Y.M.: Yolov7: Trainable bag-of-freebies sets new state-of-the-art for real-time object detectors. arXiv preprint arXiv:2207.02696 (2022)
24. Wang, C.Y., Liao, H.Y.M., Wu, Y.H., Chen, P.Y., Hsieh, J.W., Yeh, I.H.: Cspnet: a new backbone that can enhance learning capability of CNN. In: Proceedings of the IEEE/CVF Conference on Computer Vision and Pattern Recognition Workshops, pp. 390–391 (2020)
25. Wu, D., et al.: Detection of camellia oleifera fruit in complex scenes by using yolov7 and data augmentation. Appl. Sci. **12**(22), 11318 (2022)
26. Yang, G., et al.: Face mask recognition system with yolov5 based on image recognition. In: 2020 IEEE 6th International Conference on Computer and Communications (ICCC), pp. 1398–1404. IEEE (2020)
27. Yang, J., et al.: Monitoring of paddy rice varieties based on the combination of the laser-induced fluorescence and multivariate analysis. Food Anal. Meth. **10**(7), 2398–2403 (2017). https://doi.org/10.1007/s12161-017-0809-2, http://link.springer.com/10.1007/s12161-017-0809-2
28. Yung, N.D.T., Wong, W., Juwono, F.H., Sim, Z.A.: Safety helmet detection using deep learning: implementation and comparative study using yolov5, yolov6, and yolov7. In: 2022 International Conference on Green Energy, Computing and Sustainable Technology (GECOST), pp. 164–170. IEEE (2022)

Extending OCR Model for Font and Style Classification

Vu Dinh Nguyen(✉), Nguyen Tien Dong, Dang Minh Tuan,
Ninh Thi Anh Ngoc, Nguyen Vu Son Lam, Nguyen Viet Anh,
and Nguyen Hoang Dang

CMC Applied Technology Institute (CATI), No 11, Duy Tan Street, Hanoi, Vietnam
{vdnguyen,ntdong,dmtuan1,ntangoc,nvslam}@cmc.com.vn

Abstract. Font and style classification aims to recognize which font and which style the characters in the input image belong to. The conjunction of font and style classification with traditional OCR systems is important in the reconstruction of visually-rich documents. However, the current text recognition systems have yet to take into account these tasks and focus solely on the recognition of characters from input images. The separation of these tasks makes the document reconstruction systems computationally expensive. In this paper, we propose a new approach that extends the current text recognition model to include font and style classification. We also present a dataset comprising input images and corresponding characters, fonts, and styles in Vietnamese. We evaluate the effectiveness of this extension on multiple recent OCR models, including VST [10], CRNN [8], ViSTR [1], TROCR [7], SVTR [4] . Our results demonstrate that our extension achieves decent accuracy rates of 98.1% and 90% for font and style classification, respectively. Moreover, our extension can even boost the performance of the original OCR models.

Keywords: Font Classification · Style Classification · OCR

1 Introduction

OCR, or optical character recognition, is a valuable technology that allows for the conversion of scanned documents or images into editable text. In recent years, Optical Character Recognition(OCR) domain using deep learning achieve many successes. However, current OCR models are not equipped to handle the complexities of recognizing different fonts and styles within an image.

In this work, we introduce a new task of jointly recognizing text, fonts, and styles for the character level of text line images. Along with this task, we also introduce a benchmark to evaluate the performance of models on this task. Additionally, we evaluate the ability of popular models to expand to this task using the aforementioned benchmark.

Supported by organization CATI.

N. T. Nguyen et al. (Eds.): CITA 2023, LNNS 734, pp. 193–204, 2023.
https://doi.org/10.1007/978-3-031-36886-8_16

We discuss the potential benefits of such a model, including improved accuracy and reliability of the resulting text, enhanced information extraction from visually rich document images, and aiding in the process of document reconstruction.

In addition, we argue that incorporating font and style recognition into OCR models can greatly enhance information extraction from visually rich document images. By recognizing the various fonts and styles within an image, allowing for more accurate and comprehensive information extraction. This can be particularly useful in fields such as medical imaging or financial reporting, where the ability to extract information from complex documents accurately is crucial.

Incorporating font and style recognition into OCR models can also greatly enhance the reconstruction of documents in formats such as Microsoft Documents, or PDF. This would allow for a more seamless integration of the reconstructed document into a word processing program, and make it easier for the user to edit and manipulate the text.

In order to evaluate the capability of current OCR models to recognize fonts and styles within an image, we propose the introduction of a new benchmark. This benchmark would consist of a dataset of document images that include a variety of fonts and styles, along with ground truth labels for the fonts and styles within each image with high-level fidelity in a new language. We empirically evaluate popular OCR models [1,4,7,8,10] on this benchmark.

Evaluation of this benchmark shows that it is the potential to extend model OCR for font and style classification and we achieve decent accuracy on both font(98.1%) and style classification(90%).

We summarize our contribution as follows:

1. Extending OCR model for font and style recognition task.
2. Comparing font and style recognition ability between recent OCR models.
3. Comparing long text line Vietnamese between recent OCR models.
4. Introducing dataset to evaluate font and style recognition in Vietnamese.

The remainder of the paper is organized as follows. In Sect. 2, we briefly summarize state-of-the-art methods relative to font classification. Sect. 3 presents our proposed approach. We evaluate our method in Sect. 4, and conclude the paper in Sect. 5.

2 Related Works

The Optical Font Recognition (OFR) problem has been undervalued despite its usefulness in visual document transformation. Although there are some works on font recognition, they typically address the problem independently of Optical Character Recognition (OCR) models.

For instance, [2] The authors of the paper proposed a new method for font style classification using character images, which involves utilizing distance profile features in three different directions (left, right, and diagonal) of the character image. The main purpose of this technique is to simplify the process of generic OCR systems by recognizing font styles. The distance profile features extracted from the character images are inputted into a support vector machine classifier. Or [6] employs a nearest neighbor approach with the tangent distance as the classification function. Discrimination is required for different digits and different font styles of the same character. While the nearest neighbor approach is effective in digit recognition, its performance in font detection is sub-optimal. To improve the system's performance, a discriminant model called the TD-Neuron is used to differentiate between similar classes.

These methods described are only capable of processing a single character in one go. Nevertheless, with the emergence of modern OCR systems, which typically process an entire word or line at once, the applicability of this approach may become challenging to integrate with such systems.

[11] introduce a straightforward framework that utilizes Convolutional Neural Networks (CNNs) for font classification. The CNN is trained to classify small patches of text into pre-defined font classes. To classify page or line images, the CNN predictions are averaged over densely extracted patches. This approach is capable of processing a text line; however, an independent model for font classification is still necessary. Furthermore, a single line of text may exhibit the same font throughout, while the style may differ at the character level.

In this study, we present an approach that differs from previous works in that we do not aim to introduce a new method to achieve the highest accuracy in font and style recognition. Instead, we propose a more straightforward and easily adaptable approach that utilizes features extracted from an OCR model for processing text lines. This approach can output the text and its font at the sentence level, as well as the style information at the character level.

3 Methods

In this work, we divide the proposed into three tasks: classify font and generate text and style simultaneously.

3.1 Problem Formulation

Given a set of N input images, denoted as $X = x_1, x_2, ..., x_N$. For each image x_i, a set of M text regions, denoted as $T_i = t_1, t_2, ..., t_M$. The task is to: (1) Extract the written content of each text region, denoted as $W = w_1, w_2, ..., w_M$ (2) Recognize the font of each text region, denoted as $F = f_1, f_2, ..., f_M$ and (3) Recognize the style of each character in the text regions, denoted as $S = s_{1,1}, s_{1,2}, ..., s_{M,L}$ where L is the number of characters in the text region t_M. The style of a character can be one or more Bold, Italic, or Underlined. The output of the model should be a tuple (W, F, S) for each image x_i in X.

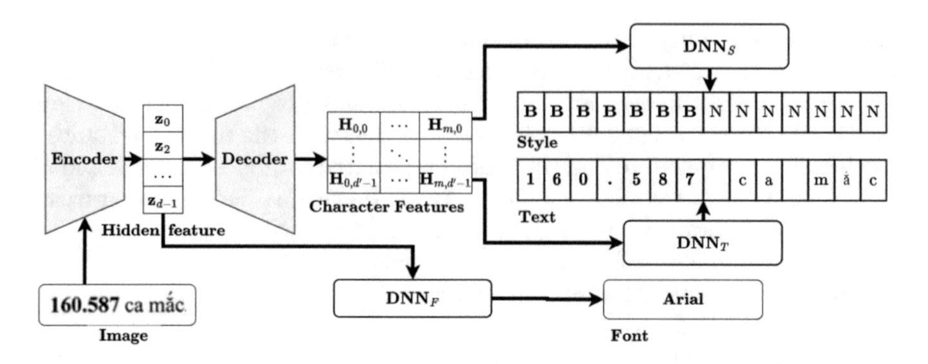

Fig. 1. Overall of architecture

3.2 Overview

The overview of our architecture is shown in Fig. 1. Here, we leverage the encoder-decoder framework as the base of our extension. This encoder-decoder framework is widely used in [1,4,7,8,10]. In this architecture, the encoder takes as input a textline image and outputs a hidden feature. This hidden feature is used as input to the decoder to output text content. We modify this architecture to support font and style classification with two additional components: a fully-connected neural network that outputs image-level font classification and another neural network to output character-level style classification. These two networks operate on the already available features and output of the encoder-decoder architecture for better computational efficiency of the extension. We describe each of these components in the following sections.

3.3 Encoder

The encoder takes as input an image $\mathbf{I} \in \mathbb{R}^{H \times W \times C}$, where $W, H \in \mathbb{N}$ are the width and height of the image respectively, $C \in \mathbb{N}$ is the number of input channels. In our implementation, we use $C = 3$ to take as input an RGB-colored image. The output of the encoder is a hidden vector $\mathbf{z} \in \mathbb{R}^d$, where $d \in \mathbb{N}$ is the hidden dimension of the vector. This hidden vector $\mathbf{z}$ is the hidden representation of the image. $\mathbf{z}$ summarizes the visual information of the image and is used by the decoder to generate the final output. In order for $\mathbf{z}$ to encode this information, we have to choose the suitable encoder architecture for aggregating the image's information. The popular choices in the literature include backbones such as VGG [9], ResNet [5], Vision Transformer (ViT) [3]. In order to thoroughly test the effect of our extension, we perform experiments with each of these backbones.

3.4 Decoder

The decoder takes hidden vector $z \in \mathbb{R}^d$ from the encoder as input. The output of the decoder is a matrix $\mathbf{H} \in \mathbf{R}^{m \times d'}$, where $m \in \mathbb{N}$ is the maximum number of characters in the images and $d' \in \mathbb{N}$ is the chosen hidden dimension of the vector. $\mathbf{H}_i$ where $i \in (1, m)$ is the index of the character in the sentence, will summarize visual feature for ith character and semantic feature with other characters in the image. To infer the character-wise features $\mathbf{H}$ from z, we use a decoder architecture in accordance with the chosen encoder. In detail, for Vision Trasformer's encoder (ViT), we apply a linear transformation on the input. For Convolutional Network-based encoders [5,9], we use either a recurrent neural network or a transformer decoder [12] to calculate character-wise features. Each row of the resulting feature matrix $\mathbf{H}$ is expected to be translatable into the corresponding character. While conventional OCR architectures leverage this feature matrix to identify the style of each character, we extend this architecture to also allow the inference of character style from the embedding.

3.5 Image-wise Font Clsasification

We assume that each image only have a single font. This is also in accordance with practical conditions. Based on this assumption, we use an image-wise feature to infer the font of the text. Intuitively, the output hidden feature $\mathbf{z}$ of the encoder should already embed image-wise information. Thus, we leverage the hidden vector $\mathbf{z}$ and use a deep neural network $\mathbf{DNN}_F$ to classify the font of the image text.

$$c_F = DNN_F(z)$$

where c_F is the likelihood of the character belonging to each font class.

3.6 Style Classification

In order to classify the style of text, it is crucial to consider the possibility of each character having a unique style, even within a single image, as illustrated in Fig. 2. To capture this variability, we extract a hidden representation for each character. Specifically, we obtain this representation from the hidden representation matrix $\mathbf{H}$ of the Decoder, using the row corresponding to the character's output feature. By sharing features between the tasks of classifying text and style, we can perform both tasks simultaneously. This joint approach allows us to leverage the interplay between text content and style, leading to improved performance on both tasks.

We define the likelihood of each character belonging to a specific character class and style class as follows:

$$ctext_i = DNN_T(\mathbf{H}_i)$$

$$cstyle_i = DNN_S(\mathbf{H}_i)$$

Here, $i \in (1, m)$ is the index of each character feature, where m is the total number of characters in the sentence. $\mathbf{H}_i \in \mathbf{R}^{d'}$ denotes the feature of the i-th character in the sentence. $ctext_i$ and $cstyle_i$ are the likelihoods of the character belonging to each character and style class, respectively. The deep neural network (DNN) DNN_T and DNN_S are used to classify the character and style, respectively.

3.7 Loss Function

The output from our model are $\mathbf{p_{o,f}}$ font probability for each image, and $\mathbf{p_{o,c}}$ character probability, $\mathbf{p_{o,s}}$ style probability for each character.
Loss of extending model will summarize of three losses: Loss of font classificaiton, text recognition and font classification.

$$L = L_F + L_T + L_S$$

where L_F is cross-entropy loss:

$$L_F = -\sum_{c=1}^{I} y_{o,f} \log(p_{o,f})$$

(I is number of font classes).
L_T, L_S are sum of cross-entropy loss over all character:

$$L_T = \sum_{i=1}^{M} -\sum_{c=1}^{J} y_{o,c} \log(p_{o,c})$$

$$L_S = \sum_{i=1}^{M} -\sum_{c=1}^{K} y_{o,s} \log(p_{o,s})$$

(i is index of character in sentence, M is maximum of character)

4 Experiments

4.1 Experiment Setting

Five OCR models were extended and compared using the same DNN. The DNN used here has two settings:

- a neural network with two fully connected layers, where the first layer is followed by ReLU activation.Feature dimension is 1024 for font classification and 256 for style and character recognition..

- a SVM classifier.

To reduce the time required for training with two different settings, we adopted a two-stage training approach. In the first stage, we trained the OCR model alone. In the second stage, the model was initialized with the weights obtained in the first stage and was then trained with the extension layers(font and style recognition) using random initialization.

The optimizer used for all five models is Adam with a learning rate of 10e−4. All five models were trained on T4, and the best accuracy for each model was recorded.

We evaluated 5 model VST [10], CRNN [8], ViSTR [1], TROCR [7], SVTR [4] and DNN we use simple two fully connected layer for experiment in 4 metrices:

$$\text{Sequence accuracy} = \frac{\text{number of correct sentence}}{\text{total number of sentence}}$$

$$\text{Character accuracy} = 1 - \frac{\text{levenshtein distance}}{\text{total character of sentence}}$$

$$\text{Style accuracy} = 1 - \frac{\text{levenshtein distance}}{\text{total character of sentence}}$$

$$\text{Font accuracy} = \frac{\text{number of correct font sentence}}{\text{total number of sentence}}$$

The evaluation of time complexity, as represented in Table 3, involves the computation of an average across all test datasets. This computation is performed repeatedly for a total of five runs, and the reported result is the average of these five runs

4.2 Benchmarks

In order to evaluate the capability of current OCR models to recognize fonts and styles within an image, we propose the introduction of a new benchmark. This benchmark would consist of a dataset of document images that include a variety of fonts and styles, along with ground truth labels for the fonts and styles within each image with high-level fidelity in Vietnamese.

The proposed benchmark for evaluating the font and style recognition capabilities of OCR models would be designed to have high fidelity at the character level. This means that the dataset would include a wide range of fonts and styles, and the ground truth labels for each image would provide detailed information about the fonts and styles of the individual characters within the image. This level of fidelity would allow OCR models to be tested on their ability to recognize and classify the fonts and styles of individual characters, rather than just at the level of the overall document. This would provide a more detailed and accurate assessment of the model's performance, and could help to identify specific challenges or limitations in its font and style recognition capabilities.

In addition, the character-level fidelity of the proposed benchmark would make it more applicable to a wider range of real-world scenarios. Many documents, particularly those containing complex layouts or formatting, may have varying fonts and styles within a single page or even within a single line of text. An OCR model with the ability to recognize and classify fonts and styles at the character level would be better equipped to handle these situations and produce more accurate results.

In conjunction with high fidelity at the character level, the proposed benchmark for evaluating the font and style recognition capabilities of OCR models would also be designed to include a new language, Vietnamese. To the best of our knowledge, there has been little previous work on developing OCR models for Vietnamese, and as such, this language presents a unique challenge for current OCR technology.

Testing Dataset. Consists of 2900 images and text labels. Each image contains one sentence one of 29 font class: *pala, uvntintuc, uvnbaihoc, uvngiongsong, arial, uvnchinhluan, uvnbachtuyet, uvnlaxanh, uvnnhan, uvnthoinay, uvngiomay, uvnvietsach, uvnhongha, uvnsachvo, cour, uvnsaigon, uvnmangcau, uvnthaygiao, uvnnhatky, times, uvnmaychup, uvngiadinhhep, uvntintuchep, uvnanhhai, uvngiadinh, uvnhonghahep, uvncatbien, uvnbachdang, calibri.* Each character in setence could have Bold, Underline, or Italic style. 2

Training Dataset. The dataset for training consists of 1 million images (Figure 3), each image has one line of text with lengths ranging from 1 to 80 characters. This dataset is synthesized from 29 fonts from the benchmark dataset and is augmented by transforms such as: rotated, shift, warp, Gaussian noise, salt noise, and blur to enhance the model.

4.3 Experiment Result and Discussion

Table 1. Experiment result OCR only with different types of extension. (O) means no extension, (NN) means Neural Network-based extension and (SVM) means SVM-based extension

Model	Sequence Accuracy			Character Accuracy		
	O	NN	SVM	O	NN	SVM
CRNN[8]	0.682	**0.701**	0.685	0.891	**0.902**	0.895
ViSTR[1]	0	0	0	**0.200**	0.172	0.160
SVTR[4]	**0.210**	0.202	0.208	0.801	**0.803**	0.801
TRocr[7]	0.788	**0.802**	0.795	0.961	**0.9701**	0.966
VST[10]	0.760	**0.780**	0.700	0.965	**0.967**	**0.967**

Table 2. Font and Style classification accuracy of each extension type.

Model	Style Accuracy		Font Accuracy	
	NN	SVM	NN	SVM
CRNN [8]	**0.010**	0.009	**0.812**	0.748
ViSTR [1]	**0.012**	0.011	**0.011**	**0.011**
SVTR [4]	**0.020**	0.01	**0.051**	0.06
TRocr [7]	**0.902**	0.850	**0.961**	0.80
VST [10]	**0.907**	0.881	**0.981**	0.892

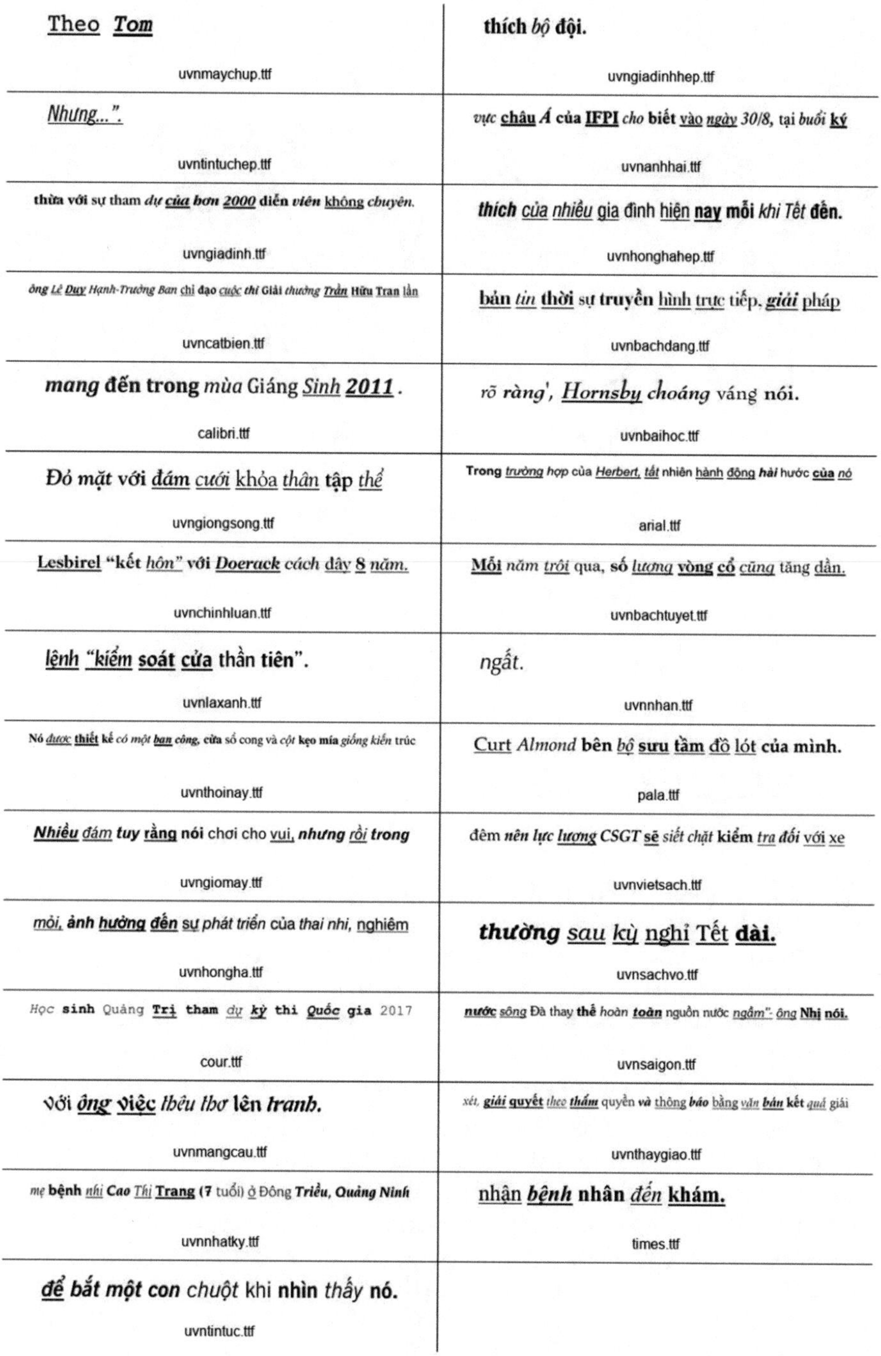

Theo Tom uvnmaychup.ttf	thích bộ đội. uvngiadinhhep.ttf
Nhưng...". uvntintuchep.ttf	vực châu Á của IFPI cho biết vào ngày 30/8, tại buổi ký uvnanhhai.ttf
thừa với sự tham dự của hơn 2000 diễn viên không chuyên. uvngiadinh.ttf	thích của nhiều gia đình hiện nay mỗi khi Tết đến. uvnhonghahep.ttf
ông Lê Duy Hạnh-Trưởng Ban chỉ đạo cuộc thi Giải thưởng Trần Hữu Tran lần uvncatbien.ttf	bản tin thời sự truyền hình trực tiếp, giải pháp uvnbachdang.ttf
mang đến trong mùa Giáng Sinh 2011 . calibri.ttf	rõ ràng', Hornsby choáng váng nói. uvnbaihoc.ttf
Đỏ mặt với đám cưới khỏa thân tập thể uvngiongsong.ttf	Trong trường hợp của Herbert, tất nhiên hành động hài hước của nó arial.ttf
Lesbirel "kết hôn" với Doerack cách đây 8 năm. uvnchinhluan.ttf	Mỗi năm trôi qua, số lượng vòng cổ cũng tăng dần. uvnbachtuyet.ttf
lệnh "kiểm soát cửa thần tiên". uvnlaxanh.ttf	ngất. uvnnhan.ttf
Nó được thiết kế có một ban công, cửa sổ cong và cột kẹo mía giống kiến trúc uvnthoinay.ttf	Curt Almond bên bộ sưu tầm đồ lót của mình. pala.ttf
Nhiều đám tuy rằng nói chơi cho vui, nhưng rồi trong uvngiomay.ttf	đêm nên lực lượng CSGT sẽ siết chặt kiểm tra đối với xe uvnvietsach.ttf
mỏi, ảnh hưởng đến sự phát triển của thai nhi, nghiêm uvnhongha.ttf	thường sau kỳ nghỉ Tết dài. uvnsachvo.ttf
Học sinh Quảng Trị tham dự kỳ thi Quốc gia 2017 cour.ttf	nước sông Đà thay thể hoàn toàn nguồn nước ngầm"- ông Nhị nói. uvnsaigon.ttf
với ông việc thêu thơ lên tranh. uvnmangcau.ttf	xét, giải quyết theo thẩm quyền và thông báo bằng văn bản kết quả giải uvnthaygiao.ttf
mẹ bệnh nhi Cao Thị Trang (7 tuổi) ở Đông Triều, Quảng Ninh uvnnhatky.ttf	nhận bệnh nhân đến khám. times.ttf
để bắt một con chuột khi nhìn thấy nó. uvntintuc.ttf	

Fig. 2. Example of Benchmarks

Fig. 3. Example of Synthesized Data

Table 3. Inference time

Model	extended(neural network) OCR (s/iter)	extended(SVM) OCR (s/iter)	OCR only (s/iter)
CRNN [8]	0.521	0.523	0.522
VisTR [1]	0.012	0.011	0.010
SVTR [4]	0.019	0.019	0.018
TrOCR [7]	1.352	1.361	1.348
VST [10]	1.523	1.511	1.522

RQ1. The impact of Font and Style Recognition Module on the Original OCR Result. The overall performances of each architecture before and after adding the extension are summarized in Table 1. Overall, the results hint that using font-and-style supervision along with our extension also improves the performance of the original text recognition tasks: NN-based extension managed to improve 3/5 models on sequence accuracy and 4/5 on character accuracy. For SVM-based extension, the improvement ratio is 2/5 and 3/5, respectively.

RQ2. Font Classification Performance. For font classification task, based on the results presented in Table 2, models that feature separate encoder and decoder components, such as CRNN [8], TrOCR [7], and VST [10], demonstrate superior performance because their encoders can focus on extracting visual features from the image, while their decoders can memorize the semantics of the text. In contrast, models such as ViSTR [1] and SVTR [4] combine their encoder and decoder components in a single block, which may not be as effective in extracting visual features for font recognition.

RQ3. Style Recognition Performance. Regarding style recognition, it is observed that models trained with CTC loss perform worse on this task, such as

CRNN [8], VisTR [1], and SVTR [4]. In contrast, models that have transformer decoders, which already have an attention mechanism, perform well and do not require CTC loss for training.

RQ4. Performance Comparison Between Different Types of Extension. When comparing the performance of neural network setting and SVM setting (Table 2), it can be observed that neural network models generally demonstrate superior performance over SVM models in both extending tasks. Hence, for both font classification and style classification in extended OCR, neural network models are preferred over SVM models.

RQ5. Inference Speed. After conducting a comparative analysis of our proposed model with extended OCR and OCR-only approaches, it was found that the processing times (refer Table 3) were similar. Therefore, adopting our approach would not significantly increase the processing time of the system. Consequently, this approach reduces the time complexity of the font classification model and eliminates the need to manage and train a different OCR model for each font.

To the best of our knowledge, no previous studies have compared OCR models for processing Vietnamese long text lines. In this study, we trained five models with the same input size of (56,672) and evaluated their performance on both font and style classification tasks and OCR effectiveness in processing Vietnamese long text lines. Our experimental results demonstrate that TrOCR [7] outperforms other models and remains the most powerful OCR model for processing Vietnamese long text lines.

5 Conclusion

Based on the results of our experiments, it is evident that there is potential to extend OCR models for font and style classification, as we achieved decent accuracy on both tasks (98.1% for font classification and 90% for style classification). This suggests that we could replace the old OCR system, which performs text and font recognition separately, with the new OCR system that can handle both tasks effectively.

Acknowledgements. This research is funded by CMC Applied Technology Institute (CATI)

References

1. Atienza, R.: Vision transformer for fast and efficient scene text recognition. In: CoRR abs/2105.08582 (2021). arXiv: 2105.08582. https://arxiv.org/abs/2105.08582
2. Bharath, V., Rani, N.S.: A font style classification system for English OCR. In: 2017 International Conference on Intelligent Computing and Control (I2C2), pp. 1–5 (2017). https://doi.org/10.1109/I2C2.2017.8321962

3. Dosovitskiy, A., et al.: An image is worth 16×16 words: transformers for image recognition at scale. In: 11929 (2020). arXiv: 2010, http://arxiv.org/abs/2010.11929
4. Du, Y., et al.: SVTR: scene text recognition with a single visual model. In: IJCAI International Joint Conference on Artificial Intelligence 1.c, pp. 884–890. (2022). issn: 10450823. arXiv: 2205.00159, https://doi.org/10.24963/ijcai.2022/124
5. He, K., et al.: Deep residual learning for image recognition. In: CoRR abs/1512.03385 (2015). arXiv: 1512.03385, http://arxiv.org/abs/1512.03385
6. La Manna, S., Colia, A.M., Sperduti, A.: Optical font recognition for multi-font OCR and document processing. In: Proceedings. Tenth International Workshop on Database and Expert Systems Applications. DEXA 99, pp. 549–553 (2008). https://doi.org/10.1109/dexa.1999.795244
7. Li, M., et al.: TrOCR: Transformer-based Optical Character Recognition with Pre-trained Models. In: CoRR abs/2109.10282 (2021). arXiv: 2109.10282, https://arxiv.org/abs/2109.10282
8. Shi, B., Bai, X., Yao, C.: An end-to-end trainable neural network for image-based sequence recognition and its application to scene text recognition. In: CoRR abs/1507.05717 (2015). arXiv: 1507.05717, http://arxiv.org/abs/1507.05717
9. Simonyan, K., Zisserman, A.: Very deep convolutional networks for large-scale image recognition. In: 3rd International Conference on Learning Representations, ICLR 2015 - Conference Track Proceedings (2015), pp. 1–14. arXiv: 1409.1556
10. Tang, X., et al.: Visual-semantic transformer for scene text recognition. In: arXiv: 2112.00948, http://arxiv.org/abs/2112.00948 (2021)
11. Tensmeyer, C., Saunders, D., Martinez, T.: Convolutional neural networks for font classification. In: CoRR abs/1708.03669 (2017). arXiv: 1708.03669, http://arxiv.org/abs/1708.03669
12. Vaswani, A., et al.: Attention is all you need. In: CoRR abs/1706.03762 (2017). arXiv: 1706.03762, http://arxiv.org/abs/1706.03762

STG-SimVP: Spatiotemporal GAN-Based SimVP for Video Prediction

Trieu Duong Le[1,2] and Duc Dung Nguyen[1,2](✉)

[1] Ho Chi Minh City University of Technology (HCMUT), 268 Ly Thuong Kiet Street, District 10, Ho Chi Minh City, Vietnam
nddung@hcmut.edu.vn

[2] Vietnam National University Ho Chi Minh City, Linh Trung Ward, Ho Chi Minh City, Vietnam

Abstract. The goal of video frame synthesis is from given frames. In other words, it tends to predict single or several future frames (video prediction - VP) or in-between frames (video frame interpolation - VFI). This is one of the challenging problems in the computer vision field. Many recent VP and VFI methods employ optical flow estimation in the prediction or interpolation process. While it helps, it also makes the model more complex and harder to train. This paper proposes a new model for solving the VP task based on a generative approach without utilizing optical flow. Our model uses simple methods and networks like CNN to reduce the hardware pressure, it can learn movement fast within the first few training epoch and still predict high-quality results. We perform experiments on two popular datasets Moving MNIST and KTH. We observe the proposed model and SimVP model on four metrics: MSE, MAE, PSNR, and SSIM in the training process. The experiments show that our model can capture spatiotemporal correlations better than previous models.(The code is available at github.com/trieuduongle/STG-SimVP.)

Keywords: Video Prediction · Video Frame Extrapolation · GAN

1 Introduction

The applications of video frame synthesis are widely adopted in many fields, such as video compression [2], enhancing frame rate in videos [14], autonomous driving, weather forecast, planning actions for robots, traffic prediction [5,22,24]. For example, self-driving cars usually will have cameras to capture the continuous sequence of images, then they will have techniques to generate future videos from the provided sequence. Based on that, these cars will use generated videos with other information to make more accurate decisions for the next actions.

Several models are built to solve this problem based on many different architectures: RNN-RNN-RNN [9,26,28]; CNN-RNN-CNN [8,10]; Transformer [15,18,30] but these models have weak point that they require a lots training resources because their architectures are too complex. Therefore, researchers are also developing models only using CNN [6,21,32] to reduce computational costs.

N. T. Nguyen et al. (Eds.): CITA 2023, LNNS 734, pp. 205–216, 2023.
https://doi.org/10.1007/978-3-031-36886-8_17

In this paper, we aim to propose a new lightweight and efficient GAN model which is built upon SimVP [6]. Our main contributions are summarized as follows:

- We propose a new model to solve the VP task with a generative approach.
- Introduce a spatiotemporal discriminator module to make the model can learn the motion from input images better.
- Utilize all the potential of SimVP when using it as a generator without any modification in its model.

2 Related Work

2.1 Video Frame Synthesis

Video frame synthesis aims to synthesize new frames based on existing ones. All introduced networks can be divided into 2 main categories:

- "Optical flow-based" approach [11,16,29,31]: This traditional solution is to estimate the optical flows from input images and combine them with features that they gain from input to synthesize new frames. Models which use this approach will have the output depending on the accuracy of optical flow, if the estimated optical flow is incorrect, the output will have a significant of artifacts.
- "Optical flow-free" approach [6,13]: These models will synthesize frames directly without optical flow estimation. By that way, this approach can reduce the computational cost and help the model scale in complex scenarios easier.

2.2 Generative Adversarial Networks - GAN

GAN was first introduced by Ian Goodfellow [7]. This model includes 2 small networks called a generator and a discriminator. The generator will try to generate new data that looks the same as the ground truth as much as possible. On the other hand, the discriminator will learn a way to classify the input data coming from the generator or they are our ground truth.

GAN models have shown their important roles in numerous image applications: image super-resolution, image inpainting, text to image... And video frame synthesis problem also has promising results when applying the GAN technique [1,4,23]

2.3 SimVP

The strategy of SimVP [6] is that the model must be simple, take less time for training, good performance. Therefore, SimVP is built upon CNN blocks, highway networks, and MSE loss to predict future frames. The Generator block in Fig. 1 and Fig. 2 are the architecture of SimVP. This model includes 3 modules:

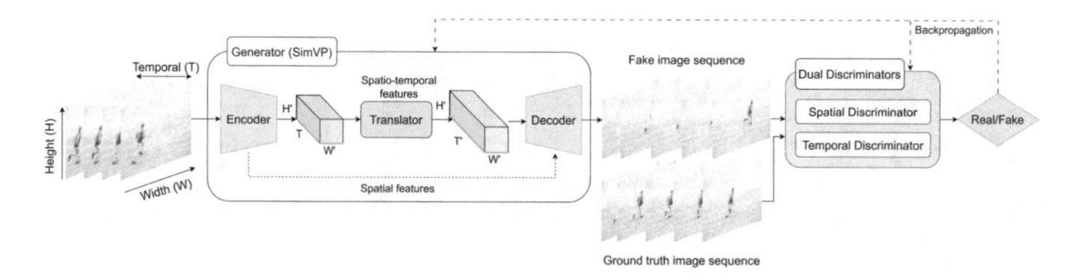

Fig. 1. The overall of our STG-SimVP. Firstly, our generator (which is SimVP [6]) will take an input of T continuous frames and predict T' frames. Then the predicted images and our ground truth frames will be used to train our discriminator module which includes 2 discriminators: a Spatial Discriminator D_s, and a Temporal Discriminator D_t. The Spatial Discriminator will decide whether the input frames are real or fake and Temporal Discriminator distinguishes whether the input images are in the correct sequence or not.

- Encoder: Used to learn and extract spatial features of the input images.
- Translator: Will learn temporal evolution.
- Decoder: Combine information that it has from Encoder and Translator to predict future frames.

Although this model is simple, it has already outperformed complex models such as PredRNN [28], PhyDNet [8], CrevNet [33]. Based on this result, SimVP can scale up for complex situations easier than other models.

3 Proposed Methods

3.1 STG-SimVP

Our proposed model - known as STG-SimVP is shown in Fig. 1. The STG-SimVP is built up on the GAN network where Generator will inherit the whole SimVP [6] architecture and a dual discriminator which will be described in detail in Sect. 3.2. The Generator will extract features and learn temporal evolution from those things then that information will be combined and predict future frames. We will use these predicted frames and the ground truth as input of Dual Discriminators, and this block will try to find real/fake sequences and finally will backpropagate to Generator and Discriminator.

3.2 Dual Discriminators

Usually, a GAN model will build up by a Generator and only one discriminator. However, in this work, we proposed a GAN model which uses dual discriminators:

- Spatial Discriminator (D_s): This network will learn a way to separate the real and fake frames.

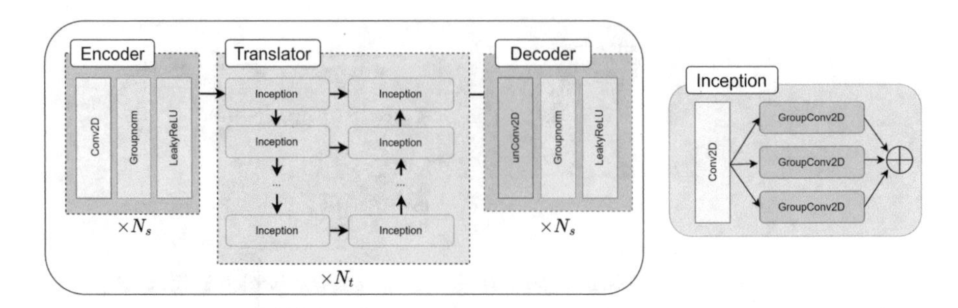

Fig. 2. The detailed architecture of SimVP includes: Encoder is the combination of N_s (Conv2D + Groupnorm + LeakyReLU) blocks, Decoder consists of N_s (unConv2D + Groupnorm + LeakyReLU) blocks and Translator has N_t Inception modules. The Inception module is built by a Conv2D with 1×1 kernel and GroupConv2D operators.

- Temporal Discriminator (D_t): the purpose of this network is to learn and decide whether the provided sequence is fake or real.

The discriminator network is based on T-PatchGAN [3]. This network (see Fig. 3) was designed to have the ability to learn spatial and temporal features to improve the quality of output. According to Ya-Liang Chang [3], this network is a combination of 6 convolutional layers and each layer will have kernel size is $3 \times 5 \times 5$ and stride is $1 \times 2 \times 2$.

The original idea of a discriminator is that the output of the discriminator will be a scalar value and it will become a binary classification problem. However, T-PatchGAN will divide images into smaller blocks and apply binary classification for these blocks. For real images, the network will be trained to expect all these blocks to be 1, otherwise, they will be 0. With this approach, T-PatchGAN has a large reception field on images and it can learn spatiotemporal feature better.

Our dual discriminators are using the T-PatchGAN network with a small modification in which D_s will use Conv2D and D_t will use Conv3D.

3.3 Objective Function

Our objective function which is used in the training process is given as follows

$$\mathcal{L} = \mathcal{L}_{image} + \mathcal{L}_{gan} \tag{1}$$

where $\mathcal{L}_{image}$ is the photometric loss and $\mathcal{L}_{gan}$ is the GAN loss.

Photometric Loss

Adopting Mean Square Error (MSE) as the photometric loss

$$\mathcal{L}_{image} = MSE(Y, \hat{Y}) = \frac{1}{n}\sum_{i=1}^{n}(Y_i - \hat{Y}_i)^2 \tag{2}$$

where Y_i is the target image, $\hat{Y}_i$ is the generated image, and n is the number of samples.

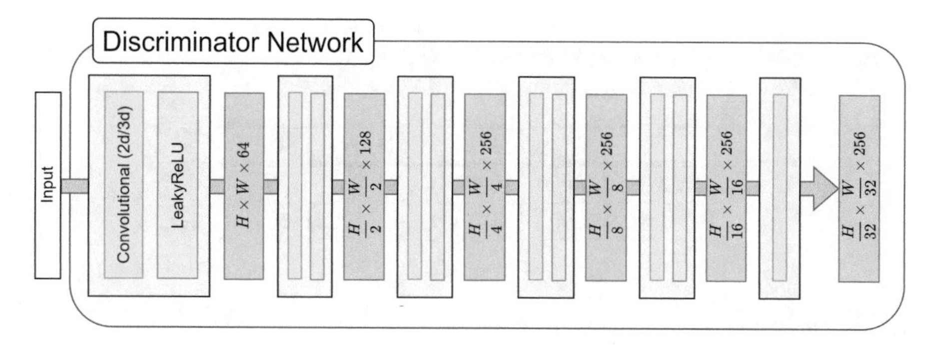

Fig. 3. The architecture of discriminator. Each convolutional layer will have kernel size $3 \times 5 \times 5$ and stride $1 \times 2 \times 2$. The output of each block is shown in purple. The final output which has shape $\frac{H}{32} \times \frac{W}{32} \times 256$ will be expected to be 1 if the input is real, otherwise the expectation value will be 0. The Conv2D layer will be used for D_s. Meanwhile, we will use the Conv3D layer for D_t.

GAN Loss

The loss for optimizing the generator is defined as the following

$$\mathcal{L}_{gan} = \lambda_s D_s(G(z)) + \lambda_t D_t(G(z)) \tag{3}$$

where G is generator that takes frame images z and λ_s, λ_t are weights of the spatial discriminator D_s, temporal discriminator D_t, respectively.

Besides, we use the Eq. 4 for training our dual discriminators.

$$\mathcal{L}_D = \mathcal{L}_D^{real} + \mathcal{L}_D^{fake} \tag{4}$$

$$\mathcal{L}_D^{real} = \mathbb{E}_{x \sim p_{\text{data}}(x)}[MSE(1, D(x))] \tag{5}$$

$$\mathcal{L}_D^{fake} = \mathbb{E}_{z \sim p_z(z)}[MSE(0, D(G(z)))] \tag{6}$$

where x is the groundtruth image in the future which is selected from the training dataset, z are input images, $MSE(1, D(x))$ means that the discriminator can classify the ground truth as real labels, and $MSE(0, D(G(z)))$ means that the predicted future images should be marked as fake labels.

From the provided objective function $\mathcal{L}$, we can expect that our generator can update weights better when our dual discriminators keep giving feedback on predicted frames of the generator throughout the training process.

4 Experiments

Due to the limitation of hardware that we are having, we only make the comparison between our STG-SimVP with the baseline.

Table 1. Dataset overview. Each dataset will include N_{train} training samples, N_{test} testing samples, each sample will have a shape (C, H, W). The input of the model will be T images and the expected output will be T' images.

	N_{train}	N_{test}	(C, H, W)	T	T'
MMNIST	10000	10000	(1, 64, 64)	10	10
KTH	5200	3167	(1, 128, 128)	10	20

4.1 Datasets

Two datasets *Moving MNIST* and *KTH* to train and evaluate models. Table 1 shows the overview of our datasets.

Moving MNIST. [22]: This is the common used dataset. This dataset contains 10000 videos, each video will include 20 frames, each frame will have a size 64 × 64 and each video will describe the counterclockwise movement of two digits independently around frames. The first 10 frames will be the input of the model, next 10 frames are the groundtruths (Fig. 4).

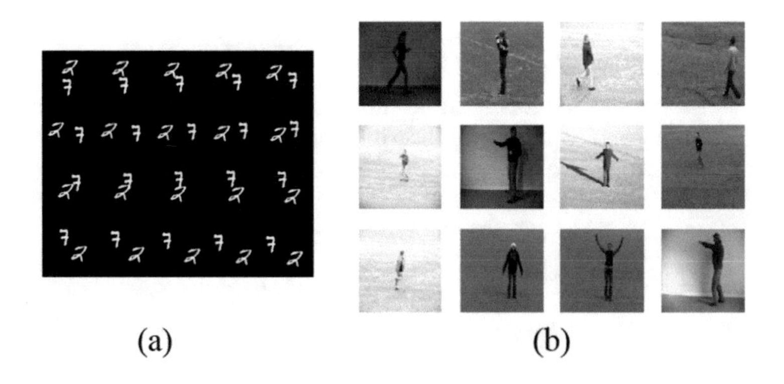

Fig. 4. Some samples of 2 used datasets: (a) MMNIST [22], (b) KTH [19]

KTH. [19] This is one of the most standard datasets. It has six action types: walking, jogging, running, boxing, hand waving, and hand clapping. Each action will be performed by 25 different individuals in four environments: outdoor, outdoor with scale variation, outdoor with different clothes, and indoor. The dataset has 600 videos with 25 fps and each frame is 160 × 120 pixels.

4.2 Training Details

The experiment environment is set up on Google Colab Pro where the provided machine is 1 GPU NVIDIA A100-SXM4 with 40 GB RAM. Our generator and SimVP can configure the layer numbers N_s, hidden dimensions C_s of the spatial

Encoder (or Decoder), and the layer numbers N_t and hidden dimensions C_t of the Translator's encoder (or decoder). For a fair comparison, both models will use the same hyperparameters:

- MMNIST dataset: $C_s = 64$, $C_t = 512$, $N_s = 4$, $N_t = 3$, learning rate $= 0.01$, training batch size $= 32$, testing batch size $= 32$, training epochs $= 250$.
- KTH dataset: $C_s = 32$, $C_t = 128$, $N_s = 3$, $N_t = 4$, learning rate $= 0.01$, training batch size $= 16$, testing batch size $= 8$, training epochs $= 100$.

4.3 Comparisons With State-of-the-Art Methods

With settings in Sect. 4.2, we will compare our model to SimVP, since SimVP uses the same settings as us and it also outperformed many other SOTA models. Following previous works, we use 4 main metrics to report the quantitative results of our model and SimVP: MSE, MAE, PSNR, and SSIM. These metrics are reported in Tables 2, 3 and two Figs. 5, 6. The results indicate a comparable performance in terms of quantitative metrics, and in some cases, ours does better.

Table 2. Quantitative results on MMNIST (250 epochs).

	MSE	MAE	PSNR	SSIM
SimVP	**38.5546**	103.1917	37.5569	0.9109
Ours	39.1384	**102.6510**	**37.6298**	**0.91111**

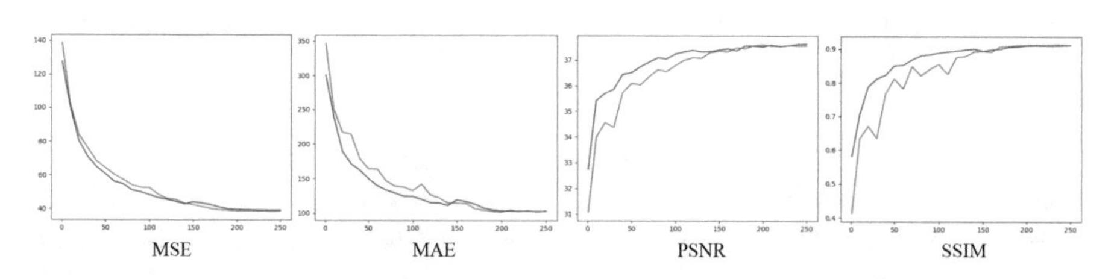

Fig. 5. Metric results for the proposed model and SimVP after 250 epochs on MMNIST dataset. The red line expresses the proposed model result while SimVP uses green color.

However, we observed the training process of the KTH dataset in detail and we have come up with conclusions:

- In the first 50 epochs, the proposed model (STG-SimVP) can decrease values better than SimVP. It shows that the dual discriminators are playing an important role at the beginning of the training process, they gave good feedback about spatiotemporal information, then the generator can learn things fast. Inspecting more in Table 4 and Fig. 6, even 4 metrics do not have a big

Table 3. Quantitative results on KTH dataset (100 epochs). For a fair comparison, the results of our model, we retrain SimVP, and other results are based on Zhangyang Gao's collection [6].

	MSE	MAE	PSNR	SSIM
MCnet [25]	–	–	25.95	0.804
ConvLSTM [20]	–	–	23.58	0.712
fRNN [17]	–	–	26.12	0.771
PredRNN [28]	–	–	27.55	0.839
PredRNN++ [26]	–	–	28.47	0.865
E3d-LSTM [27]	–	–	29.31	0.879
STMFANet [12]	–	–	29.85	0.893
SimVP [6]	53.8878	**454.5872**	**33.1617**	0.8986
Ours	**52.3847**	460.9462	33.0859	**0.8998**
Ours (without D_s)	62.3233	482.9688	32.8576	0.87599
Ours (without D_t)	54.2064	466.1641	33.0390	0.8991

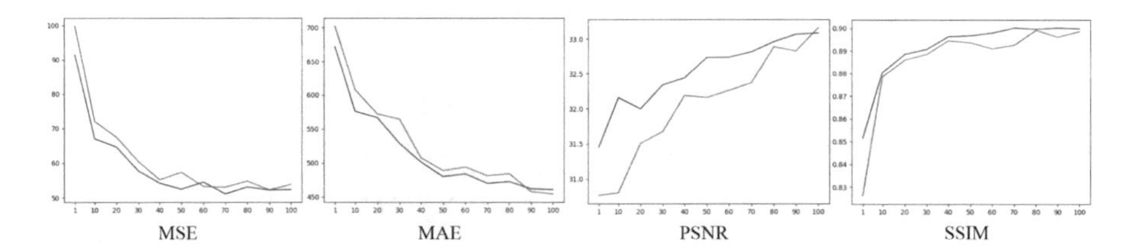

Fig. 6. Visualize four quantitative metrics after training 100 epochs on the KTH dataset. The red line is our model while the green is SimVP. (Color figure online)

difference, but STG-SimVP are expressing that it can learn the movement better than SimVP. After training, we investigated that the blurry black block in the image is what the model predicted for the human in the future. In epoch 1, SimVP cannot find that black block, and at epoch 10, it starts to predict the movement but the movement cannot go so far. Finally, at epoch 20, it can predict the movement better but the predicted black block does not have the human shape which needs longer in the vertical axis. On the other hand, STG-SimVP at the beginning started to predict the black block and right after that, it can learn the movement and predict the human shape really better.

- From epoch 50, when the training process occurs long enough, 2 models have similar performance results.
- In the last observation at epoch 100, SimVP has a black block deeper than our model but it still cannot show which part is the head, body, or legs while ours can express clearer that the head and legs will be light block, the body will be black block. We present the detailed result in Table 4

Table 4. The results of our model and SimVP at epochs 1, 10, 20, and 100. The input of the model will be 10 images and it will predict 20 future frames. From observation, the movement and human shape which is a blurry black block in each frame are learned from our model faster than SimVP.

	Epoch	Predicted 20 frames
SimVP	1	
	10	
	20	
	100	
Ours	1	
	10	
	20	
	100	
Ground Truth		
Input		

5 Ablation Study

In this section, we conduct 2 ablation studies to investigate our proposed model. We will verify the effectiveness of the spatial discriminator D_s and temporal discriminator D_t by removing one of them from the discriminator module during training. Quantitative results are reported in Table 3 and our observation throughout the training process is available in Table 5, it is showing that our STG-SimVP model cannot miss D_s or D_t:

- Missing D_s: it can learn the movement but the human in predicted frames always has blurry because the temporal discriminator only has the ability to learn temporal information as we defined.
- Missing D_t: it lost the ability to learn temporal consistency fast. Besides, on a few last predicted frames, D_s cannot give good feedback to the generator because the architecture of D_s is Conv2D layers, so it cannot learn temporal information well. Therefore, it cannot do its job well and it also makes the generator update weights incorrectly.

Table 5. Predicted frames at epoch 1, 10, 20, 100 of STG-SimVP without D_s or D_t. The conclusion is that when it is missing D_s, the model still can learn the temporal consistency but the predicted person in frames is blurry. On the other hand, STG-SimVP without D_t will learn temporal information slower and it cannot reconstruct images well.

	Epoch	Predicted 20 frames
No D_s	1	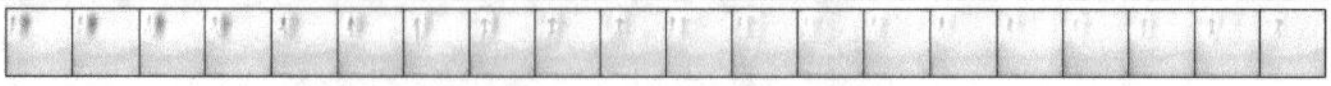
	10	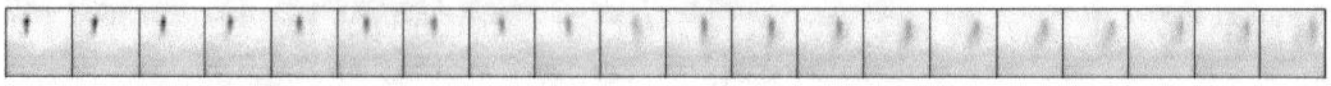
	20	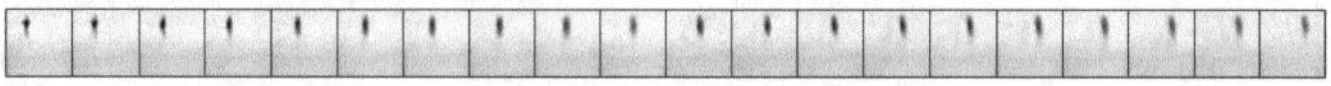
	100	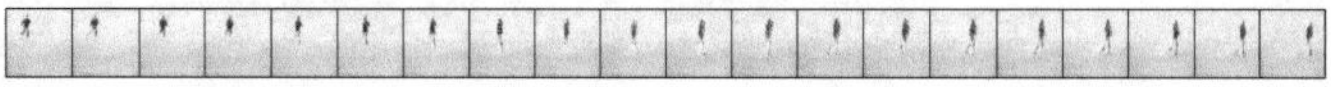
No D_t	1	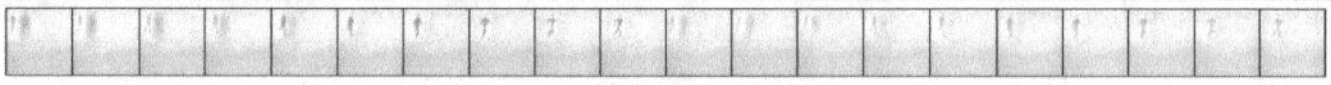
	10	
	20	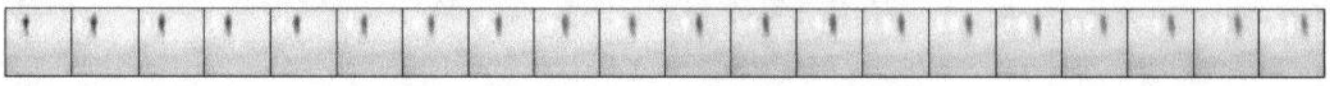
	100	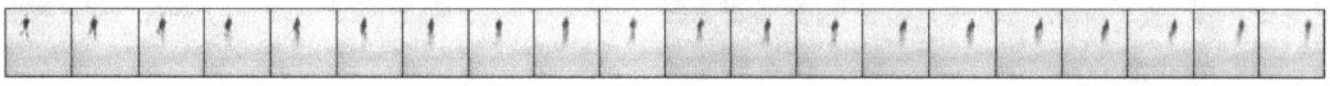

6 Limitations

The proposed model has proved that it can achieve good performance with a lightweight architecture. However, it still has some limitations.

- Number of predicted future frames cannot be changed after the training process. It means that if we train the model to predict 10 future frames, then if we want to predict more than 10 frames we need to retrain the model. However, we can continue using the predicted frames as input to predict future frames once again, but it is not an efficient way.
- STG-SimVP still does not have a chance to make an evaluation with colorful images. We need to do this soon to validate that the "strong generalization" feature is still kept

7 Conclusion

In this paper, we introduced a generative model which consists of one generator and dual discriminators for the video frame prediction task. We showed that on two datasets Moving MNIST and KTH, our model can learn spatiotemporal information faster than SimVP only after a few epochs without increasing the computing cost. In addition, we proved that our dual discriminators are important for boosting the learning process while spatial discriminator can classify images are real or fake, and temporal discriminator keep improving the temporal consistency information for predicted frames.

Acknowledgements. We acknowledge Ho Chi Minh City University of Technology (HCMUT), VNU-HCM for supporting this study.

References

1. Aigner, S., Körner, M.: Futuregan: Anticipating the future frames of video sequences using spatio-temporal 3D convolutions in progressively growing gans. arXiv preprint arXiv:1810.01325 (2018)
2. Bégaint, J., Galpin, F., Guillotel, P., Guillemot, C.: Deep frame interpolation for video compression. In: DCC 2019-Data Compression Conference, pp. 1–10. IEEE (2019)
3. Chang, Y.L., Liu, Z.Y., Lee, K.Y., Hsu, W.: Free-form video inpainting with 3D gated convolution and temporal patchgan. In: Proceedings of the IEEE/CVF International Conference on Computer Vision, pp. 9066–9075 (2019)
4. Clark, A., Donahue, J., Simonyan, K.: Adversarial video generation on complex datasets (2019)
5. Finn, C., Goodfellow, I., Levine, S.: Unsupervised learning for physical interaction through video prediction. Adv. Neural Inf. Process. Syst. **29** (2016)
6. Gao, Z., Tan, C., Wu, L., Li, S.Z.: Simvp: simpler yet better video prediction. In: Proceedings of the IEEE/CVF Conference on Computer Vision and Pattern Recognition (CVPR), pp. 3170–3180 (2022)
7. Goodfellow, I.: Nips 2016 tutorial: Generative adversarial networks. arXiv preprint arXiv:1701.00160 (2016)
8. Guen, V.L., Thome, N.: Disentangling physical dynamics from unknown factors for unsupervised video prediction. In: Proceedings of the IEEE/CVF Conference on Computer Vision and Pattern Recognition, pp. 11474–11484 (2020)
9. Hsieh, J.T., Liu, B., Huang, D.A., Fei-Fei, L.F., Niebles, J.C.: Learning to decompose and disentangle representations for video prediction. Adv. Neural Inf. Process. Syst. **31** (2018)
10. Hu, A., Cotter, F., Mohan, N., Gurau, C., Kendall, A.: Probabilistic future prediction for video scene understanding. In: Vedaldi, A., Bischof, H., Brox, T., Frahm, J.-M. (eds.) ECCV 2020. LNCS, vol. 12361, pp. 767–785. Springer, Cham (2020). https://doi.org/10.1007/978-3-030-58517-4_45
11. Jiang, H., Sun, D., Jampani, V., Yang, M.H., Learned-Miller, E., Kautz, J.: Super slomo: high quality estimation of multiple intermediate frames for video interpolation. In: Proceedings of the IEEE Conference on Computer Vision and Pattern Recognition, pp. 9000–9008 (2018)
12. Jin, B., et al.: Exploring spatial-temporal multi-frequency analysis for high-fidelity and temporal-consistency video prediction. In: Proceedings of the IEEE/CVF Conference on Computer Vision and Pattern Recognition, pp. 4554–4563 (2020)
13. Kalluri, T., Pathak, D., Chandraker, M., Tran, D.: Flavr: flow-agnostic video representations for fast frame interpolation. In: Proceedings of the IEEE/CVF Winter Conference on Applications of Computer Vision, pp. 2071–2082 (2023)
14. Liu, Z., Yeh, R.A., Tang, X., Liu, Y., Agarwala, A.: Video frame synthesis using deep voxel flow. In: Proceedings of the IEEE International Conference on Computer Vision, pp. 4463–4471 (2017)
15. Lu, L., Wu, R., Lin, H., Lu, J., Jia, J.: Video frame interpolation with transformer. In: Proceedings of the IEEE/CVF Conference on Computer Vision and Pattern Recognition, pp. 3532–3542 (2022)

16. Lu, W., Cui, J., Chang, Y., Zhang, L.: A video prediction method based on optical flow estimation and pixel generation. IEEE Access **9**, 100395–100406 (2021)
17. Oliu, M., Selva, J., Escalera, S.: Folded recurrent neural networks for future video prediction. In: Proceedings of the European Conference on Computer Vision (ECCV), pp. 716–731 (2018)
18. Rakhimov, R., Volkhonskiy, D., Artemov, A., Zorin, D., Burnaev, E.: Latent video transformer. arXiv preprint arXiv:2006.10704 (2020)
19. Schuldt, C., Laptev, I., Caputo, B.: Recognizing human actions: a local SVM approach. In: Proceedings of the 17th International Conference on Pattern Recognition, 2004. ICPR 2004. vol. 3, pp. 32–36. IEEE (2004)
20. Shi, X., Chen, Z., Wang, H., Yeung, D.Y., Wong, W.K., Woo, W.C.: Convolutional LSTM network: a machine learning approach for precipitation nowcasting. Adv. Neural Inf. Process. Syst. **28** (2015)
21. Shouno, O.: Photo-realistic video prediction on natural videos of largely changing frames. arXiv preprint arXiv:2003.08635 (2020)
22. Srivastava, N., Mansimov, E., Salakhudinov, R.: Unsupervised learning of video representations using LSTMS. In: International Conference on Machine Learning, pp. 843–852. PMLR (2015)
23. Tulyakov, S., Liu, M.Y., Yang, X., Kautz, J.: Mocogan: decomposing motion and content for video generation. In: Proceedings of the IEEE Conference on Computer Vision and Pattern Recognition, pp. 1526–1535 (2018)
24. Villegas, R., Pathak, A., Kannan, H., Erhan, D., Le, Q.V., Lee, H.: High fidelity video prediction with large stochastic recurrent neural networks. Adv. Neural Inf. Process. Syst. **32** (2019)
25. Villegas, R., Yang, J., Hong, S., Lin, X., Lee, H.: Decomposing motion and content for natural video sequence prediction. arXiv preprint arXiv:1706.08033 (2017)
26. Wang, Y., Gao, Z., Long, M., Wang, J., Philip, S.Y.: Predrnn++: towards a resolution of the deep-in-time dilemma in spatiotemporal predictive learning. In: International Conference on Machine Learning, pp. 5123–5132. PMLR (2018)
27. Wang, Y., Jiang, L., Yang, M.H., Li, L.J., Long, M., Fei-Fei, L.: Eidetic 3D LSTM: a model for video prediction and beyond. In: International Conference on Learning Representations (2019)
28. Wang, Y., Wu, H., Zhang, J., Gao, Z., Wang, J., Philip, S.Y., Long, M.: Predrnn: a recurrent neural network for spatiotemporal predictive learning. IEEE Trans. Pattern Anal. Mach. Intell. **45**(2), 2208–2225 (2022)
29. Wei, H., Yin, X., Lin, P.: Novel video prediction for large-scale scene using optical flow. arXiv preprint arXiv:1805.12243 (2018)
30. Weissenborn, D., Täckström, O., Uszkoreit, J.: Scaling autoregressive video models. arXiv preprint arXiv:1906.02634 (2019)
31. Wu, Y., Wen, Q., Chen, Q.: Optimizing video prediction via video frame interpolation. In: Proceedings of the IEEE/CVF Conference on Computer Vision and Pattern Recognition, pp. 17814–17823 (2022)
32. Xu, Z., Wang, Y., Long, M., Wang, J., KLiss, M.: Predcnn: predictive learning with cascade convolutions. In: IJCAI, pp. 2940–2947 (2018)
33. Yu, W., Lu, Y., Easterbrook, S., Fidler, S.: Efficient and information-preserving future frame prediction and beyond (2020)

Tritention U-Net: A Modified U-Net Architecture for Lung Tumor Segmentation

Nguyen Hung Le[1,2], Duc Dung Nguyen[1,2], Tuong Nguyen Huynh[3], and Thanh Hung Vo[1,2](✉)

[1] Ho Chi Minh City University of Technology (HCMUT), 268 Ly Thuong Kiet Street, District 10, Ho Chi Minh City, Vietnam
{hung.le040502,nddung,vthung}@hcmut.edu.vn
[2] Vietnam National University Ho Chi Minh City, Linh Trung Ward, Ho Chi Minh City, Vietnam
[3] Industrial University of Ho Chi Minh City, Ho Chi Minh City, Vietnam
huynhtuongnguyen@iuh.edu.vn

Abstract. Lung tumor segmentation in computed tomography (CT) images is a critical task in medical image analysis. It aids in the early detection and diagnosis of lung cancer, which is one of the primary causes of cancer deaths around the world. However, because of the variable sizes, uncertain shapes of lung nodules, and complex internal lung structure, lung tumor segmentation is a difficult problem. In this study, we propose a novel Tritention U-Net as an efficient model for solving that problem. It is integrated with the Tritention Gate on the contracting path between the encoder and decoder to highlight task-relevant salient features. The proposed Tritention U-Net model is trained and evaluated using the Medical Image Decathlon dataset - Task06_Lung, which requires the model to segment a small portion of a lung image. Our model achieved a Dice score of 91.80% and was compared to well-known models to demonstrate the improvement.

Keywords: Medical Image Processing · Lung tumor segmentation · Deep learning · Tritention U-Net

1 Introduction

Nowadays, medical image processing is playing an increasingly significant role in cancer diagnosis and treatment. Image segmentation is an essential part of medical image processing because it provides useful information for diagnosis, clinical studies, and treatment planning. With the advancement of computer vision, a variety of medical image segmentation techniques are being developed to alleviate the shortage of skilled labor. Additionally, in tumor segmentation,

Supported by Ho Chi Minh City University of Technology (HCMUT), VNU-HCM.

N. T. Nguyen et al. (Eds.): CITA 2023, LNNS 734, pp. 217–227, 2023.
https://doi.org/10.1007/978-3-031-36886-8_18

these techniques can aid in the early detection and analysis of tumors, which will increase the patient's chances of survival and enable effective treatment.

The traditional approach to segmenting images was first developed using digital image processing combined with optimization algorithms. These algorithms, such as the snake algorithm [12] and region growing [10], generate segment maps by establishing initial regions and comparing pixel values. These algorithms provide a local view of the image's features and focus on pixel differences and gradients at the local level. Much later, algorithms that provide a global view of the input image, such as Otsu's algorithm [18], adaptive thresholding [21], and clustering algorithms [16], was developed.

In recent years, fully convolutional neural networks with a U-shaped architecture have become one of the most widely used structures in medical image segmentation. The typical U-shaped network, U-Net [19], used an encoder-decoder architecture with skip connections, which has achieved distinguished performance in medical imaging applications on segmentation tasks. With that success, many Unet-based networks, such as ResUNet [7], UNet++ [25], and UNet3+ [9], have been proposed and have gained good results in many medical imaging domains. However, because of the locality of the convolutional operation, learning long-range dependencies is difficult for a CNN-based model. To deal with this problem, attention-based models such as Attention U-Net [17] and MsAUNet [3] are proposed, which emphasize relevant features while reducing irrelevant ones. However, these methods still have limitations in modeling long-term dependencies. To address this issue, transformers that have demonstrated excellent performance on a wide range of tasks related to natural language processing, such as TransUnet [5], Swin UNETR [8], and Swin-Unet [2], are integrated into the CNN-model. These models can capture global context information and effectively model long-range dependencies, making them suitable for various medical imaging tasks.

Computed tomography (CT) images are high-resolution diagnostic images that use X-rays and computer processing to create detailed images of the internal organs of the body. They are widely used to diagnose and guide the treatment of cancer. However, a radiologist must examine a CT scan with 150–500 slices for an accurate diagnosis, which is a hard and time-consuming process. Furthermore, in lung cancer diagnosis, it can be challenging to distinguish between nodules and internal lung structure, particularly when nodules are close to blood vessels, the pulmonary hila, diaphragms, or even the airways [15]. Another challenge is that lung nodules are irregular in shape and small in comparison to the surrounding noise. This makes it difficult to detect and diagnose lung cancer in its early stages, which can result in treatment delays and poorer patient outcomes.

To address this issue, we propose Tritention U-Net, an end-to-end model for automatically detecting and segmenting lung nodules from CT scans and being suitable for the segmentation of many types of lung nodules. To improve nodule segmentation accuracy, our model uses the Tritention Gate, a novel attention-based mechanism that captures both localization and semantic information. We evaluate the performance of Tritention U-Net on the public dataset Medical Segmentation Decathlon (MSD) [22] with Task06_Lung and archive high performance compared with other well-known models.

The rest of this paper is as follows. Section 2 is our approach. Experimental results are given in the Sect. 3. And finally, Sect. 4 is the conclusion and the extended direction.

2 Tritention U-Net Architechture (TU-Net)

2.1 Architechture Overview

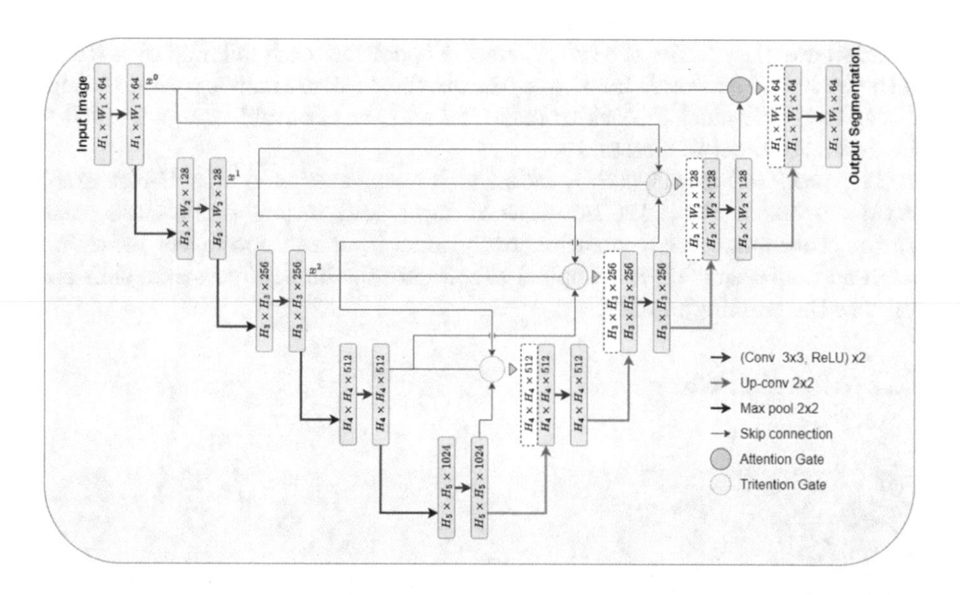

Fig. 1. Proposed Tritention Unet Architecture.

The general architecture of the proposed Tritention U-Net is presented in Figure. 1. It uses a U-Net-based backbone incorporated with the three proposed Tritention Gates (TAs) and one Attention Gate (AG) [17], which is used in the shallowest skip connection. For more detail, the U-Net encoder takes an image as input and produces corresponding output features. In the skip connections, these features are passed through TAs and an AG to highlight salient features that are relevant to the task. After that, the U-Net decoder is used to merge the feature outputs to create a combination of high-level semantic features and lower-level fine-grained features. The network's output is a mask that corresponds to the current task.

2.2 Backbone U-Net Architecture

In this paper, we construct our tritention model over the U-Net architecture. U-Net is a fully convolutional network commonly used for image segmentation tasks. In its architecture, U-Net consists of two paths: the contracting path and the expansive path. In the contracting path, two 3×3 convolutions ("same" padding) are applied repeatedly, and after each one, a rectified linear unit activation (ReLU) and a 2×2 max pooling operation with stride 2 are applied to downsample the feature. The number of feature channels is doubled after each downsampling step, while the spatial dimensions of the feature map are cut in half. At each step of the expansive path, an upsampling operation is performed, followed by a 2×2 transpose convolution, which halved the number of feature channels. After that, the feature vectors are concatenated with the corresponding feature vectors from the skip connection. Two 3×3 convolutions ("same" padding) are placed after the concatenation operation, each followed by a ReLU. In the backbone network's final layer, the obtained feature map is passed through a final 1×1 convolution block to make the number of output channels equal to the desired number of classes.

The proposed model uses a modified implementation of the U-Net architecture to take a 128×128 image as an input and output a 128×128 mask. On the other hand, between each convolutional layer and non-linear activation, batch normalization [11] was applied to mitigate the internal covariate shift and stabilize the training process.

2.3 Tritention Gate

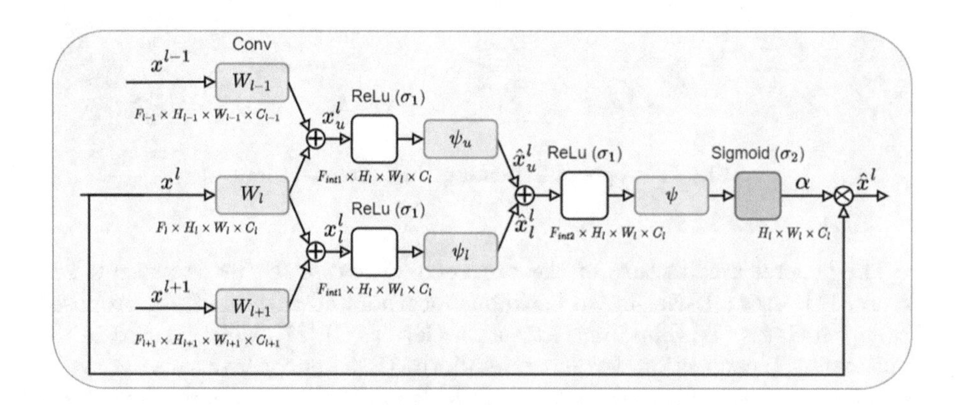

Fig. 2. Tritention Gate.

The feature map in the Fully Convolutional Network [13] for segmentation is gradually downsampled to capture more semantic information. When going deeper, however, spatial location information is lost, reducing the localization accuracy in the upsampled path. Some networks [4,14,19] use skip connections to combine the upsampled output with the corresponding feature from the downsampled path to address this issue. Furthermore, in order to improve the performance of these networks, Attention Gate (AG) is incorporated into the model to highlight salient features passed through the skip connection. We propose Tritention Gate (TG) based on this concept. Figure. 2 represent Tritention Gate architecture. TG, like AG, aims to identify significant regions in order to maintain only the activations relevant to the given task while gradually suppressing feature responses in irrelevant background regions. However, with the addition of a signal from the upper skip connection, TG can obtain more localization information to enhance the accuracy of the identified regions.

The TG generates output by element-wise multiplying the coefficient α and the input feature map x^l. If a region in the feature map x^l is important, its corresponding coefficient value in α will be close to 1 to keep the relevant signal; otherwise, it will be close to 0 to prune irrelevant signals. The Eq. 1 can be used to express the multiplication at the output.

$$\hat{x}^l = \alpha \cdot x^l \tag{1}$$

The input to TG is x^{l-1}, x^l, and x^{l+1}, which are feature maps derived from successive upper, current, and successive lower skip connections. To map input tensors to the same dimensional space $\mathbb{R}^{F_{int1}}$, the linear transformations W_{l-1}, W_l, and W_{l+1} are computed. Following that, two successive tensors of three output tensors are concatenated to fuse features from different levels for efficient feature extraction. The extracted features ${x_u}^l$, which contains more localization information, and ${x_l}^l$, which contains more semantic information, are then passed through a ReLU activation σ_1 and a convolutional layer. Next, the two obtained tensors $\hat{x_u}^l$ and $\hat{x_l}^l$ are combined and passed through another ReLU activation. The output tensor then undergoes a 1×1 convolution followed by a sigmoid activation function σ_2 to produce a coefficient α that ranges from 0 to 1 to identify significant regions of the image. Tritention can be formulated Eq. 2, 3 and 4.

$$\hat{x_u}^l = \psi_u^T(\sigma_1(W_{l-1}^T x^{l-1} + W_l^T x^l + b_{l-1})) + b_{\psi_u} \tag{2}$$

$$\hat{x_l}^l = \psi_l^T(\sigma_1(W_{l+1}^T x^{l+1} + W_l^T x^l + b_{l+1})) + b_{\psi_l} \tag{3}$$

$$\alpha = \sigma_2(\psi^T(\sigma_1(\hat{x_u}^l + \hat{x_l}^l)) + b_\psi) \tag{4}$$

where $\sigma_1(x) = max(x)$ and $\sigma_2(x) = (1 + exp^{-x})^{-1}$ represent ReLU and Sigmoid activation, respectively. The TG contains the linear transformations $W_{l-1} \in \mathbb{R}^{F_{l-1} \times F_{int1}}$, $W_l \in \mathbb{R}^{F_l \times F_{int1}}$, $W_{l+1} \in \mathbb{R}^{F_{l+1} \times F_{int1}}$, $\psi_u \in \mathbb{R}^{F_{int1} \times F_{int2}}$, $\psi_l \in \mathbb{R}^{F_{int1} \times F_{int2}}$, $\psi \in \mathbb{R}^{F_{int2} \times F_1}$ and bias terms $b_{l-1} \in \mathbb{R}^{F_{int1}}$, $b_{l+1} \in \mathbb{R}^{F_{int1}}$, $b_{\psi_u} \in \mathbb{R}^{F_{int2}}$, $b_{\psi_l} \in \mathbb{R}^{F_{int2}}$, $b_\psi \in \mathbb{R}$.

3 Experimental and Results

3.1 Dataset

Table 1. Description of the MSD Dataset - Task06_Lung. The slice quantity is shown in the "mean ± standard deviation" format.

Characteristics	Description
Target	Lung and tumours
Modality	CT
Thickness	< 1.5 mm
Size	96 3D volumes (64 Training + 32 Testing)
Image Dimension	512×512
Slice Quantity	279.4 ± 105.6
Source	The Cancer Imaging Archive

The dataset Medical Segmentation Decathlon (MSD) Task06_Lung [22] was used to compare the proposed Tritention U-Net to the standard U-Net and its variant, Attention U-Net. The overview of the dataset is provided in Table 1. This lung dataset includes non-small cell lung cancer cases from patients at Stanford University (Palo Alto, CA, USA) and was made accessible to the public through TCIA [6]. There are 96 CT scans (64 samples for training and 32 for testing) with corresponding ground truth segmentation masks of 512×512 resolution. The annotation files were annotated by an experienced thoracic radiologist using OsiriX [20]. Some of the data samples are shown in Figure. 3.

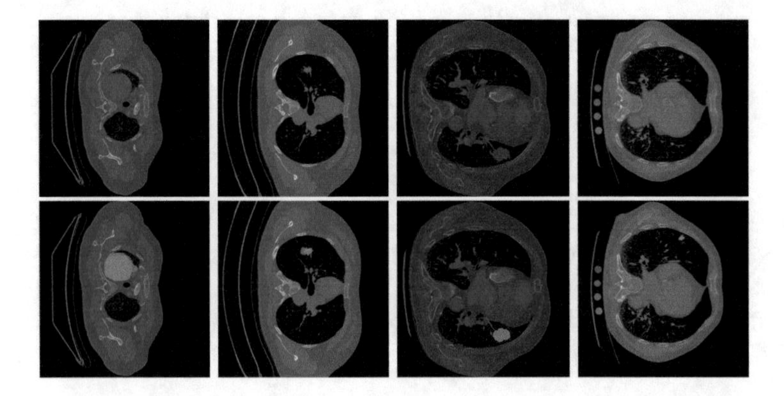

Fig. 3. Illustrate some data samples from the dataset. The first row contains the input images, and the second row contains the appropriate segmentation results (orange mask) for the above images. (Color figure online)

3.2 Experimental Setup

Data preprocessing: There are two approaches to dealing with volumetric input data in the image segmentation task. The first method involves segmenting each 2D slice of the volume separately, whereas the second method involves segmenting the entire volume [24]. In this paper, due to computation resource limitations, we used the 2D approach, which requires fewer computations for training and has a faster inference speed than the 3D approach.

In the data preprocessing process, we first divide the 3D CT scan data into appropriate slices. For example, a volume (512, 512, 250) CT scan with 250 slices was divided into 250 individual volume images (512, 512). Following that, we scale each obtained image to size 128 × 128 and only take images whose corresponding mask contains a segmentation label. After this stage, a total of 1,613 images are created and partitioned into a training subset and a test subset with a ratio of 90/10, respectively.

Evaluation Metrics: In this paper, the Dice Similarity Coefficient (DSC) is used as the main metric to compare segmentation performance between our proposed model and others. Aside from that, Precision and Recall have been utilized as additional assessments. All of the above metrics are formulated in Eqs. 5, 6, and 7.

$$DSC = \frac{2|Gt \cap Pr|}{|Gt| + |Pr|} \tag{5}$$

$$Precision = \frac{|Gt \cap Pr|}{|Pr|} \tag{6}$$

$$Recall = \frac{|Gt \cap Pr|}{|Gt|} \tag{7}$$

where Gt is a set of ground truth pixels and Pr is a set of model predict pixels.

Implementation Details: In this experiment, the Tritenion UNet is trained using the Adam optimizer with an initial *learning rate* of 0.001, *beta_1* of 0.9, *beta_2* of 0.999, and *epsilon* of 1e-07. Besides that, the input image is 128 × 128 and the *batch size* is 16. Moreover, to prevent overfitting, an early stopping strategy [23] is used, in which the model is stopped after twenty epochs with no improvement. The experiment was achieved using Python 3.6 and Tensorflow 2.9.0 [1]. All training and evaluating were done on the Google Colab, which has an integrated NVIDIA Tesla T4 GPU with 16GB. In addition, all models use batch normalization to minimize the number of training epochs required to train.

3.3 Results

Table 2 shows the comparison of the proposed Tritention U-Net with the standard U-Net [19] and Attention U-Net [17] on the MSD dataset for Task06_Lung. The experiment results show that the proposed Tritention-Unet archive performs best in terms of DSC (91.80%) and Precision (92.76%). When compared

Table 2. Segmentation Results obtained on the MSD Dataset - Task06_Lung

Method	Dice (%)	Precision (%)	Recall (%)
U-Net	90.28	86.24	**94.71**
Attention U-net	89.41	85.63	93.53
Tritention U-net	**91.80**	**92.76**	90.86

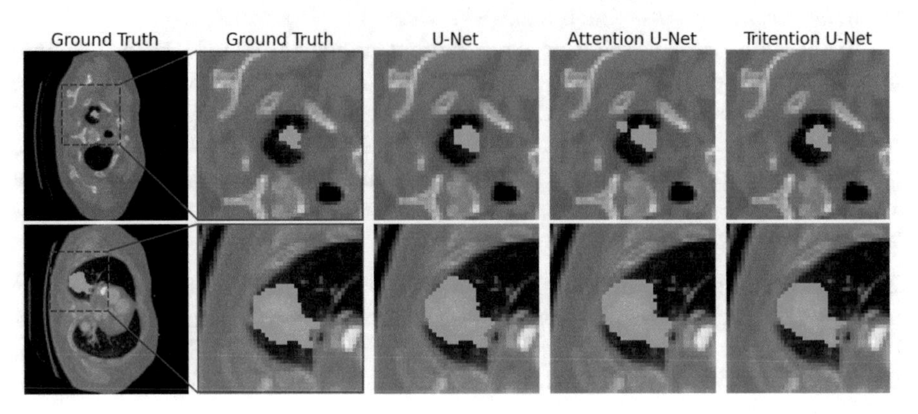

Fig. 4. The segmentation result of the different approaches on the MSD Dataset - Task06_Lung

to U-Net and Attention U-Net, our methods had no better Recall (90.86%). However, by integrating Tritention Block in U-Net, we were able to achieve a balance between Recall and Precision value, preventing the model from having the under-segmentation problem shown in Figure. 4.

Figure 5(a) shows a histogram of the dice score values and the total number of slices to aid in evaluating the output of our proposed Tritention Unet model on the test set. As shown in Fig. 5(a), the majority of slices had a dice score value greater than 90%. However, there are a few slices where the dice score is close to zero. This issue arises because some slices of the dataset contain very small tumors, and after the resizing phase of data preprocessing, the tumors have too few pixels (128×128) and contain too little information for the model to learn. Figure 5(b) depicts some samples with small tumors.

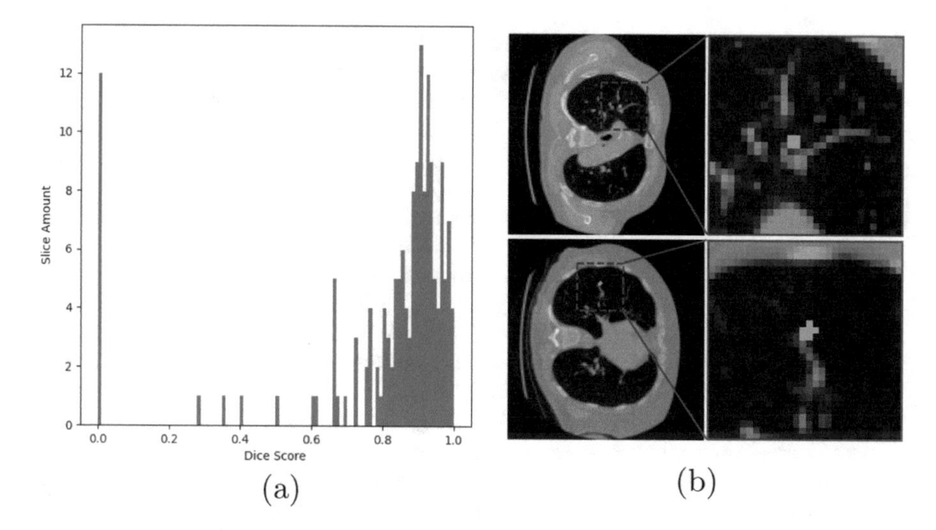

Fig. 5. (a) The visualization of the Dice Score Distributions of Tritention U-Net on MSD Task06. (b) The ground-truth lung segmentations for the tiny tumor case (on the left) and its corresponding zoom-in (on the right).

4 Conclusion

Lung tumor segmentation is an important role in the diagnosis and treatment progress. It is a time-consume task for any expert medical doctor. Automatic segmentation will be useful for medical doctors in early decision-making. However, there are many challenges for this task, including various sizes, uncertain shapes, and complex internal lung structures. In this paper, we present a novel Tritention gate model for lung tumor segmentation. The model used the U-Net backbone to encode and decode features at multiple levels and integrated Tritention Gate to highlight salient features that are passed through the skip connections. After evaluation and visualization of the results, the proposed approach showed encouraging precision in lung nodule segmentation, with a 91.80% Dice similarity coefficient for the Medical Image Decathlon dataset - Task06_Lung.

Acknowledgements. We acknowledge Ho Chi Minh City University of Technology (HCMUT), VNU-HCM for supporting this study.

References

1. Abadi, M., et al.: TensorFlow: Large-scale machine learning on heterogeneous systems (2015). https://www.tensorflow.org/, software available from tensorflow.org
2. Cao, H., et al.: Swin-unet: Unet-like pure transformer for medical image segmentation. In: Computer Vision-ECCV 2022 Workshops: Tel Aviv, Israel, 23–27 October 2022, Proceedings, Part III, pp. 205–218. Springer (2023). https://doi.org/10.1007/978-3-031-25066-8_9

3. Chattopadhyay, S., Basak, H.: Multi-scale attention u-net (msaunet): a modified u-net architecture for scene segmentation. arXiv preprint arXiv:2009.06911 (2020)
4. Chaurasia, A., Culurciello, E.: LinkNet: Exploiting encoder representations for efficient semantic segmentation. In: 2017 IEEE Visual Communications and Image Processing (VCIP). IEEE (Dec 2017). https://doi.org/10.1109/vcip.2017.8305148, https://doi.org/10.1109%2Fvcip.2017.8305148
5. Chen, J., et al.: Transunet: Transformers make strong encoders for medical image segmentation. arXiv preprint arXiv:2102.04306 (2021)
6. Clark, K., et al.: The cancer imaging archive (tcia): maintaining and operating a public information repository. J. Digit. Imaging **26**, 1045–1057 (2013)
7. Diakogiannis, F.I., Waldner, F., Caccetta, P., Wu, C.: Resunet-a: A deep learning framework for semantic segmentation of remotely sensed data. ISPRS J. Photogramm. Remote. Sens. **162**, 94–114 (2020)
8. Hatamizadeh, A., Nath, V., Tang, Y., Yang, D., Roth, H.R., Xu, D.: Swin unetr: Swin transformers for semantic segmentation of brain tumors in mri images. In: Brainlesion: Glioma, Multiple Sclerosis, Stroke and Traumatic Brain Injuries: 7th International Workshop, BrainLes 2021, Held in Conjunction with MICCAI 2021, Virtual Event, 27 September 2021, Revised Selected Papers, Part I, pp. 272–284. Springer (2022). https://doi.org/10.1007/978-3-031-08999-2_22
9. Huang, H., et al.: Unet 3+: A full-scale connected unet for medical image segmentation. In: ICASSP 2020–2020 IEEE International Conference on Acoustics, Speech and Signal Processing (ICASSP), pp. 1055–1059. IEEE (2020)
10. Ikonomatakis, N., Plataniotis, K., Zervakis, M., Venetsanopoulos, A.: Region growing and region merging image segmentation. In: Proceedings of 13th International Conference on Digital Signal Processing, vol. 1, pp. 299–302 (1997). https://doi.org/10.1109/ICDSP.1997.628077
11. Ioffe, S., Szegedy, C.: Batch normalization: Accelerating deep network training by reducing internal covariate shift. In: International Conference On Machine Learning, pp. 448–456. pmlr (2015)
12. Kass, M., Witkin, A., Terzopoulos, D.: Snakes: Active contour models. Int. J. Comput. Vision **1**(4), 321–331 (1988)
13. Long, J., Shelhamer, E., Darrell, T.: Fully convolutional networks for semantic segmentation (2014). https://doi.org/10.48550/ARXIV.1411.4038, https://arxiv.org/abs/1411.4038
14. Milletari, F., Navab, N., Ahmadi, S.A.: V-net: Fully convolutional neural networks for volumetric medical image segmentation. In: 2016 Fourth International Conference on 3D Vision (3DV), pp. 565–571 (2016). https://doi.org/10.1109/3DV.2016.79
15. Naidich, D.: Lung cancer detection and characterization: challenges and solutions. In: Multislice CT: A Practical Guide Proceedings of the 6th International SOMATOM CT Scientific User Conference Tuebingen, September 2002, pp. 215–222. Springer (2004). https://doi.org/10.1007/978-3-642-18758-2_17
16. Nameirakpam, D., Singh, K., Chanu, Y.: Image segmentation using k -means clustering algorithm and subtractive clustering algorithm. Proc. Comput. Sci. **54**, 764–771 (2015). https://doi.org/10.1016/j.procs.2015.06.090
17. Oktay, O., et al.: Attention u-net: Learning where to look for the pancreas. arXiv preprint arXiv:1804.03999 (2018)
18. Otsu, N.: A threshold selection method from gray-level histograms. IEEE Trans. Syst. Man Cybern. **9**(1), 62–66 (1979). https://doi.org/10.1109/TSMC.1979.4310076

19. Ronneberger, O., Fischer, P., Brox, T.: U-net: Convolutional networks for biomedical image segmentation (2015). https://doi.org/10.48550/ARXIV.1505.04597, https://arxiv.org/abs/1505.04597
20. Rosset, A., Spadola, L., Ratib, O.: Osirix: An open-source software for navigating in multidimensional dicom images. J. Digital Imaging Official J. Soc. Comput. Applicat. Radiol. **17**, 205–16 (2004). https://doi.org/10.1007/s10278-004-1014-6
21. Roy, P., Dutta, S., Dey, N., Dey, G., Chakraborty, S., Ray, R.: Adaptive thresholding: A comparative study. In: 2014 International Conference on Control, Instrumentation, Communication and Computational Technologies (ICCICCT), pp. 1182–1186 (2014). https://doi.org/10.1109/ICCICCT.2014.6993140
22. Simpson, A.Let al.: A large annotated medical image dataset for the development and evaluation of segmentation algorithms (2019). https://doi.org/10.48550/ARXIV.1902.09063, https://arxiv.org/abs/1902.09063
23. Yao, Y., Rosasco, L., Caponnetto, A.: On early stopping in gradient descent learning. Constr. Approx. **26**(2), 289–315 (2007)
24. Zhang, Y., Liao, Q., Ding, L., Zhang, J.: Bridging 2d and 3d segmentation networks for computation efficient volumetric medical image segmentation: An empirical study of 2.5d solutions (2020). https://doi.org/10.48550/ARXIV.2010.06163, https://arxiv.org/abs/2010.06163
25. Zhou, Z., Rahman Siddiquee, M.M., Tajbakhsh, N., Liang, J.: UNet++: a nested u-net architecture for medical image segmentation. In: Stoyanov, D., et al. (eds.) DLMIA/ML-CDS -2018. LNCS, vol. 11045, pp. 3–11. Springer, Cham (2018). https://doi.org/10.1007/978-3-030-00889-5_1

Multi-modal with Multiple Image Filters for Facial Emotion Recognition

Thong T. Huynh[1,2], My M. Nguyen[1,2], Phong T. Pham[1,2], Nam T. Nguyen[1,2], Tien L. Bui[1,2], Tuong Nguyen Huynh[3], Duc Dung Nguyen[1,2], and Hung T. Vo[1,2](✉)

[1] Ho Chi Minh City University of Technology (HCMUT), 268 Ly Thuong Kiet Street, District 10, Ho Chi Minh City, Vietnam
{thong.huynhthanh,my.nguyen060902,phong.phamthanh,nam.nguyenbknetid, tien.builam,nddung,vthung}@hcmut.edu.vn

[2] Vietnam National University Ho Chi Minh City, Ho Chi Minh City, Vietnam

[3] Industrial University of Ho Chi Minh City, Ho Chi Minh City, Vietnam
huynhtuongnguyen@iuh.edu.vn

Abstract. The need to understand people, especially their behaviors and feelings, is growing significantly in today's quickly-moving world. Despite the remarkable progress of science and technology in general and artificial intelligence in particular, facial emotion recognition remains challenging. This paper proposes a unique method for enhancing the accuracy of emotion recognition models. Through image analysis, the hair area and other facial areas have similar pixels but different intensities. However, to recognize emotions on the face, people only need to focus on facial features. Therefore, areas with the same pixels are not very helpful in accurately recognizing emotions. To solve the above problem, we conducted to eliminate or blur pixels that are the same as on the facial image. In addition, we also demonstrate that using the multi-model approach can support the learning process by allowing the sub-models to collaborate and increase accuracy. The experiments showed that the proposed approach offers a valuable contribution to the field of facial emotion recognition and has a significant improvement compared to previous approaches.

Keywords: Facial emotion recognition · Human-computer interaction · Convolutional network

1 Introduction

Facial Emotion Recognition (FER) is a large field in Human-Computer Interaction (HCI) based on two subjects that are emotional psychology and artificial intelligence. Human emotion may be expressed by speech or non-verbal, such

Supported by Ho Chi Minh City University of Technology (HCMUT), VNU-HCM.
Thong T. Huynh and My M. Nguyen contributed equally.

N. T. Nguyen et al. (Eds.): CITA 2023, LNNS 734, pp. 228–239, 2023.
https://doi.org/10.1007/978-3-031-36886-8_19

as transformations on the face or tone of voice, which are detected by sensors. In 1967, Albert Mehrabian - an American psychologist known for his research on the influence of body language and tone of voice in conveying messages [8]. He pointed out that when communicating with others, nonverbal factors such as gestures, facial expressions, and tone of voice play an important role besides language factors. Especially, he showed that 55 percent of emotions were expressed by face, 38 percent by voice, and the rest by speech.

Facial expression recognition has a wide range of applications across various fields, including education, where it can be used to understand learners' responses and engagement with the content of teaching sessions. In examination settings, it can be used to track and predict cheating behaviors by candidates. Besides, marketers can benefit from understanding how buyers react to their product advertisements. Facial emotion recognition can also be applied in the field of security where it can assist in detecting suspicious behavior and prevent potential hazards. The medical field can use FER to automate the care process as well as analyze the mental health of the patients. Finally, in the recruitment process, evaluating the quality of candidates can be more easily achieved with the assistance of this technology.

The importance of facial emotion recognition has attracted significant attention from numerous researchers. Many approaches have been proposed to address this problem, ranging from traditional machine-learning techniques to more advanced deep-learning models.

D. Lakshmi *et al.* [7] developed a way to identify facial emotions using a modified version of Local Binary Pattern (LBP) [10] and Histogram of Oriented Gradients features. They used Viola-Jones face detection to locate facial features and applied a high-pass filter to enhance the identified region. They combined the features from the identified areas using Deep Stacked AutoEncoders and used Multi-class Support Vector Machine to classify emotions. The results showed that their method successfully distinguished emotions in the CK+ and JAFFE datasets. Zia Ullah *et al.* [11] proposed a new technique for facial emotion recognition using a combination of deep learning and traditional methods. The technique uses super-resolution, facial emotion recognition, and classification phases. An Improved Deep CNN is used to increase image pixel density during super-resolution. The Viola-Jones algorithm is used for face detection and traditional methods like Texton, Bag of Words, GLCM, and Improved LGXP features are used for feature extraction. RNN and Bi-GRU neural networks are used for classification, and a Score level fusion voting mechanism is used to improve accuracy. The proposed technique achieves 95% accuracy, outperforming traditional approaches on various databases. Zhang *et al.* [14] introduced a new facial expression recognition approach using Local Binary Pattern (LBP) [10] and Local Phase Quantization (LPQ) [5] based on the Gabor face image. First, the Gabor filter is adopted to extract features of the face image among five scales and eight orientations to capture the significant visual properties. Then the Garbor image is encoded by the LBP and LPQ, respectively. Considering the size of the combined feature was too large, two algorithms Principal

Component Analysis (PCA) [1] and Linear Discriminant Analysis (LDA) [13] are used to reduce its dimension. Finally, the multi-class SVM classifiers based on the JAFFE database were used in the experiment.

In recent years, transformer architecture [12] has emerged as a powerful method. Many researchers have leveraged its ability for self-attention to create models that perform better in various tasks, including computer vision. Aayushi Chaudhari *et al.* [2] used the ResNet-18 model [4] and transformers to classify facial expression recognition. The experiment underwent associated procedures, including face detection, cropping, and feature extraction using a deep learning model combined with fine-tuned transformer. The purpose of this study was to examine the performance of the Vision Transformer and compare it to their cutting-edge models on hybrid datasets. Roberto Pecoraro *et al.* [9] proposed a self-attention module that can be easily integrated into virtually every convolutional neural network named Local multi-Head Channel (LHC). There are two principal concepts on which the method is based. First, using the self-attention pattern in computer vision, applying the channel-wise application is considered more effective than the traditional approach of using spatial attention. Second, in facial expression recognition, where images have a consistent structure, local attention is referred to as a potentially better method than the global approach in overcoming the limitations of convolution. Compared to the previous state-of-the-art in the FER2013 dataset, LHC-Net is evaluated with significantly less complexity and effect on the underlying architecture regarding computational expense.

In this paper, we propose a new approach to improve emotion recognition model accuracy by eliminating or blurring the same pixels on facial areas. We also illustrate employing a combination of multiple models to enhance accuracy. The experimental results demonstrate that our method outperforms previous approaches.

The rest of this paper is as follows. Section 2 presents our proposed method in detail. Then, experimental results are given and discussed in Sect. 3. Finally, Sect. 4 gives the conclusion and future works.

2 Multiple Filter Levels for FER

2.1 Multiple Model

In machine learning, ensemble learning is a popular approach that combines multiple models in order to improve the predictive accuracy and robustness of a single model. By aggregating the outputs of several models, ensemble learning can reduce overfitting, and capture a wider range of patterns in the data. The architecture of multiple models in this article is built according to the architecture in Fig. 1.

In the first step, the model receives an image from the data set. The image is processed by the Processor, and the model makes predictions using a set of images that includes an original image and one or more processed images. In this study, there are two approaches were proposed, such as Dropping pixels -

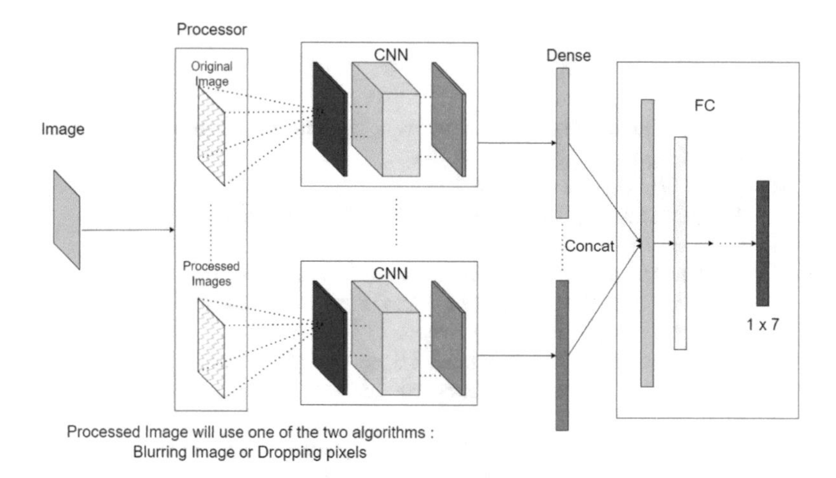

Fig. 1. Multiple model architecture

Processor, and Blurring images - Processor. Both approaches employ the original image to create special images, and all of them are used for the model. Each image is input to a separate CNN to extract unique features for each image type. The CNNs are the same for each sub-module and are based on the VGG16, which is a well-known and popular architecture. Finally, all features are concatenated and passed through a fully connected block to make predictions.

As previously mentioned, there are two approaches based on the way to process the original input image in the processor block in Fig. 1. Multi-model following the direction of dropping pixels is named Dropping Pixels Multiple (DPM-FER). And the model is in another direction of blurring image, which is named Blurring Multiple (BM-FER).

2.2 Dropping Pixels - Processor

The idea of the algorithm used is to remove pixels that are similar and lie in a certain area (a matrix of size k). Conditions considered are similar: the disparity between the largest and the smallest pixel, and the disparity between the largest pixel and the pixel are under consideration. If the disparity between max-min is too small, we consider that image area to have unchanged pixels (for example, hair area) and give no value to the model.

Dropping pixels algorithm is implemented using a matrix that iterates all the pixels of the image, deciding whether to keep or remove that pixel from the image to reduce unnecessary details in the image. Using two loops to get the submatrix and each sub-matrix will be considered with the given condition.

Figure 2 shows how the *Dropping pixels* algorithm works. A big 4×4 matrix in Fig. 2 was iterated by a 2×2 matrix, obtaining four submatrices. To determine whether a submatrix is saved or not, take into account the conditions on each submatrix.

```
Data: img(48 × 48), k, v1, v2
Result: newImg(48 × 48)
newImg ← img.copy
// Iterating the entire images with a matrix of size k, step = k
for i ← 0 to img.shape[0] − k by k do
    for j ← 0 to img.shape[1] − k by k do
        // Considering each matrix of size k in the image
        M ← img[i : i + k, j : j + k]
        // Condition 1: This matrix means that the difference between
           2 largest and smallest pixels must be large enough
        c1 ← Max(M) − Min(M) >= v1
        // Condition 2: Each pixel is meaningful when the difference
           between it and the largest pixel is small enough
        c2 ← Max(M) − M <= v2
        // Save the result of the matrix under consideration
        newImg[i : i + k, j : j + k] ← c1 * c2 * M
    end
end
```

Algorithm 1: Dropping pixels algorithm

In this algorithm, just considering the difference between the 2 largest and smallest pixels is not enough. The $v2$ condition, the difference between the largest and pixel considered, is added to enhance the ability to adjust the granularity reduction between pixels. With the input image as the Fig. 3a, $v1$, $v2$ is the condition that determines whether the pixels are saved or not. Changing the values of $v1$, $v2$ from 0-255, *step* by 15 gives the result as Fig. 3b.

Figure 3a is the original input image, also known as the sample image. The algorithm will use this image and create image 3b by adjusting the parameters of the conditions in the algorithm.

2.3 Blurring Image - Processor

This algorithm along with the purpose of reducing the detail of the image is deployed by calculating the difference between two adjacent pixels. With such a calculation, a pixel can be counted in terms of another pixel either horizontally, vertically, or diagonally. And to be objective, the result when calculated on a pixel will be the average of the three calculations above for each pixel. The Blurring image algorithm is given in the algorithm 2.

First of all, a row and a column will be added to the right and the bottom of the pixel matrix according to the symmetry mechanism. Then, after iterating the pixels and recalculating each one's value, each pixel will subtract the three adjacent ones-the bottom, right, and bottom right corner. The new value will be the average of the three results that just has been calculated.

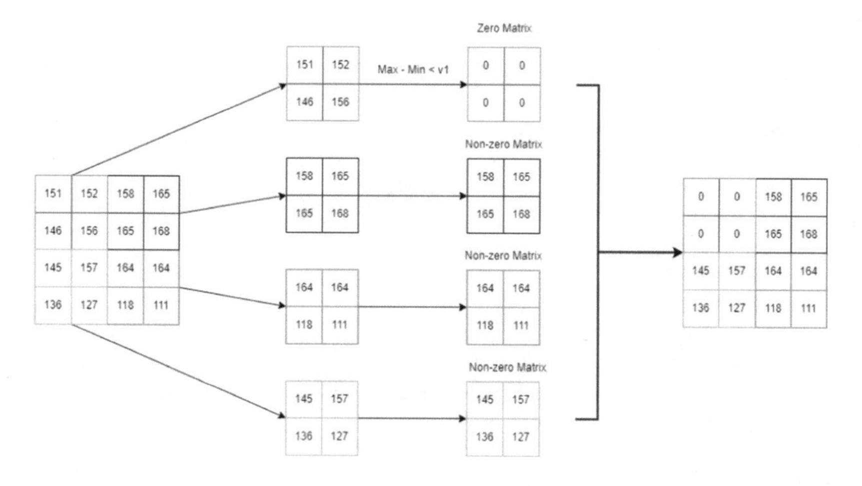

Fig. 2. Dropping pixels Sample

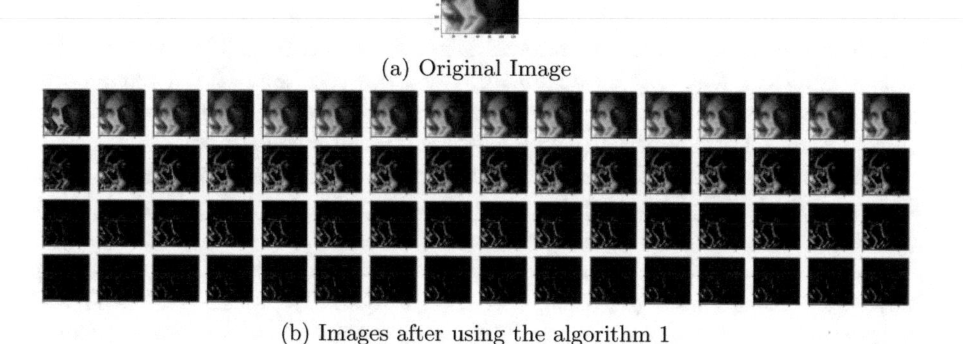

(a) Original Image

(b) Images after using the algorithm 1

Fig. 3. Illustration of image processing

3 Experimental and Results

3.1 Dataset

The ICML 2013 Workshop on Challenges in Representation Learning the Facial Expression Recognition 2013 (FER2013) dataset as a main dataset for the facial expression recognition contest [3]. FER2013 was created by Aaron Courville Pierre and Luc Carrier using the Google image search API to search for images of faces with many keywords related to different emotions. These keywords were also combined with words related to ethnicity, age, gender, etc. The authors also used OpenCV to place detect the human face of each image, reject the wrong label image then resize and convert it to grayscale. FER2013 is just a small subset of the work for the contest. FER2013 consists of 35,887 48 $\times$ 48-pixel grayscale images, most of which are human faces.

```
Data: img(48 × 48)
Result: newImg(48 × 48)
newImg ← img.padding(1, 1)
// Iterating the entire images
for i ← 0 to img.shape[0] − 1 by 1 do
    for j ← 0 to img.shape[1] − 1 by 1 do
        // horizontal
        h ← img[i][j] − img[i][j + 1]
        // vertical
        v ← img[i][j] − img[i + 1][j]
        // diagonal
        d ← img[i][j] − img[i + 1][j + 1]
        // Save the result of the matrix under consideration
        newImg[i][j] ← mean(h + v + d)
    end
end
```

Algorithm 2: Blurring algorithm

Figure 4 gives some sample images for each emotion category in the FER2013. Each column contains three different images for a category. The inconsistency of face position, face angle, and incomplete faces (e.g. some hand parts cover some of the face areas in fear or surprise images) is one of the main challenges in classifying emotions based on the features of each facial region.

All images in FER2013 were labeled into one of seven main categories: angry, disgust, fear, happy, sad, surprise, and neutral. Table 1 briefly describes some general information about the training set, public test set, and private test set for each of the emotion classes of FER2013. The dataset contains 4,953 Angry images, 547 Disgust images, 5,121 Fear images, 8,989 Happy images, 6,077 Sad images, and 6,198 Neutral images. The training set consists of 28,709 images, the public test set consists of 3,589 images and the private test set consists of 3,589 images. The human accuracy on this dataset was about $65 \pm 5\%$ [3].

3.2 Experimental Setup

Spatial Transformer Networks. The spatial transformer network (STN) was introduced by Jaderberg *et al.* [6]. It is used as a small CNN module inserted directly into our models, following the input immediately, and just before the CNN module. In this study, two options for using STN, such as using only one STN for all input, and each STN for each input which is an original image or a processed images. The output of STNs is used to input of CNNs and continue with the flow in Sect. 2.1. Based on each test, various types of STN are used, for example, DPM-FER + STN is to use 1 STN as shown in Fig. 5a, DPM-FER + 2STN is to use 2 STNs as shown in Fig. 5b. Details of how to use STN are in Fig. 5.

STN is a popular approach to increase the spatial invariance of the model. The main idea is pretty simple: its convolution layers capture features of the

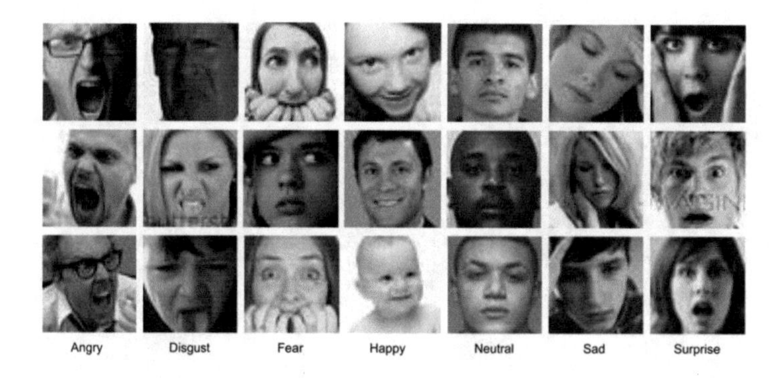

Fig. 4. Some sample from FER2013 dataset.

Table 1. Information about FER2013

Category	Training	Public Test	Private Test
Angry	3,995	467	491
Disgust	436	56	55
Fear	4,097	496	528
Happy	7,215	895	879
Sad	4,830	653	594
Surprise	3,171	415	416
Neutral	6,965	607	626
Total	28,709	3,589	3,589

image and learn how to perform an affine transformation to the original image directly from the backpropagation of the main model. The output image is the original image with canonical orientation, leading to better classification performance. It's a powerful technique to overcome the invariability of CNN.

Dropping Pixels Multiple Model - DPM. The model is trained on the Fastai[1] platform, using data with multiple dimensions and *learning_rate*. According to each training with *fit_one_cycle*, the learning rate is determined based on the available *lr_find* function in Fastai.

The trained model uses an original image with a size of 48×48 and the image which is processed with algorithm 1 has a size of 96×96, the *batch_size* used is 128, *learning_rate* change through each training phase. Firstly, each branch of the model is trained with its corresponding images. The model's CNNs will load those trained weights and fine-tune them so that the model can converge faster.

The example of input data of the DPM is shown in Fig. 6a. Image data samples for each submodel (Fig. 6a), the images are processed using algorithm 1

[1] Ref: https://docs.fast.ai/learner.html.

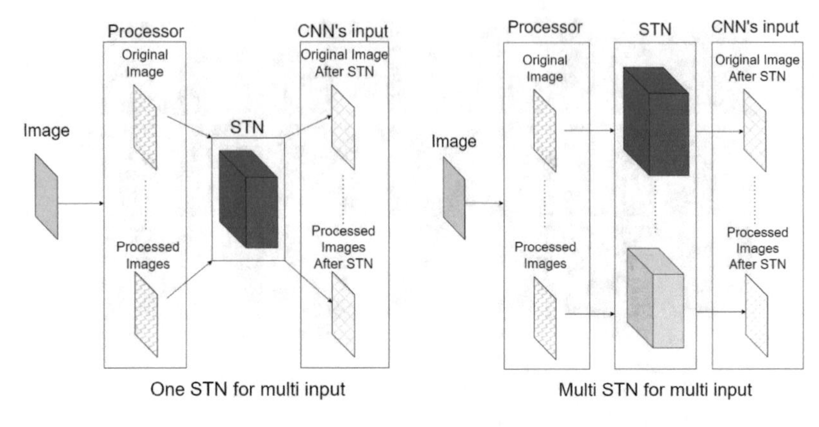

(a) One STN for multi images (b) Multi STNs for multi images

Fig. 5. Using STN for image input

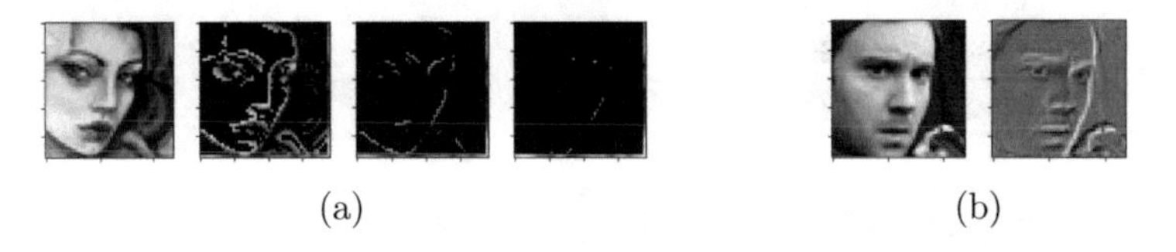

(a) (b)

Fig. 6. (a) Example of input data for the DPM (b) Example of input data for the BM

with the parameters described in the table 2. The selected parameters are the parameter sets for each level used to test the model.

Table 2. Table of input image parameters for the DPM

img	k	v1	v2
Image 1	0	0	255
Image 2	2	15	30
Image 3	2	30	30
Image 4	2	45	30

Table 3. Baseline comparison table of results

Model	Public Test(%)	Private Test(%)	TTA PV(%)	TTA PB(%)
VGG16	65.27	66.31	64.44	66.01
DPM-FER (2 CNN) + 2STN	66.45	67.87	66.01	67.40
BM-FER + STN	68.23	69.30	70.08	70.87

Blurring Multiple Model - BM. The BM-FER is also configured similarly to the DPM. Because the two models follow two different approaches, their input data will differ slightly. The input data will consist of two images: the original image and the image which is transformed by algorithm 2. As a result, both images in this model are the same size: 48×48. The example of the input data of the BM is given in Fig. 6b. On the left of Fig. 6b is the original image and on the right is the image processed with the algorithm 2.

3.3 Results

Table 3 shows the comparison of the approach in the article compared to the conventional VGG16. Overall, both our approaches outperform the baseline network (VGG16). BM-FER has significant improvement over the base network, with an accuracy of 70.08% on the private test and 70.87% on the public test with TTA. The baseline VGG16 only got about 65.27% on the public test set and 66.31% on the private test set. The margin of improvement is more than 7% in related. Compared to the baseline, the DPM-FER model produced results that were a little better than VGG16: 66.45% on the public test set and 67.87% on the private test set. The results also suggest that TTA is one of the useful methods for models, as both two of our approaches could get better performance in prediction with it enabled.

Table 4. Table comparing results of different configurations

Model	Public Test(%)	Private Test(%)	TTA PV(%)	TTA PB(%)
DPM-FER (4 CNN)	66.06	66.82	—	—
DPM-FER (2 CNN)	65	—	—	—
DPM-FER (2 CNN) + STN	66.02	68.18	65.85	67.09
DPM-FER (2 CNN) + 2STN	66.45	67.87	66.01	67.40
BM-FER	68.00	68.65	69.12	70.30
BM-FER + STN	68.14	69.07	69.37	70.63
BM-FER + STN + (64×64)	68.23	69.30	70.08	70.87

Some different configures of DPM-FER and BM-FER also conduct and results are given in the table 4. For DPM-FER, for the limitation of computing resources, there are only two configures, two filters, and four filters. We also test whether STN blocks are in shared mode. The results show that there is little improvement when using four filters over two filters, 66.05% compared with 65% in two filters. However, four filters of image needed four different of CNNs base network, then the network size is about double, heavily. It looks like the STN block shared enabled does not have a big effect. For BM-FER architecture, we also tested with the larger image size of 64×64 beside STN enabled. Compare with the base of BM-FER, this configuration got a small performance improvement.

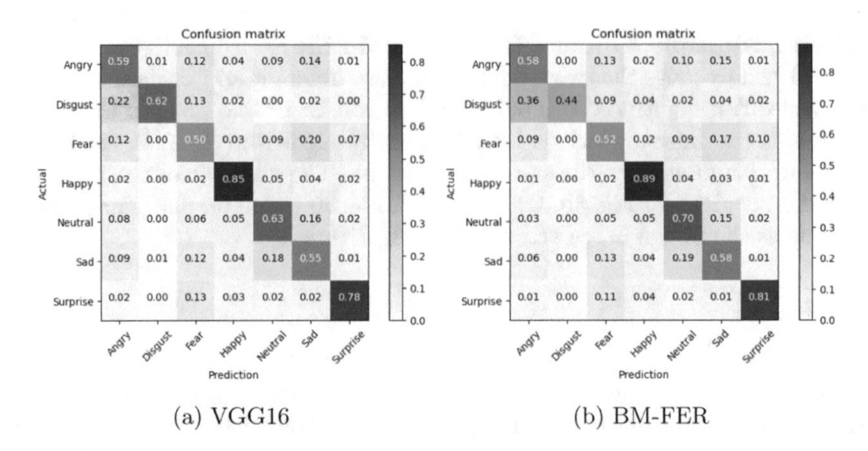

(a) VGG16 (b) BM-FER

Fig. 7. Confusion matrix of VGG16 and BM-FER on the Private Test

Figure 7 is a detailed comparison between two models VGG16 and BM-FER based on their confusion matrix on the Private Test set. For some classes, BM-FER has a large improvement compared to VGG16, such as happy, and neutral. It also decreases the confusion in some pairs of emotions such as fear-angry, fear-sad, and surprise-fear. However, there are some classes that VGG16 is better recognized, such as disgust.

4 Conclusion

In conclusion, facial emotion recognition is one of the significant attention fields in researching things related to human behaviors and emotions. The development of artificial intelligence has experienced a significant leap in recent years, especially with the emergence of cutting-edge deep-learning models. As a result, many researchers and scientists have proposed different methods to optimize this topic's output result. However, it is still a challenging task that has numerous obstacles in various aspects.

After realizing that certain areas with comparable values may not have meaningful contributions to improve the accurate predictions of the model, we decided to focus on regions with greater importance weight and omit those with similar to solve the above problem. Simultaneously, we also leveraged the mutual support of sub-models during the self-learning process by combining them together. By integrating these concepts, we have introduced two model types, DBM-FER and BM-FER, in this paper. The experimental outcomes demonstrate a notable enhancement, achieving the best results of 70.78% and 67.40%, respectively, for BM-FER and DPM-FER on the testing dataset, compared to 66.01% attained by a well-known CNN network, VGG16.

Acknowledgements. We acknowledge Ho Chi Minh City University of Technology (HCMUT), VNU-HCM for supporting this study.

References

1. Bro, R., Smilde, A.K.: Principal component analysis. Anal. Methods **6**(9), 2812–2831 (2014)
2. Chaudhari, A., Bhatt, C., Krishna, A., Mazzeo, P.L.: Vitfer: Facial emotion recognition with vision transformers. Appl. Syst. Innov. **5**(4), 80 (2022)
3. Goodfellow, I.J., et al.: Challenges in representation learning: A report on three machine learning contests. In: Neural Information Processing: 20th International Conference, ICONIP 2013, Daegu, Korea, November 3-7, 2013. Proceedings, Part III 20. pp. 117–124. Springer (2013)
4. He, K., Zhang, X., Ren, S., Sun, J.: Deep residual learning for image recognition. In: Proceedings of the IEEE Conference On Computer Vision And Pattern Recognition, pp. 770–778 (2016)
5. Heikkila, J., Ojansivu, V.: Methods for local phase quantization in blur-insensitive image analysis. In: 2009 International Workshop on Local and Non-Local Approximation in Image Processing, pp. 104–111 (2009)
6. Jaderberg, M., Simonyan, K., Zisserman, A., et al.: Spatial transformer networks. In: Advances in Neural Information Processing Systems, vol. 28 (2015)
7. Lakshmi, D., Ponnusamy, R.: Facial emotion recognition using modified hog and lbp features with deep stacked autoencoders. Microprocess. Microsyst. **82**, 103834 (2021)
8. Mehrabian, A.: Silent Messages: A Wealth Of Information About Nonverbal Communication (includes An Updated Bibliography). Wadsworth, Belmont, CA (1981)
9. Pecoraro, R., Basile, V., Bono, V.: Local multi-head channel self-attention for facial expression recognition. Information **13**(9), 419 (2022)
10. Pietikäinen, M.: Local Binary Patterns. Scholarpedia **5**(3), 9775 (2010)
11. Ullah, Z., Qi, L., Hasan, A., Asim, M.: Improved deep cnn-based two stream super resolution and hybrid deep model-based facial emotion recognition. Eng. Appl. Artif. Intell. **116**, 105486
12. Vaswani, A., Shazeer, N., Parmar, N., Uszkoreit, J., Jones, L., Gomez, A.N., Kaiser, Ł., Polosukhin, I.: Attention is all you need. In: Advances in Neural Information Processing Systems. Vol. 30 (2017)
13. Xanthopoulos, P., Pardalos, P.M., Trafalis, T.B., Xanthopoulos, P., Pardalos, P.M., Trafalis, T.B.: Linear discriminant analysis. Robust data mining, pp. 27–33 (2013)
14. Zhang, B., Liu, G., Xie, G.: Facial expression recognition using lbp and lpq based on gabor wavelet transform. In: 2016 2nd IEEE International Conference on Computer and Communications (ICCC), pp. 365–369 (2016)

Few-Shots Novel Space-Time View Synthesis from Consecutive Photos

Van Quan Mai[1,2] and Duc Dung Nguyen[1,2](✉)

[1] Ho Chi Minh City University of Technology (HCMUT), 268 Ly Thuong Kiet Street, District 10, Ho Chi Minh City, Vietnam
nddung@hcmut.edu.vn

[2] Vietnam National University Ho Chi Minh City, Linh Trung Ward, Ho Chi Minh City, Vietnam

Abstract. Despite the remarkable result of Neural Scene Flow Fields [10] in novel space-time view synthesis of dynamic scenes, the model has limited ability when a few input views are provided. To enable the few-shots novel space-time view synthesis of dynamic scenes, we propose a new approach that extends the model architecture to use shared priors learned across scenes to predict appearance and geometry at static background regions. Throughout the optimization, our network is trained to rely on the image features extracted from a few input views or from the learned knowledge for reconstructing unseen regions based on the camera view direction. We conduct multiple experiments on NVIDIA Dynamic Scenes Dataset [23] that demonstrate our approach results in a better rendering quality compared to the prior work when a few input views are available.

Keywords: NeRF · View synthesis · Few-shot view reconstruction

1 Introduction

With the rapid development of camera technology on mobile devices in recent years, we can easily capture meaningful moments with highly qualified photos. However, it requires many shoots to select moments that have the best facial expressions or actions. As a result, many redundant images remain in the phone memory or cloud storage. The novel space-time view synthesis topic has recently gained impressive progress thanks to using neural networks to learn representations of 4D space-time dimension. By applying these models, we aim to produce a short photo-realistic space-time video with motion parallax from a few consecutive images. However, most prior approaches use monocular videos as input to train their network. It introduces some issues when only a few input photos are available.

Our work introduces a novel approach that allows our framework to effectively learn and present dynamic scenes using only a limited number of input views. Unlike the previous models that do not use any image features, our model

N. T. Nguyen et al. (Eds.): CITA 2023, LNNS 734, pp. 240–249, 2023.
https://doi.org/10.1007/978-3-031-36886-8_20

leverages shared priors learned across scenes as an additional input to compensate for the lack of input photos when learning the presentation of the target scene.

Specifically, our model composes of two main components: dynamic and static. While we inherit the architecture of the time-variant (dynamic) network from prior work [10], which captures scene dynamics, we extend the time-invariant (static) network to make use of image features to predict unseen background regions when the camera is moving. First, using a convolution neural network, we extract an image feature grid from each image input. Then for each point on a camera ray and view direction, we interpolate image features by projecting each point to the coordination of input views. Finally, the location of the query point, view direction, and image features are fed to a neural network to output density, color, and the blending weight that specifies how to combine output from the dynamic and static network.

We conduct extensive evaluations on human and non-human scenes of the NVIDIA Dynamic Scenes Dataset [23]. Our experiments show that using scene priors yields better rendering quality when few views are provided in comparison with the baseline model.

2 Related Work

Novel View Synthesis: Many rendering methods [2,3,5,8,25] propose representing a scene in 3D space using a Point cloud or Triangle mesh to render new views. These methods ensure rendered image quality, but entirely building a Point Cloud or Triangle Mesh requires special equipment such as 3D scanners or photogrammetry techniques. High memory usage is also a drawback of these representation methods since almost every point in 3D space must be stored. Emerging in recent years, Neural Radiance Field (NeRF) [17] proposes to use a continuous function modeled by an MLP neural network to learn the geometry and appearance of a scene in 3D space. NERF has shown impressive view synthesis results that outperformed state-of-the-art methods [13,16,24]. Since NeRF learns each scene presentation separately, it requires many views of the scene from different viewpoints. In addition, the training process also requires much time and computational power. Yu et al. [29] proposed using an encoder to extract spatial features of input images and use these priors to predict the color and density of a point on the ray camera. The model can synthesize novel views from one or a few input views thanks to learned priors across scenes. The above methods are optimized to model a static scene with fixed appearance and geometry over time and thus are limited to model dynamic scenes with temporal changes.

Frame Interpolation: This method aims to interpolate an arbitrary frame between two or more adjacent frames. Frame interpolation increases the video quality by increasing the number of frames per second. S. Meyer et al. [15] proposed a phase-based method that represents motion in the phase shift of individual pixels; however, it is only applicable to limited-range motions. Kernel-based methods [18,19] use convolution neural networks to perform motion estimation and motion compensation in a single step. The pixel colors at a target frame are generated by performing convolution operations on the input frames with the predicted kernels. Increasing the kernel size to model large-range motions is necessary, thereby increasing the number of learned parameters. Recently, the Motion-based approach [6,9,12] has brought impressive results. This method models optical flows between two frames and uses image-warping techniques to interpolate the target frame. Many models following this approach can model non-linear [11,28] or asymmetry [20] motions from one frame to the next. Liying Lu et al. [14] proposed the VFIformer model combining the Transformer [27] with Cross-Scale Window-based Attention to model large-range movements, which has gained impressive interpolation quality compared to other proposed methods [6,9,21]. Although these models achieve promising results in interpolating in-between frames, they work only in 2D space.

Space-Time View Synthesis: Scenes captured from mobile phones are usually not static but change over time due to illumination and motion changes. Most prior approaches [1,26] in this domain require multiple views and time-synchronized input videos and have several drawbacks in modeling complicated scene geometry and appearance. Yoon et al. [23] proposed combining the depth map of a single view and the depth from multiple views. However, this method requires human-annotated foreground masks and a heavy pre-processing and training pipeline. Building on the successes of the NeRF, Park et al. [22] proposed to model an additional continuous volumetric deformation field that augments NeRF by warping each observed point into a canonical 5D NeRF. Although it can represent complex dynamic scenes, the model has limited ability to model large-range, asymmetric motions. Most closely related to ours, Li et al. [10] proposed modeling a time-variant continuous function to present the appearance, geometry, and 3D scene motion of dynamic scenes. These dense scene flow fields enable the model to interpolate novel views at arbitrary space and time. Even though the above methods can learn dynamic scene presentations, they use monocular videos as input to train their network. Instead, we show that our model can leverage learned priors across scenes to model unseen regions when a few input photos are available.

3 Proposed Methods

In this section, we briefly revisit the architecture of Neural Scene Flow Fields [10]. We then discuss our approach to extend the baseline model to learn dynamic scene presentations from a few input photos for space-time view synthesis at novel camera angles.

3.1 Background: Neural Scene Flow Fields (NSFF)

NSFF is built upon NeRF for static scene presentations. The model adds the notion of time to estimate the color and density of a point in 3D space besides location and view direction. Moreover, NSFF can predict forward and backward scene flows as dense 3D vector fields that point to neighbor frames and use this information to optimize the presentation of a dynamic scene.

NSFF optimizes both time-variant presentation (dynamic) and time-invariant presentation (static). For given 3D point x at time i and view direction d, the dynamic network predicts not only the color c_i and density α_i but also the forward and backward scene flows $\mathscr{F}_i$ which are 3D offset vectors that point to the position of x at time $i-1$ and $i+1$. The dynamic model F_θ^{dy} is formulated as follows:

$$(c_i, \alpha_i, \mathscr{F}_i) = F_\theta^{dy}(x, \boldsymbol{d}, i) \tag{1}$$

To cover for large temporal gaps and leverage observations from all input frames, NSFF models the static presentation F_θ^{st} with its own MLP:

$$(c, \alpha, v) = F_\theta^{st}(x, \boldsymbol{d}) \tag{2}$$

where v is the blending weight field that linearly combines the color and density predicted from the dynamic and static network.

Finally, the color $\hat{C}$ of an image pixel at time i at a specific target viewpoint is calculated by volume rendering blended color and density of 3D points along camera ray r:

$$\hat{C}_(r) = \int_{t_n}^{t_f} T(t)\alpha(r(t))c(r(t), \boldsymbol{d})dt \tag{3}$$

where $T(t) = \exp(-\int_{t_n}^{t} \alpha(r(s))ds)$

3.2 NSFF with Encoder

Given enough input frames from different time steps and viewpoints, NSFF can learn dynamic presentations for a scene, yet it cannot share knowledge across scenes. Therefore, when few inputs are available, the model tends to overfit the training dataset and does not generalize well to render unseen viewpoints. Inspired by prior work [29], we propose extending NSFF by conditioning the time-invariant model on spatial image features. We choose to focus on the time-invariant part (static) rather than the time-variant (dynamic) scene presentations since the dynamic network is designed to model 3D motions, which are specialized from scene to scene. The static model, however, learns the presentation of background regions that are consistent across input frames and can leverage prior knowledge of the world to reconstruct unseen areas when the viewpoint changes.

Particularly, our static model is comprised of two main components: 1) a fully-convolutional encoder E which extracts spatial image features from an input view into a pixel-aligned feature grid, and 2) a convolution network f which predicts color, density, and blending weight. We show the entire model architecture in Fig. 1.

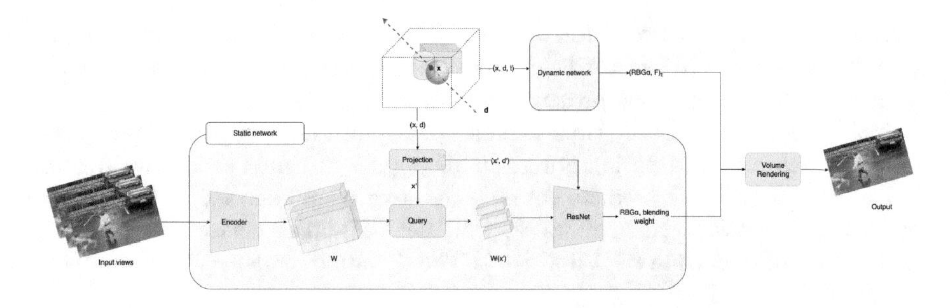

Fig. 1. To predict the color of an image pixel at time t, we first use an Encoder (E) to encode input views to pixel-aligned feature grids (W). Then, we query image features for each input view at projected coordinate x' to get $W(x')$. Next, the projected coordinate x', view direction d', and image features $W(x')$ is fed to a ResNet to predict color, density, and blending weight. At the same time, we use the dynamic network from NSFF [10] to predict color, density, and scene flows of point x at the time t. Finally, results from the static and dynamic networks are combined and volume is rendered to output the final color.

Given n input views, we first use the encoder E to extract features from each input image I_i. This results in a feature grid of view i: $W_i = E(I_i)$. Then for a target camera ray r with view direction d, we project a query point x along with the view direction d into the coordination system of each input view i using known intrinsic as follows:

$$x_i = P_i x, d_i = R_i d \tag{4}$$

where $P_i = [R_i t_i]$ is the change of basis transformation matrix of input view i which transforms from world coordinate to input view space.

Using projected coordinate x_i, we query the features from the feature grid of view i and then apply bi-linear interpolation between the pixel-wise features to extract the image feature $W_i(x_i)$. Then the image features of n input views are fed to the convolution network f along with the projected coordinates and view directions to predict color, density, and blending weight as follows:

$$(c, \alpha, v) = f(V_1, V_2, ..., V_n) \tag{5}$$

where V_i is a concatenation vector of x_i, d_i and $W_i(x_i)$.

Notably, the camera view direction can be interpreted as a nature attention weight that the model can rely on to determine the importance of image features from multiple input views. If the camera view direction is aligned with an input view direction, the model can leverage its image features exclusively. In contrast, if none of the input view directions are similar to the camera view direction, the model must rely on the learned priors.

4 Experiments

4.1 Implement Details

For the dynamic network, following NSFF architecture, we use positional encoding to transform the inputs and feed them to a network with 8 fully-connected layers, each with 256 channels and a skip connection that concatenates the input to the fifth layer's activation.

For the static network, we use the ResNet34 backbone pre-trained on ImageNet [4] for the encoder to extract a feature pyramid from input images. Then extracted features, projected positions, and view directions are fed through a ResNet that consists of 5 blocks with the image feature as a residual at the beginning of each ResNet block. We found during our experiments that the static network's loss has not converged as quickly as the dynamic network due to its complexity. It results in the blending weights tending to choose pixel colors in the dynamic network exclusively in the first few iterations, and the static network stops reducing its loss. Therefore, rather than optimizing the two networks simultaneously on the rendering loss of the combined predicted results, we warm up the static model by optimizing it with its rendering loss for the first 1k iterations.

Similar to the prior work [10], we train and evaluate our model on the NVIDIA Dynamic Scenes Dataset [23]. The dataset consists of 8 human and non-human motion scenes recorded by 12 synchronized cameras. Training the full model for each scene using the Adam optimizer [7] with a learning rate of 0.0005 takes around 10 h on 1 NVIDIA A100-SXM4 with 40 GB RAM provided by the Google Colab (Fig. 2).

Fig. 2. Scene samples in NVIDIA Dynamic Scenes Dataset [23].

4.2 Evaluations

To evaluate our model, we use the same evaluation dataset prepared by Li et al. [10] that contains 11 held-out images per time instance at different camera viewpoints for each scene. We train our model and NSFF using three input views per scene and compare the rendering quality with three standard error metrics: structural similarity index measure (SSIM), peak signal-to-noise ratio (PSNR), and perceptual similarity through LPIPS [30], on each scene in the NVIDIA Dynamic Scenes Dataset [23]. The quantitative results in Table 1 show that our model has outperformed the NSFF model on the three error metrics. We also provide our quantitative results in Fig. 3. Our model produces fewer artifacts and better render quality than the NSFF model at static background regions.

Table 1. Observation of 3 metrics PSNR, SSIM, and LPIPS on the NVIDIA Dynamic Scenes Dataset. At a specific scene, values giving better results will be marked by **bold**.

Scene	PSNR(↑)		SSIM(↑)		LPIPS(↓)	
	NSFF	Ours	NSFF	Ours	NSFF	Ours
Balloon1-2	12.405	**14.5**	0.306	**0.36**	0.514	**0.47**
Balloon2-2	14.72	**16.337**	0.424	**0.453**	0.47	**0.352**
DynamicFace-2	10.412	**11.636**	0.18	**0.217**	**0.594**	0.635
Jumping	16.5	**18.646**	0.566	**0.636**	0.385	**0.299**
Playground	13.6	**14.236**	**0.313**	0.3	**0.485**	0.512
Skating-2	17.15	**18.907**	0.592	**0.653**	0.376	**0.34**
Truck-2	**17.93**	17.604	0.571	**0.589**	0.263	**0.254**
Umbrella	18.079	**18.738**	**0.463**	0.45	0.391	**0.27**
Mean	15.1	**16.326**	0.427	**0.457**	0.435	**0.392**

5 Limitations

The task of novel space-time view synthesis of dynamic scenes is challenging especially when limited input views are available. Although we result in a better performance compared to the prior work, there are several limitations to our approach.

- Despite the generative nature of our static network, we are not able to train it on the whole dataset due to limitations in computing resources. However, only using pre-trained weights of the encoder in the ImageNet [4] outperformed the baseline model.
- Inheriting the limitations of prior works, our model's training and rendering time are high. Given the complexity of our static network, training the model requires a heavier pipeline and is more time-consuming compared to the baseline model.

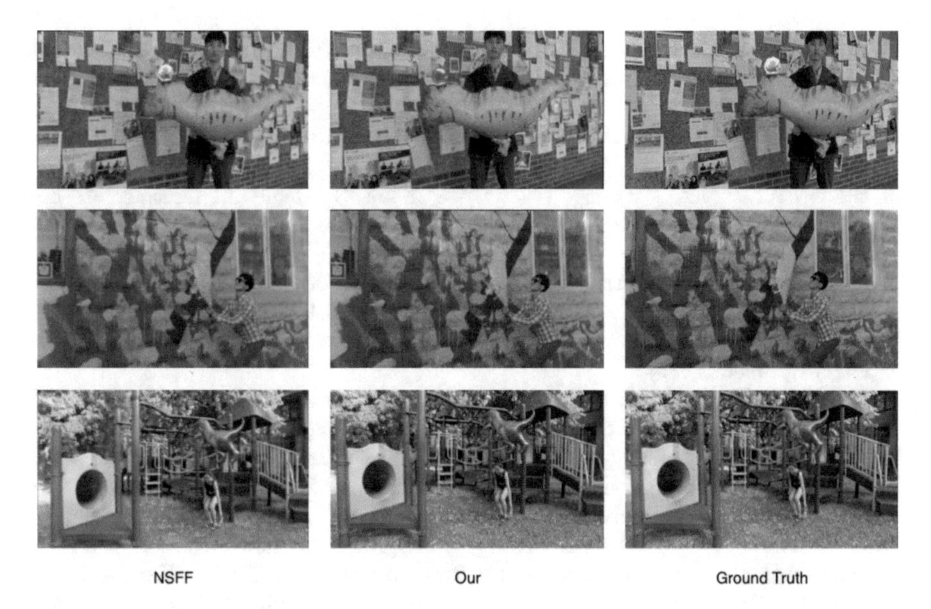

Fig. 3. Qualitative comparison between our model and NSFF. Our model reduces fewer artifacts than the NSFF model.

6 Conclusion

This paper presented a new approach to synthesizing novel space-time views of dynamic scenes from a few input images. We have shown that our model can leverage the learned priors to predict pixel colors at unseen regions. It resulted in a better performance in the NVIDIA Dynamic Scenes Dataset than the baseline model. For future work, we aim to train our model on the whole dataset and optimize the network architecture to reduce the training and rendering time. Furthermore, we aim to apply learned priors to the dynamic network to predict scene flow fields.

Acknowledgements. We acknowledge Ho Chi Minh City University of Technology (HCMUT), VNU-HCM for supporting this study.

References

1. Bansal, A., Vo, M., Sheikh, Y., Ramanan, D., Narasimhan, S.: 4D visualization of dynamic events from unconstrained multi-view videos. In: 2020 IEEE/CVF Conference on Computer Vision and Pattern Recognition (CVPR), pp. 5365–5374 (2020)
2. Buehler, C., Bosse, M., McMillan, L., Gortler, S.J., Cohen, M.F.: Unstructured lumigraph rendering. In: Proceedings of the 28th Annual Conference on Computer Graphics and Interactive Techniques (2001)
3. Chaurasia, G., Duchene, S., Sorkine-Hornung, O., Drettakis, G.: Depth synthesis and local warps for plausible image-based navigation. ACM Trans. Graph. **32**(3) (2013)

4. Deng, J., Dong, W., Socher, R., Li, L.J., Li, K., Fei-Fei, L.: ImageNet: a large-scale hierarchical image database. In: 2009 IEEE Conference on Computer Vision and Pattern Recognition, pp. 248–255 (2009)
5. Hedman, P., Alsisan, S., Szeliski, R., Kopf, J.: Casual 3D photography. ACM Trans. Graph. **36**(6) (2017)
6. Huang, Z., Zhang, T., Heng, W., Shi, B., Zhou, S.: Real-time intermediate flow estimation for video frame interpolation. In: Avidan, S., Brostow, G., Cissé, M., Farinella, G.M., Hassner, T. (eds.) Computer Vision – ECCV 2022: 17th European Conference, Tel Aviv, Israel, 23–27 October 2022, Proceedings, Part XIV, pp. 624–642. Springer, Heidelberg (2022). https://doi.org/10.1007/978-3-031-19781-9_36
7. Kingma, D.P., Ba, J.: Adam: a method for stochastic optimization. CoRR abs/1412.6980 (2014)
8. Klose, F., Wang, O., Bazin, J.C., Magnor, M., Sorkine-Hornung, A.: Sampling based scene-space video processing. ACM Trans. Graph. **34**(4) (2015)
9. Lee, H., Kim, T., Chung, T.Y., Pak, D., Ban, Y., Lee, S.: AdaCof: adaptive collaboration of flows for video frame interpolation. In: 2020 IEEE/CVF Conference on Computer Vision and Pattern Recognition (CVPR), pp. 5315–5324 (2020)
10. Li, Z., Niklaus, S., Snavely, N., Wang, O.: Neural scene flow fields for space-time view synthesis of dynamic scenes. In: Proceedings of the IEEE/CVF Conference on Computer Vision and Pattern Recognition (CVPR), pp. 6498–6508, June 2021
11. Liu, Y., Xie, L., Siyao, L., Sun, W., Qiao, Yu., Dong, C.: Enhanced quadratic video interpolation. In: Bartoli, A., Fusiello, A. (eds.) ECCV 2020. LNCS, vol. 12538, pp. 41–56. Springer, Cham (2020). https://doi.org/10.1007/978-3-030-66823-5_3
12. Liu, Z., Yeh, R.A., Tang, X., Liu, Y., Agarwala, A.: Video frame synthesis using deep voxel flow. In: 2017 IEEE International Conference on Computer Vision (ICCV), pp. 4473–4481 (2017)
13. Lombardi, S., Simon, T., Saragih, J., Schwartz, G., Lehrmann, A., Sheikh, Y.: Neural volumes: learning dynamic renderable volumes from images. ACM Trans. Graph. **38**(4) (2019)
14. Lu, L., Wu, R., Lin, H., Lu, J., Jia, J.: Video frame interpolation with transformer. In: Proceedings of the IEEE/CVF Conference on Computer Vision and Pattern Recognition (CVPR), pp. 3532–3542, June 2022
15. Meyer, S., Wang, O., Zimmer, H., Grosse, M., Sorkine-Hornung, A.: Phase-based frame interpolation for video. In: 2015 IEEE Conference on Computer Vision and Pattern Recognition (CVPR), pp. 1410–1418 (2015)
16. Mildenhall, B., et al.: Local light field fusion: practical view synthesis with prescriptive sampling guidelines. ACM Trans. Graph. **38**(4) (2019)
17. Mildenhall, B., Srinivasan, P.P., Tancik, M., Barron, J.T., Ramamoorthi, R., Ng, R.: NeRF: representing scenes as neural radiance fields for view synthesis. In: Vedaldi, A., Bischof, H., Brox, T., Frahm, J. (eds.) Computer Vision - ECCV 2020 - 16th European Conference, Glasgow, UK, 23–28 August 2020, Proceedings, Part I. LNCS, vol. 12346, pp. 405–421. Springer, Cham (2020). https://doi.org/10.1007/978-3-030-58452-8_24
18. Niklaus, S., Mai, L., Liu, F.: Video frame interpolation via adaptive convolution. In: 2017 IEEE Conference on Computer Vision and Pattern Recognition (CVPR), pp. 2270–2279 (2017)
19. Niklaus, S., Mai, L., Wang, O.: Revisiting adaptive convolutions for video frame interpolation. In: 2021 IEEE Winter Conference on Applications of Computer Vision (WACV), pp. 1098–1108 (2021)

20. Park, J., Lee, C., Kim, C.S.: Asymmetric bilateral motion estimation for video frame interpolation. In: 2021 IEEE/CVF International Conference on Computer Vision (ICCV), pp. 14519–14528 (2021)
21. Park, J., Lee, C., Kim, C.S.: Asymmetric bilateral motion estimation for video frame interpolation. In: Proceedings of the IEEE/CVF International Conference on Computer Vision (ICCV), pp. 14539–14548, October 2021
22. Park, K., et al.: Nerfies: deformable neural radiance fields. In: 2021 IEEE/CVF International Conference on Computer Vision (ICCV), pp. 5845–5854 (2021)
23. Shin Yoon, J., Kim, K., Gallo, O., Park, H.S., Kautz, J.: Novel view synthesis of dynamic scenes with globally coherent depths from a monocular camera. In: 2020 IEEE/CVF Conference on Computer Vision and Pattern Recognition (CVPR), pp. 5335–5344 (2020)
24. Sitzmann, V., Zollhoefer, M., Wetzstein, G.: Scene representation networks: continuous 3D-structure-aware neural scene representations. In: Wallach, H., Larochelle, H., Beygelzimer, A., d' Alché-Buc, F., Fox, E., Garnett, R. (eds.) Advances in Neural Information Processing Systems, vol. 32. Curran Associates, Inc. (2019)
25. Snavely, N., Seitz, S.M., Szeliski, R.: Photo tourism: exploring photo collections in 3D. ACM Trans. Graph. **25**(3), 835–846 (2006)
26. Vagharshakyan, S., Bregovic, R., Gotchev, A.: Light field reconstruction using Shearlet transform. IEEE Trans. Pattern Anal. Mach. Intell. **40**(1), 133–147 (2018)
27. Vaswani, A., et al.: Attention is all you need. In: Proceedings of the 31st International Conference on Neural Information Processing Systems, pp. 6000–6010, NIPS 2017. Curran Associates Inc., Red Hook, NY, USA (2017)
28. Xu, X., Siyao, L., Sun, W., Yin, Q., Yang, M.: Quadratic video interpolation. Adv. Neural Inf. Process. Syst. **32** (2019). Publisher Copyright: © 2019 Neural information processing systems foundation. All rights reserved; 33rd Annual Conference on Neural Information Processing Systems, NeurIPS 2019; Conference date: 08 December 2019 Through 14 December 2019
29. Yu, A., Ye, V., Tancik, M., Kanazawa, A.: pixelNeRF: neural radiance fields from one or few images. In: Proceedings of the IEEE/CVF Conference on Computer Vision and Pattern Recognition (CVPR), pp. 4578–4587, June 2021
30. Zhang, R., Isola, P., Efros, A.A., Shechtman, E., Wang, O.: The unreasonable effectiveness of deep features as a perceptual metric. In: Proceedings of the IEEE Conference on Computer Vision and Pattern Recognition (CVPR), June 2018

Information Technology Skills Extractor for Job Descriptions in vku-ITSkills Dataset Using Natural Language Processing

Nguyen Huu Nhat Minh[1(✉)], Nguyen Ket Doan[1], Pham Quoc Huy[1], Kieu Xuan Loc[1], Hoang Nguyen Vu[1], Huy Nguyen[2], and Huynh Cong Phap[1]

[1] Vietnam-Korea Information Technology and Communication University, The University of Danang, Da Nang, Vietnam
{nhnminh,nkdoan.20it7,pqhuy.20it3,kxloc.20it10,hnvu, hcphap}@vku.udn.vn

[2] Olinno Technology Ltd., Da Nang, Vietnam
huy.nguyen@olinno.com

Abstract. The IT skills extractor is convenient and efficient for recent job recommendation systems and job seekers to find suitable jobs. In this paper, we design an efficient SpaCy pipeline for extracting IT skills based on Natural Language Processing (NLP) and Named Entity Recognition (NER) methods from the job description. The main proposed method helps to extract potential hard-soft skills and later could provide to job recommender and job seekers. As the state-of-the-art open-source NLP framework, we first construct a new IT skills dictionary based on ChatGPT and perform automatic labeling for scrapped job description dataset, named vku-ITSkills dataset. Using this dataset, the quality of labels could be improved by the Part-of-Speech (POS) function and additional rules. We then fine-tune the pre-trained RoBERTa-base model for Transformer based word embedding in training NER model to extract skills. Thereafter, we define additional logical rules to enhance the extracted results that could further find out more skills based on syntactic such as the comma rule. In this language pipeline, RoBERTa embedding, NER, and additional rules play important roles to cope with unseen and new IT skills that are non-existed in vku-ITSkills dataset and are missed from NER. Throughout the evaluation, we test the proposed pipeline with 200 job descriptions manually labeled by our team and demonstrate the efficiency of each step in the pipeline.

Keywords: IT Skills Dataset · Named Entity Recognition · Natural Language Processing

1 Introduction

Recently, the freelancer market and IT jobs keep increasing and demanding more efficient job recommenders could help job seekers to find suitable positions. Job Descriptions (JDs) often contain valuable information such as the required skills for each job position.

N. T. Nguyen et al. (Eds.): CITA 2023, LNNS 734, pp. 250–261, 2023.
https://doi.org/10.1007/978-3-031-36886-8_21

Hence, one of the promising recommendation designs for finding the most suitable jobs for job seekers by extracting the potential skills from both the job description and CVs for matching similar pairs between JDs and CVs. Unlike salary or working location entities which could be easier to extract using syntax, regular expression [1] and existing Named Entity Recognition (NER) [2], the IT skills extraction is much more complex and challenging and even easily misunderstood by humans, e.g.: *Reading* technical documents is a hard skill, but *Reading* is also a city in England. Therefore, the extractor should carefully be considered the useful context in order to produce highly accurate and efficient results.

Nowadays, the world of online job recruitment platforms could make it an easier process and more convenient for human resource teams and job seekers to connect based on job requirements and user profiles or CVs. Most of JDs can be too descriptive or too simple about the requirements of the job as well as the required skills which could make it challenging for a recommendation. Conventional job recommenders often suffer from the below situations:

- CVs may have many formats and styles, containing tables and other formatting features to make them look attractive, but this makes them difficult for syntactical extraction
- Complex context of skills
- The JDs may be too descriptive or too simplistic

Moreover, recent digital transformation keeps defining many more new job titles as well as new required IT skills. Thus, poor skills extractors could lead to poor matching between CVs and JDs. As a result, developing online job recommender engines attracts more research to improve overall performance. One of the promising research domains by leveraging Natural Language Processing (NLP) techniques since it has already demonstrated their efficiency in many recent language applications such as Grammarly, ChatGPT [3], etc. In this paper, we produce a good quality labeled JDs vku-ITSkills dataset and develop the novel processing pipeline based on one of the best open-source NLP frameworks, namely SpaCy [2]. Extracting skills could provide crucial information for successfully matching CVs to JDs, and play a major role in job recommender. In summary, we present the proposed method as follows:

i) We first construct the vku-ITSkills dataset with automatic labeling using EntityRuler in SpaCy and a novel skill dictionary generated by ChatGPT.
ii) We then build additional rules and train NER for skill extraction with the pre-trained Transformer based RoBERTa and additional rules with Part of Speech (POS).
iii) To evaluate the proposed pipeline, we demonstrate the efficiency of each step.

2 Preliminary

To generate a list of suitable jobs for a candidate, the recommender systems should know the job seekers' skills, and experiences [4]. Regarding extracted skills and experiences, the recommendation algorithms could provide the most relevant jobs for each candidate [4]. One of the conventional approaches [5] is to match these skills word-by-word from

CV and JD. However, this approach yields some disadvantages and is unreliable and inaccurate such as:

i) Matching similar words may be inaccurate due to the context of words, more than one way to present a similar skill. For example, *"A project of Java coffee using Java Language"*. Here the first *"Java"* means a type of coffee while the second *"Java"* is the IT skill. As another example, in some CVs we counter *"OPP"* as IT skills with the meaning "object-oriented programming", but in another CV, the real meaning could be "*Oriented Polypropylene*".
ii) The skills contain special characters, e.g., c++, problem-solving
iii) Some unseen skills

To resolve the above challenges, NLP based approach could provide much efficient solution by understanding word contexts [4]. NLP represents the words by using advanced neural network models for language tasks (i.e., BERT [6], RoBERTa [7], GPT3 [8], and DistilBERT [9]) using advanced word embedding technologies such as Word2Vec [10], Transformer [11]. We demonstrate the significant improvement by using the proposed NLP SpaCy pipelines in our experiment results. Accordingly, it brings more reliable and accurate results than that of conventional word-by-word matching methods. The missing skills further could be filled up with additional logical rules such as comma rule and noun phrases by utilizing POS functions in NLP. The extracted skills from job descriptions and CVs will be used as inputs for the other recommendation modules.

In this work, we leverage the NLP components for NLP tasks as shown in Fig. 1. The functions, the revelant SpaCy components, and description for each component in the proposed pipeline are shown in Table 1.

Fig. 1. The basic NLP pipeline in SpaCy [2]

Table 1. The basic SpaCy components will be adopted in the proposed pipeline [2]

Name	SpaCy Component	Description
Tokenizer	Tokenizer	Segment text into tokens
Transformer	Transformer	Token embedding for multi-task language tasks
Tagger	Tagger	Assign Part-of-Speech tags for each word such as noun, verb, adjective, pronoun, adverd, etc
Parser	DependencyParser	Assign dependency labels
EntityRuler	EntityRuler	Label the phrase by exact matching rule
NER	EntityRecognizer	Detect and label named entities

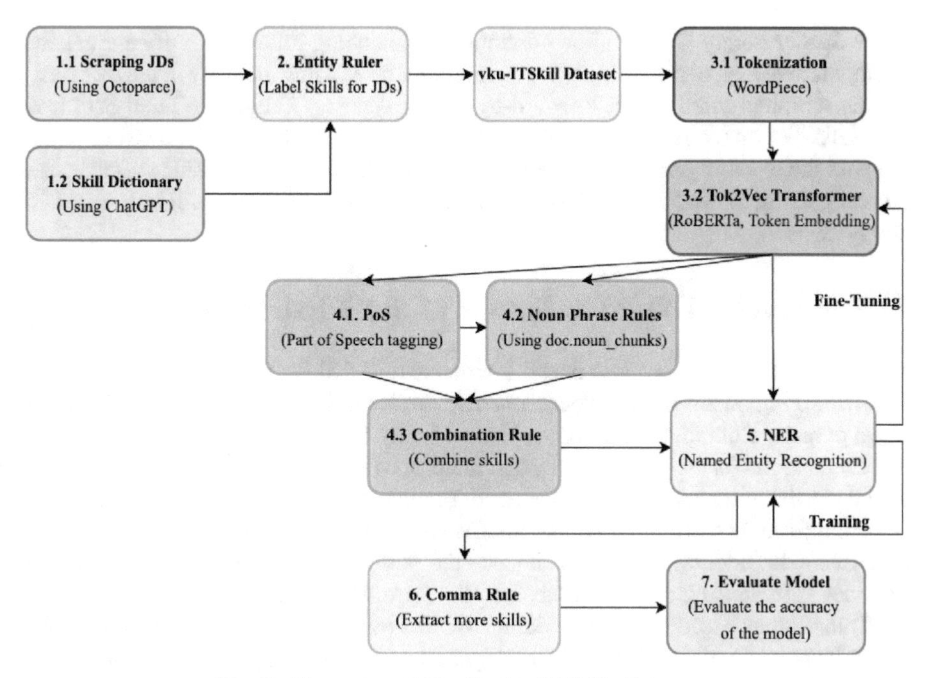

Fig. 2. The proposed pipeline for IT Skills Extractor

3 Construct IT Skills Dictionary and vku-ITSkills Dataset

One of the critical issues to produce a specialized NLP-based IT skill NER extractor is building the initial skills dictionary and labeling skills for JDs. Hence, we need to construct a novel good JDs dataset and skills dictionary for the later training IT Skills Extractor. For the JDs dataset, our first approach is utilizing the JDs dataset from Kaggle. However, the dataset is not clean and good enough quality to train the IT skills extractor. Specifically, we analyze the dataset and figure out missing special punctuations and white space, which may lower the model accuracy. Even though, it could provide training JDs but often duplicated and has some incorrect sentence formats. Then, we clean the data manually which consumes a long time to double-check. Hence, we use the python regex method [1] to improve data quality. However, the data is still not completely clean and still outdated. In the second approach, we considered the web scraping approach by using Octoparse [12] to extract data from two primary jobs recruiting platforms such as LinkedIn [13] and Indeed [14]. It is a fast, more convenient, and accurate method. However, to prepare the data for model training purposes, we continue cleaning the scrapped data. Our dataset comprises 30,000 distinct JDs within the IT industry from a variety of domains such as web and mobile development, networks, cybersecurity, artificial intelligence, machine learning, QA/QC, blockchain, and cloud computing. Thereafter, we use 25,000 JDs for training purposes.

As another crucial component, we need to build the skill dictionary with ChatGPT [3]. We utilize ChatGPT to compile a comprehensive list of 5000 IT skills frameworks, libraries, and tools from various domains in the IT industry. Entity Ruler is a component in the SpaCy library [2] that allows defining the matching rules to identify entities in a document. We then perform Entity Ruler to automatic labeling the skills in our dictionary for the scrapped training JDs by word-by-word matching. Taking the input data from our skills dictionary. Entity Ruler helps to scan through each JD if the word or phrase matches the word or phrase as "HARD-SKILL" or "SOFT-SKILL". This paper focuses on recognizing IT Skills as the Hard-Skills for IT jobs. These steps will be 1.1 and 1.2 components as shown in Fig. 2.

4 The Proposed NLP Pipeline for IT Skills Extractor

As shown in Fig. 2, we propose the NLP pipeline for IT Skills Extractor comprise of the following components of Tokenizer, Token Embedding, Part of Speech (POS) Tagging & Noun phrases, Combination of Skills, Named Entity Recognition (NER), Comma Rule, and Evaluation. As the results of steps in the proposed pipeline, the extracted skills are stored in a list of tuples containing the entity's name, the start-end indexes, and the word. Different from the normal process of SpaCy, we have combined both the NER and PoS functions with additional rules to enhance the performance. The vku-ITSkills dataset provides sufficient IT skills and labeled JDs for training NER as well as fine-tuning the Transformer based token embedding. While both the NER and PoS modules have their advantages and drawbacks, their integration procedure and proposed addition rules result in the superior performance of IT skills Extractor.

4.1 Tokenizer

In SpaCy, a tokenizer is a component of the NLP pipeline that is responsible for breaking down a text input into individual tokens, such as words or punctuation marks. To tokenize the raw text, first, it is split based on whitespace characters. Then, the tokenizer processes the text from left to right and results in the list of substrings/subwords/tokens. In doing so, the first check of the tokenizer looks for tokenizer exception rules, such as splitting *"don't"* into *"do"* and *"n't"* or keeping *"U.K."* as one token. The second check examines prefixes, suffixes, or infixes like commas, periods, hyphens, or quotes and splits them off if appropriate. Whenever a match is found, the tokenizer applies the rule and continues looping, starting with the newly split substrings. As a result, SpaCy can effectively tokenize complex, nested tokens such as those containing abbreviations and multiple punctuation marks.

WordPiece Tokenization

This tokenization technique known as WordPiece was developed by Google [15] for pretraining BERT and is being utilized in various BERT based language models such as RoBERTa [7], DistilBERT [9], etc. WordPiece begins with a limited vocabulary including special tokens used by the model and the initial alphabet. For instance, initially the word "love" is splitted into set of tokens as l, ##o, ##v, ##e. Consequently, WordPiece algorithm defines merge rules instead of choosing the most common pairs, it calculates a score for each pair of splitted tokens using the following the equation [15]:

$$Score = (freq\_of\_pair)/(freq\_of\_first\_element \times freq\_of\_second\_element) \quad (1)$$

where *freq_of_pair*: the frequency of the pair

freq_of_first_element: the frequency of first element

freq_of_second_element: the frequency of second element

After merging pairs, the Worpiece algorithm stores the longest subword with infrequent individual components from original words.

4.2 Transformers Based Token Embedding

The human written text based on human knowledge is an unnatural representation for computers to understand. As the computer language, the computer system relies on the binary or numerical representation. Thereby, word embedding techniques enable computers to understand and represent words and sentences using mathematical embedding models. The embedding could capture semantic and syntactic relationships between words or subwords (e.g., *"Softwares"* and *"Applications"*).

Transformer models use self-attention mechanisms [11], which allow the model to selectively focus and incorporate the complex relations on different parts of the input sequence. Transformer models typically consist of an encoder and a decoder where the encoder processes the input sequence using self-attention and feed-forward neural networks, producing a series of encoded representations from each input sentences. The decoder then uses the encoded representations and self-attention to generate output sequence for the trained language tasks such as a machine translation. The attention mechanism is the heart of Transformer based language models as follows:

$$Attention(Q, K, V) = softmax(\frac{QK^T}{\sqrt{d_k}})V \quad (2)$$

SpaCy also leverages the transformer based pre-trained models from Hugging Face [17]. These models are pre-trained on a large corpus of text, and then can be fine-tuned on task-specific datasets. Taking advantage of this efficient strategy, we fine-tune the RoBERTa-base [17] token embedding model with the JDs in vku-ITSkills dataset in training NER to make better language models for detecting named entities such as IT skills. RoBERTa model is based on the same architecture as BERT which is a Transformer-based neural network model. RoBERTa and BERT have differences in their training mechanisms and dataset sizes. RoBERTa has been trained on a dataset of 160 GB of text, which is more than 10 times larger than the dataset used to train BERT. Furthermore, during training, RoBERTa implements "dynamic masking" to help the model learn more resilient and adaptable word representations. RoBERTa has 355 million parameters, which is much larger than BERT's 110 million parameters. This increased size and larger training dataset allow RoBERTa model to capture more complex patterns in language and perform better on NLP tasks. Using the dynamic masking mechanism [7], tokens are masked randomly from a distribution at each training step. This design helps to prevent the model from memorizing specific token patterns during training.

The *RobertaEmbeddings* takes three inputs such as token IDs, segment IDs, and position IDs. The first layer of embeddings is a learned embedding matrix, followed by positional embeddings and segment embeddings, which are added to capture the position of each token in the sequence and distinguish between different segments of the input sequence. The *RobertaSelfAttention* is responsible for performing multi-head self-attention on the input sequence embeddings. The multi-head attention enables multiple embeddings of inputs and performs in parallel, which can lead to better representation learning. The *RobertaSelfOutput* takes the input from self-attention and employs a series of transformations such as a linear layer, followed by a residual connection and layer normalization. The *RobertaIntermediate* is a fully connected feed-forward layer that is used as an intermediate layer in the RoBERTa encoder. The *RobertaOutput* is a feedforward neural network layer that processes the outputs of the *RobertaSelfAttention* and the *RobertaIntermediate*.

4.3 Part of Speech (POS) Tagging, Noun Phrases and Combination Rules

POS tagging is the process of labeling each word in a sentence with its corresponding part of speech, such as noun, verb, adjective, adverb, preposition, conjunction, or interjection [2]. SpaCy's POS tagger is based on a statistical model that uses deep learning techniques and contextual information to assign the most likely part-of-speech tag to each word in a sentence. The model is trained on large, annotated corpora of text, and it is highly accurate and efficient.

Noun Phrases Rules

After using POS tagging to label a word with a typing word, we propose a list of the noun phrases in the Job Description using SpaCy's "doc.noun_chunks". SpaCy's *doc.noun_chunks* uses Matchers to extract noun phrases in JDs according to rules. Hence, to enhance the quality of training data more IT skills can be found via the following rules:

R1: If a noun phrase contains technological terms such as "tool", "library", "software", "platform", "module", "framework", and so on. If the noun phrase contains words labeled as "HARD-SKILL", then that noun phrase will be labeled as IT skills.

E.g.: Tensorflow ***framework****, NLP* ***modules****, Web and Mobile* ***platform***

R2: For noun phrases labeled as "HARD-SKILL", the tokens are tagged by POS tagger such as "PRON" (e.g., your, our), "DET" (e.g., a, an, the), "ADV" (e.g., more, much), "CCONJ" (e.g., and, or, &), "NUM" (e.g., 1, one) will be eliminated.

E.g.: ***one or more*** *Machine Learning tools* $\Rightarrow$ *Machine Learning tools,*

the *Android modules* $\Rightarrow$ *Android modules.*

R3: We examine the tokens preceding word of the noun phrase "HARD-SKILL" that has the suffix "*ing*" and include it in the noun phrase. We repeat the same process by skipping the token "*and*", "*or*", "*&*", or "*/*" until cannot find "*ing*" in the preceding word.

E.g.: ***performing*** *data analysis;* ***building*** *and* ***developing*** *Web applications;* ***downloading****,* ***installing*** *and* ***using*** *Office* 365

R4: Search for noun phrases that include "*developing*", "*building*", "*deploying*", "*optimizing*", and so on. We then combine these two words and label the phrase as "HARD-SKILL". This applies to all noun phrases, including those that are not initially labeled as "HARD-SKILL". Although this rule may sound similar to rule 3, it is used for a different purpose. If we label all noun phrases after the word has the "ing" suffix can produce a significant number of incorrect outcomes.

$$E.g.\text{: } Having\ employees,\ interns,\ and\ contractors\ is\ not\ "HARD-SKILL"$$

After using SpaCy Part of Speech (POS) Tagging & Noun phrases, we had a complete list of hard skills, next step we used Combination Rule to group skills together, and label them and added new hard skills.

Combination Rule: In this rule, we merge all the IT skills next to each other as "HARD-SKILL"

$$Ex\text{: } Two\ consecutive\ IT\ skills:\ "C"\ "Programming"\ will\ be\ merged\ into\ "C\ Programming"\ and\ label\ us\ "HARD-SKILL"$$

Overall, the above additional rules preprocess the labeled data using the logical rules and heuristic design to improve the quality of labeling data before feeding in the later training process (Fig. 3).

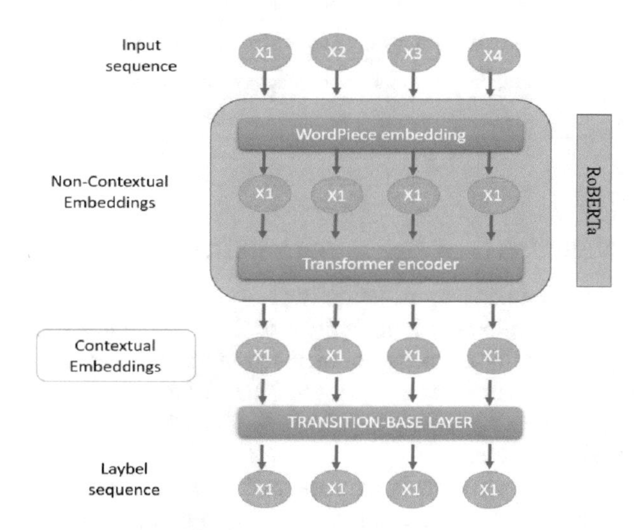

Fig. 3. SpaCy NER Model [2]

4.4 Named Entity Recognition

Named Entity Recognition (NER) [2, 4] is a crucial NLP task in dealing with identifying and categorizing named entities for a given input text. Named entities refer to useful terms or concepts which belong to predefined categories such as individuals, organizations, geographic locations, products, etc. Recently, the improved ability to automatically identify and categorize named entities using advanced language models could help to extract better valuable information from texts and documents. The language model is typically trained on a large annotated dataset of text, where each named entity is labeled with its corresponding type. The machine learning model then learns to identify patterns and features in the text that are associated with different types of named entities. These patterns and features can include the presence of certain words, phrases, or context clues.

SpaCy utilizes a transition-based parser model [16] for its Named Entity Recognition (NER) functionality. This parsing approach involves predicting the structure of the text by mapping it to a series of state transitions. By that way, it build a dependency tree for a input sentence. The neural network state prediction model for NER consists of either two or three following subnetworks: **RoBERTa** is the embedding subnetwork that maps each token into a vector representation; **Transition-based parser** represent feature-specific vector for each (token, feature) pair; **Feed-forward layers** (optional) consists of feed-forward layers to predict scores according to the entity recognition.

Table 2. Training method of proposed NLP pipelines

PIPELINE	TRAINING
Transformers	Fine-tune the pre-trained RoBERTa embedding
NER	Train a new component from scratch on our vku-ITSkills dataset
Tagger	Using the pre-trained component without updating
Parser	Using the pre-trained component without updating

4.5 Training NER Model and Fine-Tuning RoBERTa

After all the aforementioned preprocessing steps, we have a set of hard skills for each job description. SpaCy allows training NER model on a custom dataset to recognize new specific types of entities that are not included in the pre-trained models. We then start training the model to extract more skills that have been previously unidentified and unlabeled. Also, this training process can fine-tune RoBERTa token embedding model according to IT Skill NER task.

Our training process performs in a dataset of 25,000 JDs in total 30,000 JDs. We choose this number due to this reaching the maximum training capacity of our workstation PC. The specification of the training workstation: Intel Core i7 13th Generation with Nvidia RTX 3090 and 64 GB memory. Our SpaCy pipeline for training the model will comprise the following components: Transformers, NER, Tagger, and Parser. Each component of the pipeline will follows different training configuration in Table 2.

To train the models were trained with the batch size is *64*, and the *"roberta-base"* transformer model using Adam optimizer with the initial learning rate is *0.00012* and dropout parameter is *0.12*. Through various testing trials, we have determined that these parameter values are suitable for the proposed NLP pipelines.

4.6 Comma Rule

In the search for additional rules to improve the extracted or labeled skills, we explored that most pre-defined rules were unreliable and hardly generalized for most of the cases. Fortunately, we figure out a highly reliable and simple rule such as the comma rule. Using this rule, if a set of words or phrases is separated by commas, 'and', 'or', or '/', and at least one of them is identified as IT skill, then all of the remaining terms are considered skills. We have found this simple heuristic rule is extraordinarily efficient and has significantly improved our skill extraction performance in addition to the extracted skills from Named Entity Recognition (NER).

5 Experimental Results

In order to evaluate the effectiveness, and performance of the proposed pipeline, i.e., **IT_Skill_NER**, we developed two distinct sub-datasets. The first subset included 100 JDs that were different from the training 25,000 JDs, while the second subset comprised 100 job descriptions scraped from other websites such as GlassDoor, Dice, Monster, and CareerBuilder. These JDs were more diverse and spanned a wide range of domains in the IT industry, including web and mobile development, networking, cybersecurity, artificial intelligence, machine learning, QA/QC, blockchain, and cloud computing. Our objective in creating these two separate testing sub-datasets is to determine if our model was able to meet the expected outcomes on LinkedIn and Indeed job descriptions as well as the others platform. We manually label these 200 JDs and double-check with the extracted results by using our proposed NLP pipeline. In that way, we evaluate the efficacy by providing a comparative analysis of the skill-extracted performance throughout each module in the proposed NLP pipeline.

In Table 3 and 4, we list out the demonstrated results for each testing sub-dataset. Our evaluation will be based on a set of four performance metrics such as correct extraction, partially missing labeling, fully missing skills, and redundant extraction. We will assess the effectiveness of each hard skill extraction module for testing the effectiveness of EntityRuler, process; POS tagging and Noun phrases rules; Combination rules; IT_Skill_NER, and Comma rule.

Table 3. Hard skills extraction results on 100JDs from LinkedIn and Indeed

Performance Metrics	Entity Ruler	POS & Noun Phrase Rules	Combination Rule	IT_Skill_NER	Comma Rule
Correct Extraction	956 (28.3%)	2034 (66.2%)	2198 (73.2%)	2265 (75.4%)	**2551 (84.9%)**
Partially Missing Skills	1467 (43.3%)	422 (13.7%)	187 (6.2%)	**162 (5.4%)**	**162 (5.4%)**
Fully Missing Skills	962 (28.4%)	616 (20.1%)	616 (20.5%)	578 (19.2%)	**292 (9.7%)**
Redundancy	179	**69**	**69**	79	79

Table 4. Hard skills extraction results on 100JDs from GlassDoor, Dice, Monster, Career Builder.

Performance Metrics	Entity Ruler	POS & Noun Phrase Rules	Combination Rule	IT_Skill_NER	Comma Rule
Correct Extraction	1020 (29.8%)	2038 (64%)	2168 (71%)	2210 (72.8%)	**2526 (83.2%)**
Partially Missing Skills	1368 (40%)	470 (14.8%)	212 (6.9%)	**187 (6.2%)**	**187 (6.2%)**
Fully Missing Skills	1032 (30.2%)	676 (21.2%)	676 (22.1%)	638 (21%)	**322 (10.6%)**
Redundancy	179	92	92	**88**	**88**

According to the experimental results as shown in Table 3 and 4, our full proposed pipeline keep improving the overall performance and achieve the best accuracy such as *84.9%* in *100* JDs from LinkedIn, Indeed and *83.2%* in *100* JDs from GlassDoor, Dice, Monster, and CareerBuilder. Even the comma rule does not help to reduce the partially missing skills but it helps to figure out more correct extracted skills and significantly reduce fully missing skills which often appear due to unseen and novel IT skills. That could help the IT_Skill_Extractor ease for future unseen IT skills and jobs.

6 Conclusions

In this paper, we not only design a good NLP pipeline for IT skills extractor but also construct a novel vku-ITSkills JDs dataset. With other modules to optimize the main module. Building vku-ITSkills JDs dataset based on web scraping and automatic labeling from 5,000 IT skills dictionary generated by ChatGPT. The proposed pipeline with NER including the additional logical rules such noun phrase rules, combination rule, comma rules exhibit extraordinary performance for extracting IT skills. In the future, we continue developing a recommendation module using extracted skills. In doing so, we plan to improve the quality of data and develop a relevance score for important skills in job recommendation.

Acknowledgement. This work is supported by Vietnam-Korea Information Technology and Communication University.

References

1. Friedl, J.E.F.: Mastering Regular Expressions. O'Reilly Media Inc., USA (2006)
2. SpaCy: Industrial-Strength Natural Language Processing with Python and Cython. https://spacy.io/. Accessed 15 Mar 2023
3. OpenAI. ChatGPT. https://chat.openai.com/. Accessed 15 Mar 2023
4. Gugnani, A., Misra, H.: Implicit skills extraction using document embedding and its use in job recommendation. In: Proceedings of the AAAI Conference on Artificial Intelligence, vol. 34, no. 08, pp. 13286–13293 (2020)

5. Al-Otaibi, S.T., Ykhlef, M.: A survey of job recommender systems. Int. J. Phys. Sci. **7**(29), 5127–5142 (2012)
6. Devlin, J., Chang, M.W., Lee, K., Toutanova, K.: Bert: pre-training of deep bidirectional transformers for language understanding. In: Proceedings of naacL-HLT (2019)
7. Liu, Y., et al.: Roberta: A robustly optimized bert pretraining approach. arXiv preprint arXiv: 1907.11692 (2019)
8. Radford, A., et al.: Language models are unsupervised multitask learners. OpenAI Blog **1**(8), 9 (2019)
9. Sanh, V., Debut, L., Chaumond, J., Wolf, T.: DistilBERT, a distilled version of BERT: smaller, faster, cheaper and lighter. arXiv preprint arXiv:1910.01108 (2019)
10. Mikolov, T., Chen, K., Corrado, G., Dean, J.: Efficient estimation of word representations in vector space. arXiv preprint arXiv:1301.3781 (2013)
11. Vaswani, A., et al.: Attention is all you need. In: Advances in Neural Information Processing Systems, vol. 30 (2017)
12. Octoparse. https://www.octoparse.com/. Accessed 15 Mar 2023
13. LinkedIn Jobs. https://www.linkedin.com/jobs/. Accessed 15 Mar 2023
14. Indeed Jobs. https://indeed.com/. Accessed 15 Mar 2023
15. Kudo, T., Richardson, J.: Sentencepiece: a simple and language independent subword tokenizer and detokenizer for neural text processing. arXiv preprint arXiv:1808.06226 (2018)
16. Chen, D., Manning, C.D.: A fast and accurate dependency parser using neural networks. In: Proceedings of the 2014 Conference on Empirical Methods in Natural Language Processing (EMNLP), pp. 740–750 (2014)

Comprehensive Study on Semantic Segmentation of Ovarian Tumors from Ultrasound Images

Thi-Loan Pham[1,2], Van-Hung Le[3(✉)], Thanh-Hai Tran[2], and Duy Hai Vu[2]

[1] Hai Duong University, HaiDuong, Vietnam
[2] School of Electrical and Electronic Engineering (SEEE), Hanoi University of Science and Technology, Hanoi, Vietnam
[3] Tan Trao University, Tuyên Quan, Vietnam
van-hung.le@mica.edu.vn

Abstract. Ovarian cancer is one of the most mortal diseases in women. It is commonly detected by medical experts while observing ultrasound images. With the increasing advance in artificial intelligence, deep learning in particular, many medical image analysis have been applied to improve the efficiency in diagnosis and support for training young doctors. This paper introduces a comprehensive study on the segmentation of ovarian tumors from ultrasound images. We investigate state-of-the-art segmentation models such as PSPNet, U-net, DANet, Deeplabv3, and PSANet and evaluate them on a recently published dataset MMOTU. Different from the original works on the MMOTU dataset that just provided binary segmentation, we generate also the label of 8 tumor categories for each pixel. By doing so, it provides not only the size and shape of tumors but also the diseases. Experimental results show that DANet gives the highest accuracy of 71.65% in average. Overall, Chocolate cyst can be accurately segmented with IoU of 96.33% by PSANet.

Keywords: Ovarian Tumor Ultrasound Image · Semantic Segmentation · PSPNet · DANet · UNet · DeepLabv3 · PSANet and Convolutional Neural Networks

1 Introduction

Early detection and diagnosis of cancer is being one of the most critical tasks of doctors in protecting human health in recent years. According to the statistics of the American Cancer Society (ACS) [1], in 2021 around the world, about 21,410 people receive a new diagnosis of ovarian cancer and there are about 13,770 deaths from ovarian cancer. Ovarian cancer is the fifth leading cause of death in women. Despite advances in medical diagnosis, an accurate diagnosis of ovarian cancer requires a variety of complex tests. Or there may be misdiagnosis due to a lack of obvious symptoms for ovarian cancer. In the process of building an application for early ovarian cancer detection and warning, ultrasound images

N. T. Nguyen et al. (Eds.): CITA 2023, LNNS 734, pp. 262–273, 2023.
https://doi.org/10.1007/978-3-031-36886-8_22

are commonly utilized as input. Ultrasound imaging has many following advantages [2] such as no radioactivity, safer imaging, greater accuracy, sensitivity, and specificity, and lower costs than other imaging modalities such as computed tomography (CT) or magnetic resonance imaging (MRI). Currently, with the development of artificial intelligence in computer vision, Computer-aided diagnosis (CAD) has been proposed to reduce interobserver variation among various doctors and thereby motivate them to produce more reliable and efficient diagnostic results [3].

Recently, Qi et al. [4] proposed the DS2Net model for semantic segmentation of ovarian tumors on the MultiModality Ovarian Tumor Ultrasound (MMOUT) [4] dataset. The authors have shown the results of semantic segmentation based on several CNNs such as PSPNet [5,6] DANet [7], SegFormer [8], U-Net [9], TransUNet [10], BiseNetV2 [11]. However, in that work, the authors produced and compared binary segmentation results without assigning labels to each tumor category. Such results can only provide the shape, and size of the tumor without classifying the tumor as benign or malignant to initially support the diagnosis of early ovarian cancer. However, knowledge about tumor categories helps to quickly diagnose the diseases.

In this paper, we perform a follow-up study based on the MMOTU dataset for semantic segmentation of ovarian tumors. Specifically, we assign each pixel in the ultrasound images a label belonging to the label of 9 pre-defined categories (8 tumor categories in the dataset and one normal case). The output of our study is semantic segmentation results with eight labels of ovarian tumors. At the same time, quantitative and qualitative analysis and challenges of ovarian tumor detection and classification are also presented.

The structure of the paper is organized as follows. In section II, we briefly present some related works on ovarian tumor segmentation. In section III, we describe our framework for a comprehensive study of semantic segmentation of ovarian tumors with various deep-learning models. The dataset and experimental results will be reported in section IV. We finally conclude and give some ideas for future works.

2 Related Works

Detection and diagnosis of ovarian cancer using imaging is a research area of great value in medicine. However, research activities in this area are still relatively modest. We will introduce some research on ovarian tumors shortly.

Enshaei et al. [12] have evaluated the potential of an artificial intelligence model by comparing it with conventional statistical methods for ovarian cancer diagnosis. The authors created a database that included 668 cases of epithelial ovarian cancer over 10 years and collected data routinely available in the clinical setting. They also collected survival data for all patients, then built an artificial intelligence model capable of comparing various algorithms and classification of tumors in addition to traditional statistical methods such as regression.

Recently, several methods focus on computer-aided diagnosing and ovarian tumor detection [13]. Despite their notable works, there still exist the following two main weaknesses. First, recent methods only focus on single-modality images (mainly 2d ultrasound images) segmentation and recognition. There is still a lack of research on exploring the presentation potential of multi-modality ultrasound images because of lacking standard datasets. To overcome two weaknesses, Qi et al. [4] construct a Multi-Modality Ovarian Tumor Ultrasound (MMOTU) image dataset. Qi et al. [4] only performed semantic segmentation of ovarian tumors (only classified with or without ovarian tumor in the image). The ovarian tumor classification and segmentation challenges of the MMOTU dataset have not been shown.

In the study of Jin et al. [14], the authors utilized various techniques U-net, CE-Net, U-net ++, and U-net with Resnet. In this work, U-net based automatic segmentation is precise enough to delineate the target mass on the US image for patients with ovarian cancer, the classical U-net scheme and its multiple variations were used for the automatic segmentation task. Generally, the U-net is a symmetrical U-shaped model consisting of an encoder-decoder architecture [15]. The encoder on the left is downsampling used to get the feature map, similar to a compression operation, while the right-side decoder is an upsampling used to restore the encoded features to the size of the original image and to output the results. Skip-connection was added to encoder-decoder networks to concatenate the features of high- and low-level together [15]. When Resnet is used as a fixed feature encoder to deepen the layers of the network and solve the vanishing gradient, the U-net structure is changed to Unet with Resnet as the backbone (U-net with Resnet) [16].

3 Segmentation of Ovarian Tumor Ultrasound Image by CNNs

3.1 Background

Image segmentation is a very important problem in computer vision and it is applied in many fields such as medicine, autonomous vehicles, satellite image processing, agriculture, forest fire warning, etc. The problem of image segmentation and object classification in images on ovarian ultrasound images is the process of determining which pixels belong to the ovarian tumor area with which label. In the following, we will introduce some CNNs that can solve both image segmentation and object classification problems.

PSPNet: The PSPNet architecture [5] takes into account the global context of the image to provide the local level predictions and hence gives better performance on benchmark datasets like PASCAL VOC 2012 and cityscapes. The model was needed because FCN-based pixel classifiers were not able to capture the context of the whole image. The PSPNet encoder contains the CNN backbone with dilated convolutions along with the pyramid pooling module. It is the main part of this model as it helps the model to capture the global context in the image which helps it to classify the pixels based on the global information present in the image. The backbone of PSPNet is the Resnet-50 [16]. The feature map from the backbone is pooled at different sizes and then passed through a convolution layer and after which upsampling takes place on the pooled features to make them the same size as of the original feature map. Finally, the upsampled maps are concatenated with the original feature map to be passed to the decoder. This technique fuses the features of different scales hence aggregating the overall context.

U-Net: The U-Net [9] is a CNN architecture for fast and precise segmentation of images. The U-Net is an elegant architecture that solves most of the occurring issues, It uses the concept of fully connected CNNs for this approach. The U-Net intends to capture both the features of the context as well as the localization. The main idea of U-Net is to utilize successive contracting layers, which are immediately followed by the upsampling operators for achieving higher resolution outputs on the input images.

The architecture shows that an input image is passed through the model and then it is followed by a couple of convolutional layers with the ReLU activation function. The image size is reduced from 572×572 to 570×570 and finally to 568×568. The reason for this reduction is because made use of unpadded convolutions (defined the convolutions as "valid"), which results in the reduction of the overall dimensionality. Apart from the convolution blocks, the encoder block is on the left side followed by the decoder block on the right side. The encoder block has a constant reduction in image size with the help of the max-pooling layers of strides 2. Convolutional layers are repeated with an increasing number of filters in the encoder architecture. The number of filters in the convolutional layers starts to decrease, along with a gradual upsampling in the following layers to the top. The use of skip connections enables connections between the previous outputs and the layers in the decoder blocks.

DANet: Jun et al. [7] proposed a novel framework, called Dual Attention Network (DANet) for natural scene image segmentation. It introduces a self-attention mechanism to capture feature dependencies in the spatial and channel dimensions, respectively. Specifically, two parallel attention modules are on top of the dilated FCN. One is a position attention module, and the other is a channel attention module. For the position attention module, the self-attention mechanism captures the spatial dependencies between any two positions of the feature maps. For the feature at a certain position, it is updated via aggregating features at all positions with the weighted summation, where the weights are decided by the feature similarities between the corresponding two positions.

DeepLabv3: The DeepLabv3 architecture is built on top of the ResNet-101 architecture by adding the Spatial Pyramid Pooling with holes (SPP) module. The SPP module was first used in the DeepLabv2 architecture. This module applies convolutional filters with different hole sizes in parallel on the output of the ResNet, producing feature maps with different details depending on the hole sizes of the filters. These maps are then concatenated and passed through a 1×1 standard convolutional filter. In this work, the implementation of DeepLabv3 [17] is used, which includes a secondary loss function in addition to the original DeepLabv3 architecture. This loss function is added to the end-of-architecture loss function multiplied by 0.4. A multi-task cross-entropy loss function is used as the loss function and the optimization is implemented with the SGD algorithm.

PSANet: Jie et al. [18] propose a Point Set Attention Network (PSANet). In the network, to update the current pixel with an intra-class common feature representation, pixels with the same class may have similar feature representations and further promote intra-class mutual improvement. Reassembling pixels in a point set Specifically, the current pixel and pixels in its neighborhood are used as a point set to generate a context-aware mask. With the guidance of the mask feature, aggregation of the pixels of the same class as the current pixel is performed in the point set for a center pixel, with which the current pixel is updated. The motivation is that the current pixel and pixels near it often belong to the same class and share similar feature representations. Therefore, selecting pixels of the same class as the current pixel helps to correct noisy representation.

3.2 Segmentation of Ovarian Tumor Ultrasound Images

In this paper, we perform semantic segmentation of ovarian tumors, the process is illustrated in Fig. 1. This is essentially a comparative study for instance segmentation of ovarian tumors with multi-task CNNs: PSPNet [5,6] DANet [7], U-Net [9], DeepLabv3 [17], PSANet [18]. Input data is an OTU_2D sub-set ultrasound image of the ovaries. The images in the dataset with various resolutions are then normalized to a fixed size of 320×240.

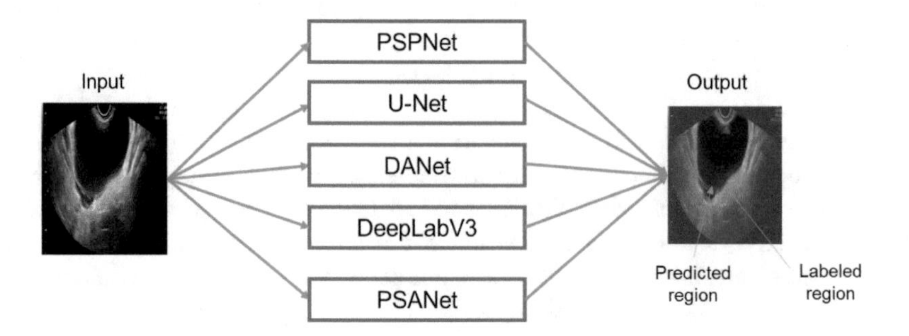

Fig. 1. The processing pipeline of the comparative study for instance segmentation of ovarian tumors.

4 Experimental Results

4.1 Data Collection

In this paper, we use a MultiModality Ovarian Tumor Ultrasound (MMOTU) [4] image dataset. The MMOTU is obtained from Beijing Shijitan Hospital, Capital Medical University with 1639 ovarian ultrasound images collected from

294 patients. It consists of two subsets with two modalities, which are OTU_2D and OTU_CEUS. The OTU_2D includes 1469 2D ultrasound images, and the OTU_CEUS includes 170 CEUS images. We only focus on the OTU_2D subset, as illustrated in Fig. 2. The width and height of images respectively range from $302 \sim 1135$ and $226 \sim 794$ pixels.

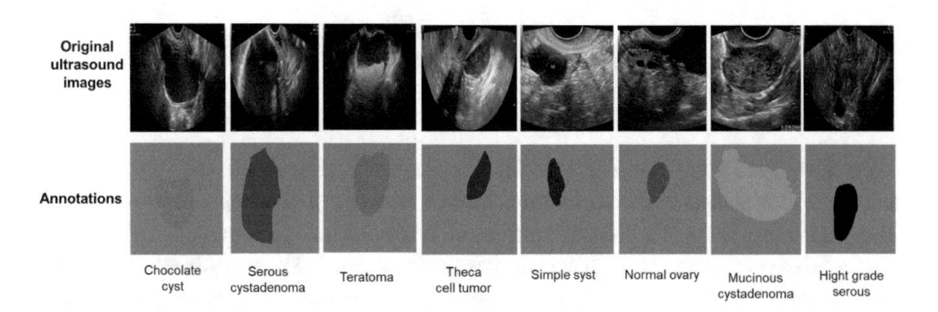

Fig. 2. Illustration of some OTU_2D images with eight labels of ovarian tumor in the dataset MMOTU. The top row is an ultrasound image. The bottom row is the annotation data, each label has a different color.

In the MMOTU dataset, the pixel-wise semantic annotation and global-wise category annotation are provided by 27 experts in Obstetrics and Gynecology department. Each image is first annotated by one expert and then checked by another expert, which guarantees the annotating quality. During annotating, experts refer to pathological reports, which makes the annotations accurate and convincing.

4.2 Evaluation Metrics

To evaluate the performance of segmentation of ovarian tumors, we utilize two metrics of evaluation. The first metric is the Intersection over Union (IoU) index [19], also known as the Jaccard index. IoU is calculated as the number of pixels in the intersection between the predicted region $P = TP + FP$ and the ground-truth/annotation region $G = TP + FN$ divided by the number of pixels in the union region of P and G, as presented in Fig. 3, Eq. 1.

$$IoU = \frac{|TP|}{|TP| + |FP| + |FN|} \tag{1}$$

where TP is the correct prediction, FP is the false prediction, and FN is the no predicted ground truth data area.

The second metric is the accuracy (ACC). ACC is calculated as the formula 2.

$$ACC = \frac{True\_Po + True\_Ne}{True\_Po + True\_Ne + False\_Po + False\_Ne} \tag{2}$$

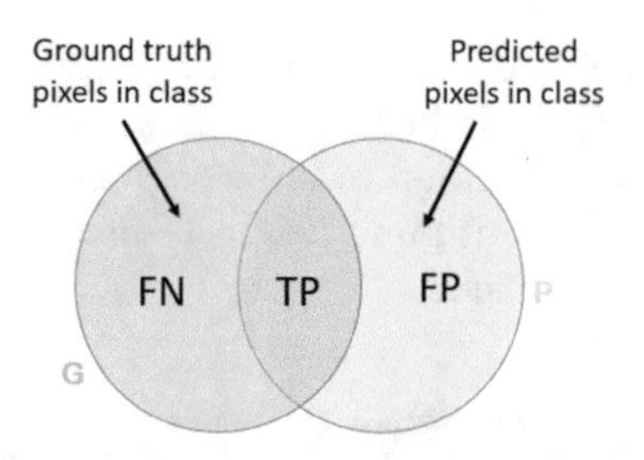

Fig. 3. Illustration of IoU calculation.

where True positive ($True\_Po$) is the number of regions that are assigned with the correct label and make the correct prediction of the ovarian tumor type label, True negative ($True\_Ne$) is the number of regions that are correctly labeled and make a false prediction of the ovarian tumor type label, False positive ($False\_Po$) is the number of regions that are mislabeled and make the correct prediction of the ovarian tumor type label, and False negative ($False\_Ne$) is the number of regions that are mislabeled and make an incorrect prediction of ovarian tumor type label.

In this paper, we deployed the studied models on a server with NVIDIA GeForce RTX 2080 Ti, 12GB GPU for fine-tuning, training, and testing. The programs were written in the Python language ($\geq$3.7 version) with the support of the CUDA 11.2/cuDNN 8.1.0 libraries. In addition, there are several libraries such as OpenCV, matplotlib, mmcls$\geq$0.20.1, numpy, packaging, prettytable, PyTorch 1.5+, etc. Especially the source code is inherited and developed on the "MMsegmentation" library [20], the pre-trained model and the code of all methods used in the evaluation of this paper are integrated with the library [20]. The instance segmentation of ovarian tumor models is fine-tuned with 20k iterations on the model trained from the "Cityscapes" [21] dataset. The IoU threshold for ovarian tumor segmentation is 0.5.

4.3 Results and Discussion

The results of ovarian tumor segmentation and label prediction based on PSPNet [5,6] DANet [7], U-Net [9], DeepLabv3 [17], and PSANet [18] with IoU and ACC measures are shown in Table 1, Table 2, respectively. In Table 1, the average IoU of the PSPNet is 56.87%, Unet is 49.7%, DANet is 58.65%, DeepLabV3 is 59.75%, PSANet is 58.66%, respectively. In Table 1, ovarian tumor segmentation results based on DeepLabv3 are the best, the average of IoU is 59.75%. In Table 2, the average result with the ACC metric of the PSPNet is 67.37%, Unet is 61.27%, DANet is 71.56%, DeepLabV3 is 69.99%, and PSANet is 70.06%, respectively. In Table 2, DANet-based label prediction for ovarian tumor region is the best, the average of ACC is 71.56% The results segmentation and label prediction results of "Mucinous cystadenoma" and "Hight grade serous" are the lowest, as shown in Table 1 and Table 2. The average segmentation result above is very low to be used as an aid in the classification of ovarian tumors when combined with IOTA

[22]. The results also show that "Chocolate cyst" is the type of ovarian tumor with the highest segmentation results. The results of the methods are all greater than 95%. This is also the simplest type of ovarian tumor and is an endometrial ovarian cyst.

Table 1. The results of ovarian tumor segmentation (*IoU*) on the OTU_2D sub-set of MMOTU dataset.

Measurement/ Labels/ Methods		**PSPNet (%)**	**U-net (%)**	**DANet (%)**	**Deeplabv3 (%)**	**PSANet (%)**
IoU	Chocolate_cyst	96.06	95.69	95.98	96.22	96.33
	Serous_cystadenoma	61.9	60.46	64.49	63.59	62.26
	Teratoma	53.13	49.32	59.53	57.31	55.36
	Theca_cell_tumor	64.39	55.8	66.35	65.8	65.09
	Simple_syst	43.13	38.73	45.08	53.57	48.22
	Normal_ovary	52.9	25.24	52.07	51.8	53.82
	Mucinous_cystadenoma	41.68	34.09	40.84	43.55	44.28
	Hight_grade_serous	41.77	38.29	44.82	46.13	43.95
	Average	56.87	49.70	58.65	**59.75**	58.66

Table 2. The results of label prediction (*ACC*) on the OUT_2D sub-set of MMOTU dataset.

Measurement/ Labels/ Methods		**PSPNet (%)**	**U-net (%)**	**DANet (%)**	**Deeplabv3 (%)**	**PSANet (%)**
ACC	Chocolate cyst	**98.84**	**98.65**	**98.93**	**98.92**	**98.75**
	Serous cystadenoma	80.84	74.3	93.56	72.81	78.28
	Teratoma	66.54	68.94	75.87	77.96	70.81
	Theca cell tumor	71.61	66.79	77.84	77	75.15
	Simple syst	52.4	49.73	58.24	61.5	56.28
	Normal ovary	55.49	34.73	61.43	58.03	60.03
	Mucinous cystadenoma	50.71	42.44	50.47	55.05	57.82
	Hight grade serous	62.51	54.59	56.17	58.66	63.38
	Average	67.37	61.27	**71.56**	69.99	70.06

Figure 4 shows the results of segmenting and correctly predicting the labels of ovarian tumors. The upper row shows the original ultrasound images and the resulting segmentation and label prediction of "Chocolate cyst", respectively. The lower row shows the original image and the resulting segmentation and

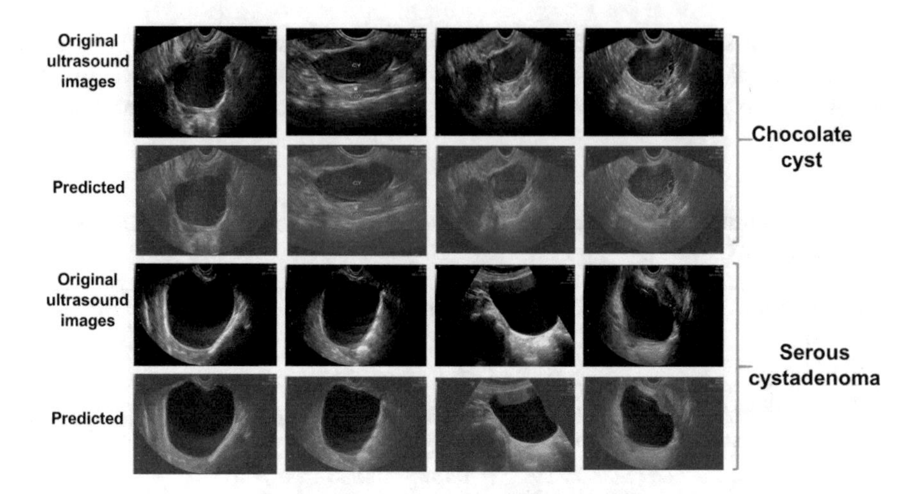

Fig. 4. Segmentation results from ovarian tumors images.

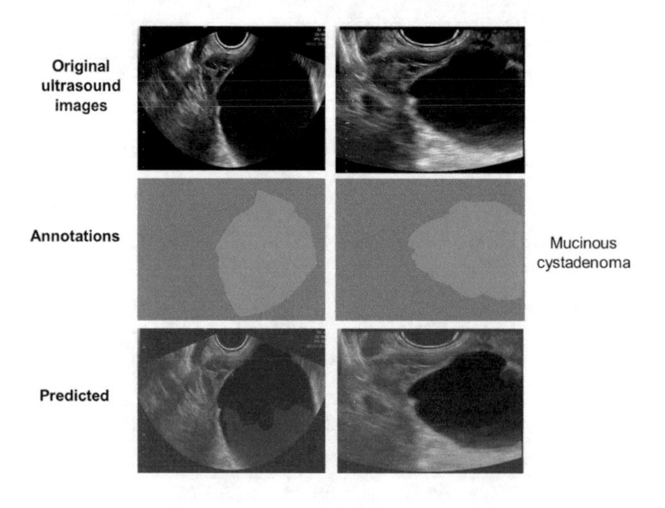

Fig. 5. Illustrating the results of false segmenting and wrong predicting the labels of ovarian tumors.

label prediction of "Serous cystadenoma". Where the ovarian tumor region is segmented and shown in the color of the original data is opacity.

Figure 5 shows the results of false segmentation and mislabelling of ovarian tumors. The tumor is a contiguous area and bears the label "Mucinous cystadenoma", segmentation results are divided into two regions and misclassified to another label.

Figure 6 shows the segmentation and label prediction results of "Serous cystadenoma" with the CNNs models: PSPNet, U-net, DANet, Deeplabv3, and PSANet. Figure 6 shows the segmentation results and the label prediction of the PSPnet and U-Net models is wrong, the PSPnet model has a small segmented and predicted "Theca cell tumor", the result of the U- net has a large area segmented and predicted as "Theca cell tumor". This also shows that segmentation results and label predictions of these two methods are lower than that of other CNN models, as shown in Table 1, and Table 2.

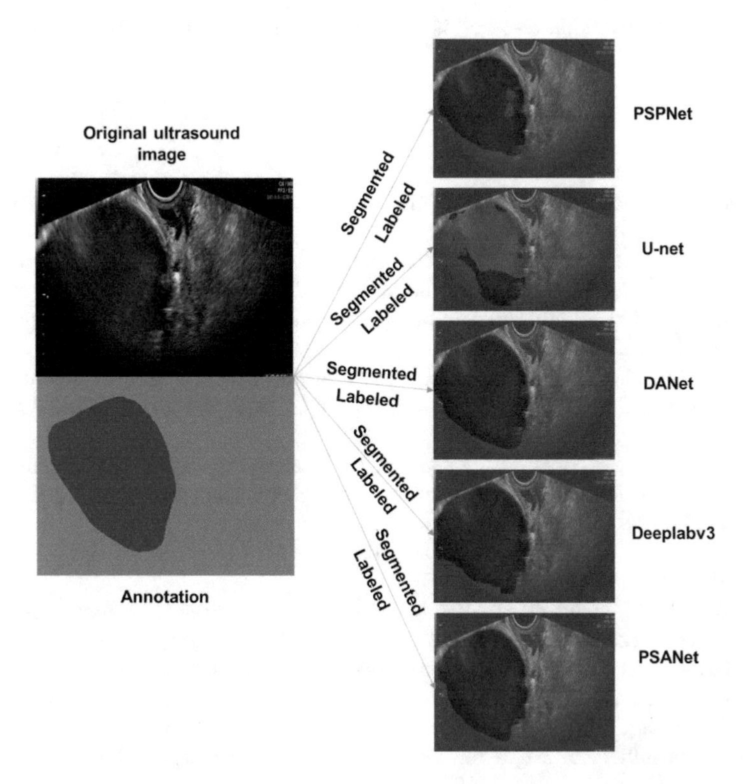

Fig. 6. Illustrated results of segmentation and prediction of ovarian tumor labeled with "Serous cystadenoma" using models: PSPNet, U-net, DANet, Deeplabv3, and PSANet.

5 Conclusion and Future Works

Ovarian cancer is the leading cause of death in women, so early diagnosis is a very important medical issue. In this paper, we performed a follow-up comparative study based on CNNs for ovarian tumor partitioning. In this study, two main problems were solved: segmentation and label prediction of eight ovarian tumor types in the OTU_2D subset of the MMOTU database. The average result for ovarian tumor segmentation was no greater than 60% and the tumor classification result was no more than 72%. These results are very low compared to the practical requirements to use artificial intelligence technology to support the detection and diagnosis of ovarian cancer. This study also shows that there are still great challenges to building a support system for the diagnosis of ovarian cancer, especially with malignant ovarian tumors. In the future, we will continue to improve the CNNs for ovarian tumor instance segmentation and test on other ovarian tumor datasets.

Acknowledgment. This research is funded by Tan Trao University in Tuyen Quang province, Vietnam.

References

1. American Cancer Society, Ovarian cancer statistics (2021). https://www.medicalnewstoday.com/articles/323798#how-common-is-it. Accessed 28 Feb 2023
2. Bierig, S.M., Jones, A.: Accuracy and cost comparison of ultrasound versus alternative imaging modalities, including CT, MR, PET, and angiography. J. Diagnostic Med. Sonography **25**(3), 138–144 (2009). https://doi.org/10.1177/8756479309336240
3. Yan, J., Lv, D., Cui, Y.: A novel segmentation approach for intravascular ultrasound images. J. Med. Biological Eng. **37**(3), 386–394 (2017), issn: 21994757. https://doi.org/10.1007/s40846-017-0233-5
4. Zhao, Q., Lyu, S., Bai, W., et al.: A multi-modality ovarian tumor ultrasound image dataset for unsupervised cross-domain semantic segmentation, CoRR, vol. abs/2207.06799 (2022)
5. Zhao, H., Shi, J., Qi, X., Wang, X., Jia, J.: Pyramid scene parsing network. In: CVPR (2017)
6. Wightman, R., Touvron, H., Jégou, H.: Resnet strikes back: an improved training procedure in timm. arXiv preprint arXiv:2110.00476 (2021)
7. Jun Fu, J.L., et al.: Dual attention network for scene segmentation. In: The IEEE Conference on Computer Vision and Pattern Recognition (CVPR) (2019)
8. Xie, E., Wang, W., Yu, Z., Anandkumar, A., Alvarez, J.M., Luo, P.: Segformer: Simple and efficient design for semantic segmentation with transformers (2021). arXiv preprint arXiv:2105.15203
9. Ronneberger, O., Fischer, P., Brox, T.: U-Net: convolutional networks for biomedical image segmentation. In: Navab, N., Hornegger, J., Wells, W.M., Frangi, A.F. (eds.) MICCAI 2015. LNCS, vol. 9351, pp. 234–241. Springer, Cham (2015). https://doi.org/10.1007/978-3-319-24574-4_28
10. Chen, J., Lu, Y., Yu, Q., et al.: Transunet: transformers make strong encoders for medical image segmentation (2021). arXiv preprint arXiv:2102.04306
11. Yu, C., Gao, C., Wang, J., Yu, G., Shen, C., Sang, N.: Bisenet v2: bilateral network with guided aggregation for real-time semantic segmentation. Int. J. Comput. Vis., 1–18 (2021)
12. Enshaei, C.N.R.A., Edmondson, R.J.: Artificial intelligence systems as prognostic and predictive tools in ovarian cancer. Ann. Surg. Oncol. **155**, 124–138 (2015)
13. Li, H., Fang, J., Liu, S., et al.: CR-Unet: a composite network for ovary and follicle segmentation in ultrasound images. IEEE J. Biomed. Health Inf., 1–10 (2019)
14. Jin, J., Zhu, H., Zhang, J., et al.: Multiple U-net-based automatic segmentations and radiomics feature stability on ultrasound images for patients with ovarian cancer. Front. Oncol. **10**, 1–8 (2021), issn: 2234943X. https://doi.org/10.3389/fonc.2020.614201

15. Navab, N., Hornegger, J., Wells, W.M., Frangi, A.F. (eds.): MICCAI 2015. LNCS, vol. 9351. Springer, Cham (2015). https://doi.org/10.1007/978-3-319-24574-4
16. He, K., Zhang, X., Ren, S., Sun, J.: Deep residual learning for image recognition. In: 2016 IEEE Conference on Computer Vision and Pattern Recognition (CVPR) (2016). https://doi.org/10.1002/chin.200650130
17. Yurtkulu, S.C., Sahin, Y.H., Unal, G.: Semantic segmentation with extended DeepLabv3 architecture. In: 27th Signal Processing and CommunicationsApplications Conference, SIU 2019, pp. 10–13 (2019). https://doi.org/10.1109/SIU.2019.8806244.
18. Zhao, H., Zhang, Y., Liu, S., Shi, J., Loy, C.C., Lin, D., Jia, J.: PSANet: point-wise spatial attention network for scene parsing. In: Ferrari, V., Hebert, M., Sminchisescu, C., Weiss, Y. (eds.) ECCV 2018. LNCS, vol. 11213, pp. 270–286. Springer, Cham (2018). https://doi.org/10.1007/978-3-030-01240-3_17
19. Berman, M., Triki, A.R., Blaschko, M.B.: The lovász-softmax loss: a tractable surrogate for the optimization of the intersection-over-union measure in neural networks. In: Proceedings of the IEEE Conference on Computer Vision and Pattern Recognition (CVPR), June 2018
20. M. Contributors, MMSegmentation: Openmmlab semantic segmentation toolbox and benchmark (2020). https://github.com/open-mmlab/mmsegmentation. Accessed 28 Feb 2023
21. Cordts, M., Omran, M., Ramos, S., et al.: The cityscapes dataset. In: CVPR Workshop on The Future of Datasets in Vision (2015)
22. IOTA, International ovarian tumor analysis. https://iotagroup.org/. Accessed 28 Feb 2023, 1999

VR VanMieu: The Interactively Virtual Temple of Literature

Đinh Ngọc Vân(✉), Ngô Hồng Giang, and Ma Thị Châu

HMI Laboratory, University of Engineering and Technology, Vietnam,
144 Xuân Thủy, Dịch Vọng Hậu, Cau Giay, Hanoi, Vietnam
19021390@vnu.edu.vn

Abstract. Historical heritage is representative stowage for our memories and provides us with a physical means to connect to past values. Recently, digital preservation of heritage has attracted many researchers. Virtual reality (VR) technology is the ability to simulate a complete and interactive virtual environment for audiences to get a more comprehensive and concrete presentation. In this paper, we propose a process of building a VR VanMieu application - a virtual heritage environment using VR technology to promote historical knowledge of the Temple of Literature.

Keywords: Virtual Reality · Vitual Reality application · The Temple of Literature

1 Introduction

A virtual world is a computer-simulated environment, and virtual reality (VR) can be defined as a simulated experience. It has helped users to solve many difficult issues that cannot be resolved in the physical world. It also enables individuals to grasp anything in the virtual world better and increases their interest in it. For instance, by simply clicking, users can visit several well-known locations like historically virtual museums. These product series are gaining popularity as a means of learning and experiencing new things. By creating such a product, users will have a new channel through which to acquire information. History plays an important part in our life. Virtual History Museum is an application that helps users discover models and pictures that partially reflect the stages that people have gone through.

Vietnam has a rich and diverse history and culture spanning more than 1000 years with thousands of historical relics. In particular, The Temple of Literature (Van Mieu in Vietnamese) is a complex on the list of special National Monuments of Vietnam, the pride of the Vietnamese people. With a rich culture and long history, the Temple of Literature relic has become an attractive destination in the tours and discovery of Hanoi tourism. The Temple of Literature is a long-standing, prominent, and diverse historical relic including many unique and detailed architectures built from the Ly Dynasty in Hanoi, Vietnam.

N. T. Nguyen et al. (Eds.): CITA 2023, LNNS 734, pp. 274–284, 2023.
https://doi.org/10.1007/978-3-031-36886-8_23

In this paper, we created a VR VanMieu application with the desire to promote the image of the Temple of Literature to people, both domestic and foreigners who have not had the opportunity to visit it directly or are looking to learn about tourist attractions when visiting Vietnam.

2 Background and Related Works

2.1 Virtual Reality

VR technology is described as a simulation technology of a realistic virtual environment that is created using interactive software and hardware and can be controlled by body movements [1]. A believable, interactive 3D computer-created world that the users can feel immersive as they really are there, both mentally and physically. The user's involvement in virtual environments is created by the immediacy provided by the VR hardware used, where no other medium is visible between the user and the visible virtual world, while multimodal interaction involving visual and tactile feedback ensures the feeling of immersion [2]. The simultaneous development of consumer-grade VR headsets, 3D game engines, and mobile devices, and the increase of computational power for processing multimedia content have allowed the emergence of consumer-grade VR technologies in various domains such as education and entertainment [3]. The literature combining works at the frontier between education, computer science, and engineering has significantly developed to propose a wide range of studies and applications related to immersive learning and cultural heritage [3, 4]. It is two-way interaction: as you respond to what you see, what you see responds to you: if you turn your head around, what you see or hear in VR changes to match your new perspective.

With VR technology, several virtual objects and their related material (multimedia data) were placed in a virtual environment in which users could explore their meaning. To evaluate a VR application's effectiveness, authors in [5] conducted an experiment comparing how respondents remember information presented in different forms of data: visual, audio, textual and kinaesthetic. They showed that the test score of the VR application experiment depends on the respondent's experience with VR. The more users use VR the higher score they get. The benefits of virtual reality go beyond just simulating the real world. For instance, through the use of virtual reality technology, we are able to recreate historic buildings that have been demolished and experience them under the dim light of oil lamps rather than more contemporary lighting sources.

There are some famous programs to create and make virtual worlds possible for users to experience, such as Unity 3D, Blender, Maya, Unreal, etc. Maya, Blender, and Unity are three great choices among them. Maya is comparable to Blender, however, it is more expensive. They have commonly been used in scene and model construction. Unity 3D is a game engine in which developers can design interactions. People can also create 3D models in Unity, then manipulate their movement, 360 rotation, check information, etc.

2.2 Unity

Unity, a power game engine, supports 2D and 3D graphics, and stands out for its ability to build games that run on multiple platforms. In Unity, the functions are written mainly in CSharp language. Unity is widely used and has a large user community and has a lot of resources stored in the Asset Store. Asset Store contains resources for graphics, particle effects, sound, etc. Users can create packages themselves, and send them to the Asset Store, packages can be free or paid depending on the owner of the product. Products. Therefore, with a large user community, Asset Store is also becoming richer and richer. The Unity package helps users share and reuse Unity projects and content. Unity Assets on the Unity Asset Store are available in a package. Packages are collections of files and data from Unity projects or project components such as assets, shaders, textures, icons, scripts, plug-ins, and others. Users can open Package Manager to view and manage all their packages. Besides, users can add the package they want by adding the package from the Asset Store to the product, then downloading the package, and then importing the package into Unity. That's the reason why we have developed the VR VanMieu application based on the Unity platform.

3 A Process of Building VR VanMieu Application

Different phases of building a virtually interactive environment application were presented and discussed by Paulauskas, L et al. [5] for researchers and virtual application developers when they build a virtual heritage application. Based on the phases, we

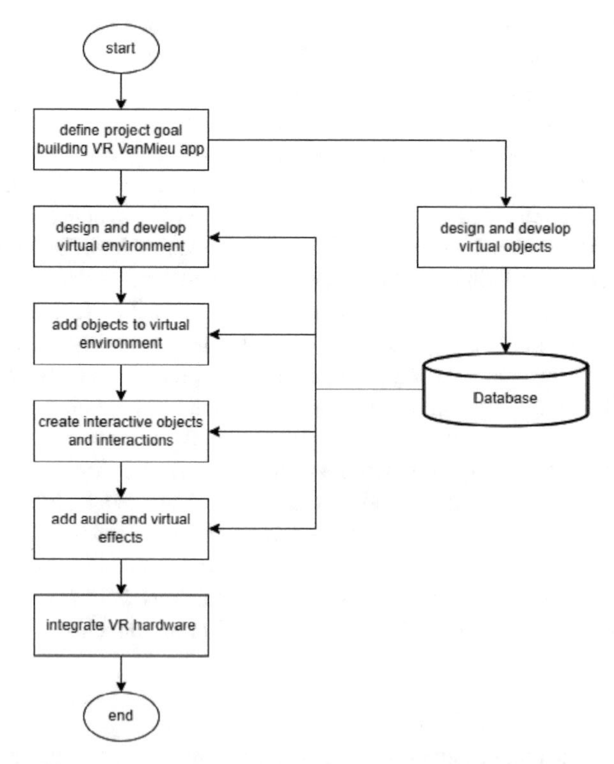

Fig. 1. A process of building the VR VanMieu application.

proposed a process of 7 phases (Fig. 1) in creating the VR VanMieu application and analyzing the experiment when the user uses the application.

The Temple of Literature is the pinnacle and symbol of Vietnamese Confucian education and is also one of the most famous temples in Vietnam. Vietnamese people are inherently studious and attach great importance to academics. Temple of Literature is the first university to train and organize a talent competition to serve the country. This is the place where many talented people have made great contributions to the country throughout history. There are also regular cultural events to promote learning. We built VR VanMieu to promote The Temple of Literature to more domestic and foreign audiences. The Temple of Literature (Fig. 2) is located in the heart of Hanoi capital, surrounded by four main streets of Dong Da district: Nguyen Thai Hoc, Ton Duc Thang, Van Mieu, and Quoc Tu Giam streets. The main gate of the Temple of Literature is located at 58, Van Mieu street.

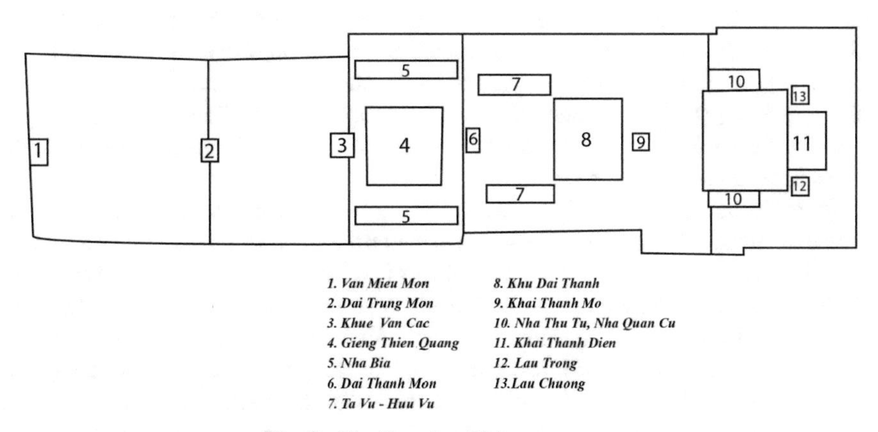

Fig. 2. The Temple of Literature map

With the desire for the application to be realistic, vivid, and easy to use. We use the kinds of data that make it possible for users to experience the most realistic and intuitive. Data types include images, 3D models, texts, and audio. Each data type has a different processing method.

Images are divided into two categories: images used to build the application and images used in the application. We went on a field trip and took pictures of the Temple of Literature. We focus on architecture and environments, these are two important things to create a virtual Temple of Literature. To create the most realistic environment, we captured all areas of the Temple of Literature in as much detail as possible. We use images of architectures and environments to create 3D models for visiting virtual reality environments. In addition, we cut patterns that are difficult or time-consuming to build to incorporate into the models.

We use images of architectures and environments to create 3D models for visiting virtual reality environments. Then, we cut patterns that are difficult or time-consuming to build to incorporate into the models. In addition, we use Unity's built-in environment packages such as plants, flowers, etc. to make the environment come to life.

Users can view information about the Temple of Literature through text. There are two types including page flipping and scrollable readers.

After we collect written information, we record that information into audio. This gives users more ways to access information. In addition, the background sound also helps the user's experience become better when visiting the Temple of Literature.

We create the Temple of Literature environment following the real map (Fig. 2) by aligning 3D models to their real-world locations. For the outdoor models, we arrange the environment to become harmonious so that users can easily see the whole view of the Temple of Literature. Besides, we put some more supporting models for our functions such as the info panel, weather switch column, etc.

4 Architecture and Campus of the Temple of Literature

The VR VanMieu application consists of elements that mimic the real-world Temple of Literature. The 3D models are constructed from real objects with high details and include many architectures such as Phuong Dinh, Nha Dai Bai, Nha Bia, Khue Van Cac, Dai Trung Mon, Den Khai Thanh, Dai Thanh Mon, Van Mieu Mon, and so on (Fig. 3, 4). Besides that, other objects are built to make the environment more lively and realític, such as a small lake, trees, separating walls, etc.

About the buildings of the Temple of Literature: We constructed the buildings from genuine schematics and then adjusted them to match the correct architectural proportions. The majority of buildings are made out of blocks that have been assembled and arranged. Other methods must be used to refine some other elements, such arched doors. Then, we add materials to structures. We come to observe reality and develop fresh, high-quality materials that, in terms of textures, are most like the genuine thing. Ultimately, we assembled the buildings based on maps and actual observations. Most of the buildings in the Temple of Literature have been built since the very last decade with that period style, some are reconstructed after being destroyed by the war. Thus, the main materials

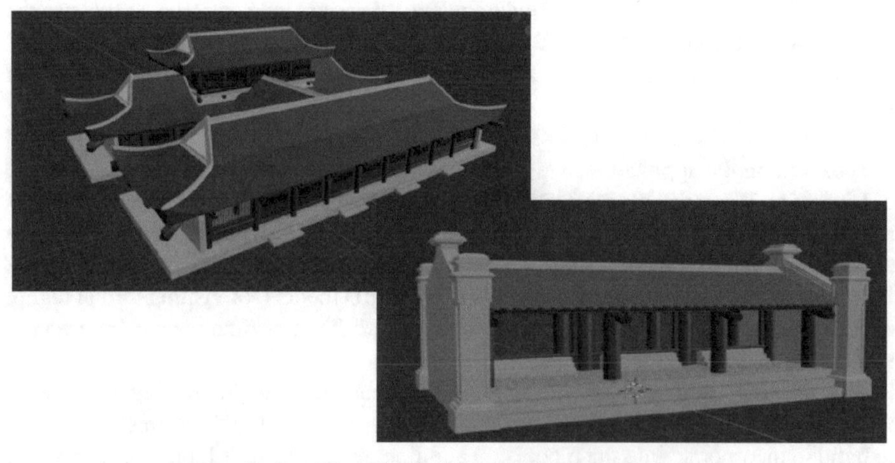

Fig. 3. Den Khai Thanh (left) Dai Thanh Mon (right)

for the buildings are old wood, red brick for the roof, and concrete color for the steps and walls.

Fig. 4. Khue Van Cac (left) Van Mieu Mon (right)

In addition to buildings, we have built Thien Quang well (Fig. 5), Sam drum (Fig. 6, Left), and Thai Hoc bell (Fig. 6, Right). Thien Quan well, i.e. "the well that illuminates the sky", is also known as Van Tri. The name Thien Quang is given to the well with the implication that people want to obtain the quintessence of the universe, enlighten intellectuals, and improve the good human qualities of people. Thien Quang well and Khue Van Cac, one square and one round, collect all the quintessence of heaven and earth at this majestic educational and cultural center. Sam drum is one of the two biggest drums in Vietnam. The reason for this name is because when beating, the drum sounds loud like thunder, the sound is resounding and majestic. It is used to serve the 1000th anniversary of Thang Long Hanoi. The Thai Hoc bell is also created to celebrate the 1000th anniversary.

Fig. 5. Thien Quang well

The body of the bell is engraved with the history of the Temple of Literature and the nation's studious spirit, and at the same time reminding generations of descendants to study, study more, study forever, study for the prosperity of the Fatherland, study for the happiness of the whole people. Unlike buildings which are largely composed of blocks, things require more techniques to create depending on the difficulty and details required.

Fig. 6. Sam drum (left), Thai Hoc bell (right)

Fig. 7. VanMieu VR environment

After constructing the building and orther models, we embed these models into the built environment and have a VanMieu VR environment (Fig. 7).

5 Interactive Design

We designed a VR VanMieu application with 3 scenarios: The Temple of Literature tour, please calligraphy from the master and participate in the quiz game. Each scenario is contained in a box with the image and name of that scenario, this means each box represents each scene. Clicking on the box will switch to the corresponding scene. Those boxes include: Visiting the Temple of Literature tour is where players visit and experience the virtual reality environment of the Temple of Literature. For the VR VanMieu, we created a menu that includes visiting the Temple of Literature as the main scenario and two secondary scenarios are quiz games and please calligraphy.

The Temple of Literature tour is where players visit and experience the virtual reality environment of the Temple of Literature. In each scenario, the developed functions include moving, rotating 360°, changing the sky according to the weather and time of day, playing quiz games, viewing the user's location via the minimap in the right corner, reading information about Temple of Literature on bulletin boards or books, listening information about Temple of Literature, please calligraphy, the user's storage bag.

The second scenario is the quiz game (Fig. 8). Users can play quiz games to check the information they have received during the visit to the Temple of Literature. Besides, users can also exchange rewards with coins in the quiz game.

Finally, users can use coins that are achieved from quiz games to exchange the calligraphy of the master. Calligraphy paintings have many different designs such as Lotus and butterfly, Peach blossom and dragonfly, etc. Calligraphic letters also have many different characters such as Hoc, Phuc, Loc, Tho.

Fig. 8. Quiz game

Humans in VanMieu VR are divided into two types: play-character and non-play characters (NPCs).

User plays the role of play-character. As a result, users will have a more authentic experience. The user has three main interactions: moving, rotating, and interacting with objects. Users can navigate in the environment to left, right, up, and down, also, when the user turns the other way, users will rotate to see the other side. We provide the function of changing between first-person and third-person perspectives to suit the needs and preferences of users. Besides, when a user interacts with objects in the environment, each object will have a different form of interaction. We have handled collisions with external objects. Therefore, users can easily interact. NPCs play the role of inserting captivation and realism to the environment. Each NPC will have different behavior such as sightseeing, walking, answering the phone, etc. With text and images, we graft to the surface of the 3D object so that the user has a more intuitive image. With page-turning text, the text will be shaped like a book, the user will turn the page to read the information. Compared to the dashboard, users scroll down to see texts. As for the sound, when clicking on the sound box model, the user can turn it on and off conveniently. Besides, when users touch the gate, it will open.

The 3D models will act as transitions to other types of data such as text or images when the user clicks.

With text and images, we graft to the surface of the 3D object so that the user has a more intuitive image. With page-turning text, the text will be shaped like a book, the user will turn the page to read the information. Compared to the dashboard, users scroll down to see texts. As for the sound, when clicking on the sound box model, the user can turn it on and off conveniently. Besides, when user touch the gate, it will open.

The Temple of Literature is added with soothing sounds to help users enjoy the process of visiting the virtual environment and some animations such as the door opening when the user approaches or the water ripples.

6 Experiment and Evaluation

6.1 Experiment

For the virtual reality Temple of Literature, the user will first see the menu. Users can choose scenarios in which they want to experience.

Here users can move and see the whole view of the Temple of Literature. Besides, players can also listen to information about the architecture of the Temple of Literature, see the minimap and change the sky according to the time of day or the weather. In addition, during the virtual reality tour of the Temple of Literature, users will be provided with information about the Temple of Literature. After that, players can play QuizGame to consolidate, recall the information when visiting and receive rewards. With the please calligraphy scenario, users can exchange the teacher's text when exchanging rewards from QuizGame. Users can view and manage what they can exchange in the user bag.

Although it is possible to access any screen, first-time players should follow the flow below the image. When that flow is done right, users will experience the virtual reality Van Mieu Quoc Tu Giam with ease and know what to do as they go through each screen one by one.

Questionnaire. We provide users with a survey (Table 1) for users to evaluate issues related to the application such as compatibility.

Table 1. Some questions in the survey about experience VanMieu VR

Question type	Question
Model	Which model do you think is the best?
	What aspects should we adjust to make the model more attractive?
Interaction	Please rate how you feel about the functions of the application
	Which function do you find the most attractive?
User experience	During your experience with the product, did you experience any symptoms?

6.2 Practical Experience

We surveyed the application with a scale of 50 people, who are students of the University of Technology. We examine 3 characteristics, which include significance, attractiveness, and convenience. With all three, we got pretty good results (Fig. 9). The app received reviews for its harmonious colors, pleasant sounds, concise information, and high historical significance. Most users in the process of experience find it quite convenient, but there are some people who are dizzy due to their location.

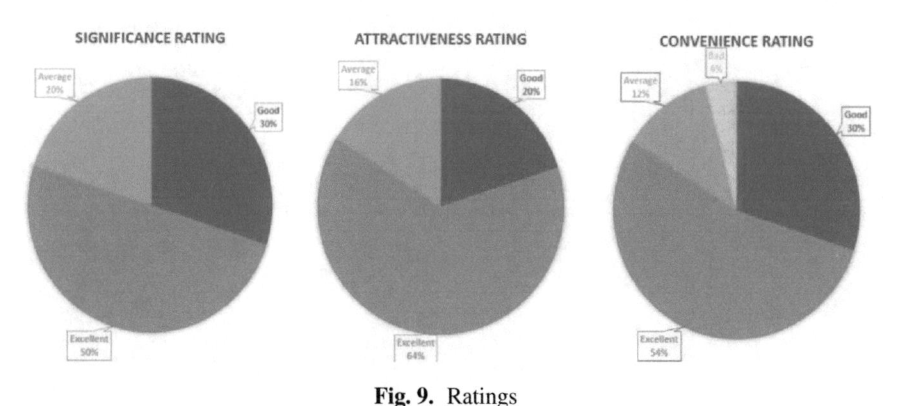

Fig. 9. Ratings

7 Conclusion

This paper presents an interactive virtual reality application about the Temple of Literature, a famous heritage of Vietnam. Here, we present the process and method of building the application, including the main function of visiting the virtual reality environment and some other functions such as quiz games, please calligraphy. Users have positive feedback after trying the product demo. This work has been supported by University of Engineering and Technology, project number CN.22.05 – "Exploring the architecture of Van Mieu Quoc Tu Giam with virtual reality glasses" (Khám phá kiến trúc quần thể Văn Miếu Quốc Tử Giám bằng kính thực tại ảo).

References

1. Coyne, L., Takemoto, J.K., Parmentier, B.L., Merritt, T., Sharpton, R.A.: Exploring virtual reality as a platform for distance team-based learning. Curr. Pharm. Teach. Learn. **10**(10), 1384–1390 (2018). https://doi.org/10.1016/j.cptl.2018.07.005. Epub 2018 Jul 21. PMID: 30527368
2. Sanfilippo, F., Blažauskas, T., Girdžiūna, M., Janonis, A., Kiudys, E., Salvietti, G.: A multimodal auditory-visual-tactile e-learning framework. In: Sanfilippo, F., Granmo, O.C., Yayilgan, S.Y., Bajwa, I.S. (eds.) Intelligent Technologies and Applications. INTAP 2021. Communications in Computer and Information Science, vol. 1616. Springer, Cham (2022). https://doi.org/10.1007/978-3-031-10525-8_10
3. Cecotti, H.: Cultural Heritage in Fully Immersive Virtual Reality. Virtual Worlds, pp. 82–102. Cultural Heritage in Fully Immersive Virtual Reality (2022). https://doi.org/10.3390/virtualworlds1010006
4. Zara, J.: Virtual reality and cultural heritage on the web, pp. 101–112. ResearchGate (2014)
5. Paulauskas, L., Paulauskas, A., Blažauskas, T., Damaševiˇcius, R., Maskeliunas, R.: Reconstruction of industrial and historical heritage for cultural enrichment using virtual and augmented reality. Technologies **11**, 36 (2023). https://doi.org/10.3390/technologies11020036

Neural Sequence Labeling Based Sentence Segmentation for Myanmar Language

Ye Kyaw Thu[1,2](✉), Thura Aung[2], and Thepchai Supnithi[1]

[1] Language and Semantic Technology Research Team, NECTEC, Khlong Nueng, Thailand
{yekyaw.thu,thepchai.supnithi}@nectec.or.th

[2] Language Understanding Laboratory, Mandalay, Myanmar

Abstract. In the informal Myanmar language, for which most NLP applications are used, there is no predefined rule to mark the end of the sentence. Therefore, in this paper, we contributed the first Myanmar sentence segmentation corpus and systematically experimented with twelve *neural sequence labeling* architectures trained and tested on both sentence and sentence+paragraph data. The word LSTM + Softmax achieved the highest accuracy of 99.95% while trained and tested on sentence-only data and 97.40% while trained and tested on sentence + paragraph data.

Keywords: Sentence Segmentation · Neural Sequence Labeling · Myanmar language · CRF · NCRF^{++} · CNN · Bi-LSTM

1 Introduction

Sentence segmentation can be defined as the task of segmenting text into sentences that are independent units and grammatically linked words. In the formal Myanmar language, sentences are grammatically correct and typically end with a "။" pote-ma. Informal language is more frequently used in daily conversations with others due to its easy flow. There are no predefined rules to identify the ending of sentences in informal usages for the machine itself. Some of the applications based on conversations, e.g, Automatic Speech Recognition (ASR), Speech Synthesis or Text-to-Speech (TTS), and chatbots, need to identify the end of sentences. To address this problem, we used the sequence labeling approach in which each unit is labeled, and the pairs of unit and label are trained using a supervised learning algorithm. In this paper, we studied the neural sequence labeling models using character and word sequence representations with Convolutional Neural Networks (CNN) and Bi-directional Long Short-Term Memory (LSTM) Recurrent Neural Networks.

2 Related Works

With the sequence labeling approach, Win Pa Pa et al. [1] examined the effectiveness of Conditional Random Fields (CRFs) for Myanmar word segmentation.

N. T. Nguyen et al. (Eds.): CITA 2023, LNNS 734, pp. 285–296, 2023.
https://doi.org/10.1007/978-3-031-36886-8_24

Furthermore, there are additional text segmentation approaches for the Myanmar language. Ye Kyaw Thu et al. [2] proposed seven different word segmentation schemes for statistical machine translation systems. However, there were no methodological sequence labeling studies for sentence segmentation in the informal Myanmar language.

Previous researchers have worked on sentence segmentation problem by using rule-based approaches (e.g., Lingua::EN::Sentence [3], which is a Perl module for English sentence segmentation) and machine learning based sequence labeling approaches like Conditional Random Fields (CRFs) [4] and Hidden Markov Models (HMM) [5].

Sadvilkar et al. introduced a multilingual rule-based sentence segmentation tool called PySBD [6] in which Myanmar sentence segmentation is available but it is only useful for formal usages because sentence segmentation is based on the sentence delimiter "။" pote-ma, which is not used in informal communications. Deep learning-based sequence labeling, also known as the neural sequence labeling approach is the current state-of-the-art approach for sequence labeling. Yang et al. [7] investigated the design challenges of building effective as well as efficient neural sequence labeling systems. For Myanmar sentence segmentation, we examined the performances of state-of-the-art neural sequence labeling models.

3 Corpus Development

This section describes the information of *mySentence* tagged corpus, as well as an overview of word segmentation and tagged text data annotation.

3.1 Corpus Information

Myanmar NLP researchers are facing many difficulties arising from the lack of resources; in particular parallel corpora are scarce [8]. For this reason, we annotated text data manually with *mySentence* tag information. The myPOS corpus version 3.0 [9] consists of 43,196 meaningful word sequences written in formal and informal formats from various domain areas and the whole corpus has already been word-segmented manually. But not all sequences are used for the experiments as sequences with only one word are ignored except for interjections.

We also collected Myanmar sentences and paragraphs from different online resources such as Facebook and Wikipedia and from the short stories available on Facebook pages [10, 11].

Table 1 shows resources of data collected to use for building mySentence corpus for sentence segmentation.

3.2 Word Segmentation

In the Myanmar language, spaces are used only to segment phrases for easier reading. There are no clear rules for using spaces in the Myanmar language. The myPOS version 3.0 corpus has been already word-segmented manually.

Table 1. Data Resources of the corpus

Data Resources	Sentence	Paragraph
myPOS (version 3.0) [9]	40,191	2,917
Covid-19 Q&A [13]	1,000	1,350
Shared By Louis Augustine Page [10]	547	1,885
Maung Zi's Tales Page [11]	2,516	581
Wikipedia	2,780	1,060
Others	93	672
Total	47,127	8,465

We used myWord word segmentation tool [12] to do word segmentation on our manually collected data and checked word segmentation results manually. We applied the word segmentation rules proposed by Ye Kyaw Thu et al. in myPOS [14] corpus. The segmented example for the Myanmar sentence (How are you, Sayar?) is shown as follows:

Unsegmented sentence: ဆရာနေကောင်းလား

Word segmented sentence: ဆရာ|နေကောင်း|လား

3.3 Corpus Annotation

After the word segmentation, we annotated the word sequences in the corpus into a tagged sequence of words. Each token within the sentence is tagged with one of the four tags: B (Begin), O (Other), N (Next), and E (End).

The beginning word which is on the left of the sentence in the Myanmar language is tagged B and the ending word of each sentence is tagged E. The three words left to the ending words are tagged N while other words in the sentence are tagged O. Tagging process was done manually for both sentences and paragraphs in the dataset.

Table 2. Statistics of tags in the mySentence corpus

Tag	Frequency	Proportion
B	47,264	7.24%
E	48,690	7.33%
N	137,592	20.46%
O	436,942	64.97%

Table 2 shows the statistics of mySentence tags in the corpus. If there are more than two /E tags in a sequence, it is considered to be a paragraph. The tagged example Myanmar sentence, (I get bored.) is shown as follows:

Untagged sentence: ကျွန်တော် ပျင်း လာ ပြီ

Tagged sentence: ကျွန်တော်/B ပျင်း/N လာ/N ပြီ/E

The tagged example Myanmar language paragraph, (I am sorry. I like drama films more.) is shown as follows:

Untagged paragraph: တောင်းပန် ပါ တယ် ကျွန်တော် က အချစ် ကား ပို ကြိုက် တယ်

Tagged paragraph:
တောင်းပန်/B ပါ/N တယ်/E ကျွန်တော်/B က/O အချစ်/O ကား/N ပို/N ကြိုက်/N တယ်/E

4 Methodology

For neural sequence labeling, Pytorch-based framework NCRF^{++} [15], a toolkit with flexible running time and a customizable configuration file, was used. It is designed for the rapid implementation of different neural sequence labeling architectures with a CRF or softmax inference layer. NCRF^{++} can be regarded as a neural version of a famous statistical CRF framework, CRF^{++}.

As shown in Fig. 1, NCRF^{++} framework supports three layers, i.e., character and word sequence representation layers with CNN and LSTM for feature extractions, and a CRF or Softmax inference layer for predicting.

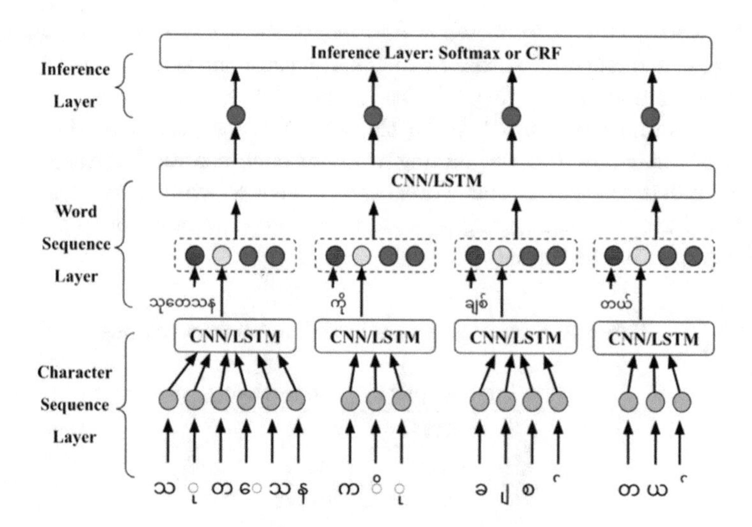

Fig. 1. NCRF^{++} for a Myanmar language sentence "သုတေသန ကို ချစ် တယ်" (I love research.) Green, red, yellow, and blue circles represent character embeddings, word embeddings, character representations, and word representations. The sparse feature embeddings are represented as grey circles.

4.1 Character Sequence Layer

Character features can be represented with character embeddings through neural network models without human-defined hand-engineering. NCRF^{++} supports different character sequence representation approaches such as character CNN, character LSTM, character GRU, and handcrafted word features. In our experiments, we used character CNN and character LSTM, which are state-of-the-art models for sequence representation.

- **Character CNN** used a CNN structure to learn character-level representations. The idea was first introduced by Santos et al. [16] to learn character representations of words for Part-of-Speech tagging.
- **Character LSTM** used a Bi-directional LSTM structure to capture the global feature of the character sequence information. The forward LSTM captures the character sequence information from left to right then right to left and concatenates the final hidden states of two RNNs as the encoder of the input character sequence, i.e., word.

4.2 Word Sequence Layer

Words from word sequences can be represented similarly to character sequences in words. Word sequence information can be captured with word embeddings through CNN and LSTM models. NCRF^{++} supports different word sequence representation approaches such as word CNN, word LSTM and word GRU. In our experiments, we used word CNN and word LSTM, which are state-of-the-art architectures for sequence representation.

- **Word CNN** used a multi-layer CNN on the word sequences to learn word-level representations. If the character sequence layer is used, the character sequence representations and word embeddings are concatenated for word representations.
- **Word LSTM** used a Bi-directional LSTM structure to capture the context information of each word, i.e., the global feature of the word sequence information. The forward LSTM captures the word sequence information from left to right and the backward LSTM in a reversed direction. And calculate the global information of the whole word sequence.

4.3 Inference Layer

The inference layer accepts the word sequence representations from the CNN and LSTM based feature extractors and learns with the assigned label (tag) to predict the correct *mySentence* tags. NCRF^{++} provides two inference functions - Softmax and CRF. In this paper, we examined both of them in order to compare the performances of the different approaches.

- **Softmax** maps the input sequence representations to the label scores, which are used to model the label probabilities of each word and support parallel decoding. In the training process, for classification, NCRF^{++} supported cross-entropy loss.
- **CRF** considers the label dependencies among the predicted segmentation tags that are inherent in the state transitions of finite state sequence models, on which exact inference over sequences can be efficiently performed. The decoding process is done with the Viterbi algorithm by searching the label sequence with the highest probability.

5 Experimental Setup

In this section, we describe data preparation, hyperparameters, and evaluation for the experiments. As shown in Fig. 1, NCRF^{++} framework provides different structure combinations on three levels: character sequence representation, word sequence representation, and inference layer with Softmax or CRF function. We trained and tested all of the combinations on both sentence-level and sentence+paragraph-level data.

5.1 Data Preparation

mySentence corpus was used to prepare two types of data - one containing sentence-only data and the other with sentence+paragraph data. And we split both types of data into training, validation, and test data as shown in Table 3. Here, sent is the abbreviation for sentence-level data and para for paragraph-level data.

Table 3. Dataset splitting for Experiments

	sent	sent+para
train	40,000	47,000
validation	2,414	3,079
test	4,712	5,512

All the training, the validation and the test files need to be in a particular format for NCRF++ to work properly. Therefore, after splitting the corpus, the format of train, validation, and test *mySentence*-tagged corpora were converted into word and tag parallel columns. Both types of data, i.e., sentence-only and sentence+paragraph data, were used for the experiment. The word distribution with Zipf's law [18] between two datasets measured with top 1,000 words for 1-g and 2-g are as shown in Fig. 2 and Fig. 3. The Zipf curves for the two datasets are almost identical and show the similarity of word distributions.

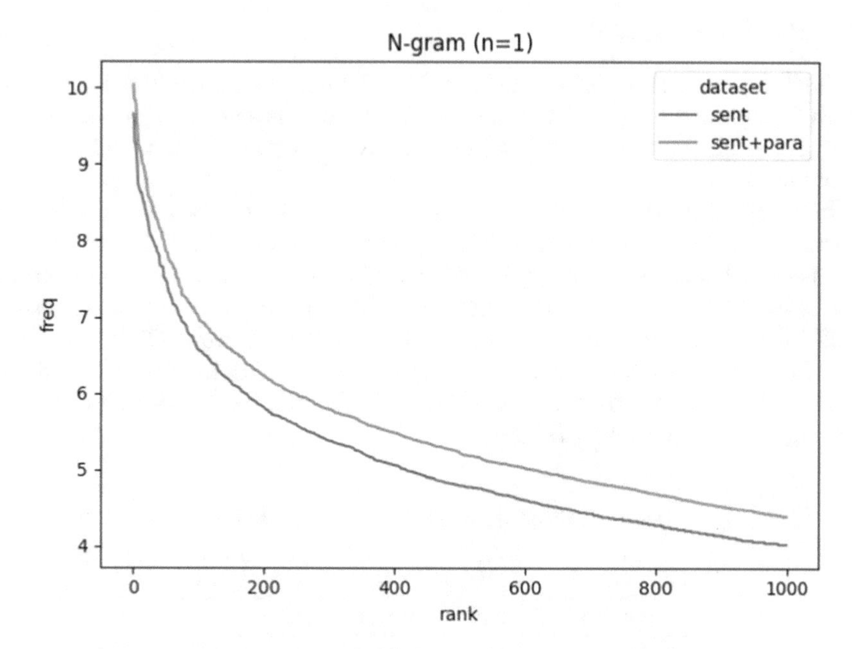

Fig. 2. Zipf's law distributions in 1-g analysis of word between sentence only and sentence+paragraph datasets.

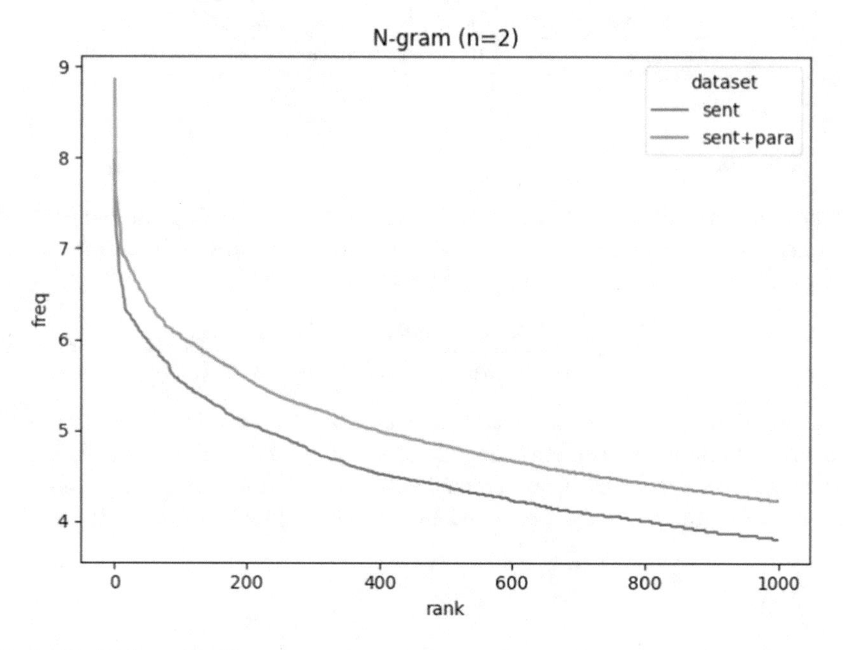

Fig. 3. Zipf's law distributions in 2-g analysis of word between sentence only and sentence+paragraph datasets.

5.2 Hyperparameters

We used character embedding size 30 and word embedding size 50 with 50 hidden layers for character representations and 200 hidden layers for word representations. For both character CNN and word CNN, 4 layers of CNN with kernel size 3 were used.

In order to prevent overfitting and underfitting the models, we used L2 regularization λ. Learning rate $\eta = 0.015$ was used for word LSTM-based models but $\eta = 0.010$ was used for word CNN-based models because a large learning rate could cause the convergence problem. The learning rate η was reduced to 0.008 for word CNN with character CNN and 0.005 for word CNN with character LSTM models respectively. Although a variety of optimizers are available in NCRF^{++}, we only used the mini-batch SGD with the batch size of 10 and learning rate decay of 0.5 (Table 4).

Table 4. Hyperparameters used in experiments

Parameter	Value	Parameter	Value
char emb size	30	word emb size	50
char hidden	50	word hidden	200
CNN layer	4	CNN kernel size	3
dropout rate	0.5	batch size	10
L2 regularization λ	1e–8	learning rate decay	0.05
Epochs	100	Optimizer	SGD

5.3 Evaluation

For evaluation, we conducted several experiments with various models on both sentence-only level and sentence+paragraph level test data. The automatic tagging performance was measured using accuracy.

$$Accuracy = \frac{No.\ of\ Correct\ mySentence\text{-}tags}{No.\ of\ predicted\ tokens\ in\ the\ test\ corpus} \quad (1)$$

In our experiments, the accuracy score is used as an evaluation metric, which measures the number of tokens tagged correctly by the model in relation to the number of tagged tokens. It can be calculated by dividing the number of correct mySentence-tags the model predicted by the total number of predictions.

6 Results and Discussion

This paper contributes the first corpus with a total size of around 55K sentences and paragraphs for Myanmar sentence segmentation. NCRF^{++} architectures were trained and tested on sentence and sentence+paragraph data.

Table 5. Accuracy % comparison of sentence-level models (c = Character, w = Word, sent = sentence and para = paragraph)

	Test Data	wCNN + Softmax	wCNN + CRF	wLSTM + Softmax	wLSTM + CRF
NoChar	sent	99.92	**99.95**	**99.95**	**99.95**
	sent+para	93.58	**93.63**	**93.63**	**93.63**
cCNN	sent	**99.95**	**99.95**	**99.95**	92.02
	sent+para	**93.63**	**93.63**	**93.63**	87.65
cLSTM	sent	**99.95**	**99.95**	**99.95**	99.91
	sent+para	**93.63**	**93.63**	**93.63**	93.59

Tables 5 and 6 show the accuracy comparison between each neural sequence labeling model on different levels of test data. The NCRF^{++} architectures trained on sentence-only data were considered sent models and those trained on sent+para data are sent+para models. Both types of models were not only tested on the sentence data but also on the sent+para test data.

The bold results show the highest accuracies achieved in each test data. According to Tables 5 and 6, the best sentence models achieved **99.95%** accuracy, and the best sentence+paragraph model; wLSTM+Softmax with no character representation achieved the highest value with **97.40%** accuracy. The cross-test results show that the best sentence models achieved **93.63%** accuracy on sentence+paragraph data, and the wCNN+Softmax model with LSTM-based character representation has the highest value with **99.66%** accuracy on sentence-only test data respectively.

Table 6. Accuracy % comparison of sentence+paragraph-level models (c = Character, w = Word, sent = sentence and para = paragraph)

	Test Data	wCNN + Softmax	wCNN + CRF	wLSTM + Softmax	wLSTM + CRF
NoChar	sent	99.41	99.49	99.44	86.44
	sent+para	96.82	96.25	**97.40**	96.61
cCNN	sent	99.26	99.27	74.81	86.44
	sent+para	96.87	96.17	74.69	83.13
cLSTM	sent	**99.66**	99.49	99.49	99.56
	sent+para	96.36	96.04	97.29	96.61

7 Error Analysis

We also did the error analysis using the SCLITE (score speech recognition system output) program from the NIST scoring toolkit (Version 2.4.11) [19]. It is used to align the hypothesis tags with error-free reference tags and calculate the word error rate (WER) [17]. This program shows the recognition rate at the sequence level and word level and also gives the confusion pairs.

For WER calculation, the SCLITE scoring method first aligns the hypothesis and reference sequences and then calculates a minimum Levenshtein distance which weights the cost of correct words (C), insertions (I), deletions (D), substitutions (S), and the number of words in the reference (N).

To know the counts of I, D, C, and S for the tag sequence "B O N N N E", at first, the output (hypothesis) sequence is compared to the reference sequence. Then, WER is calculated based on the counts.

Scores: (#C #S #D #I) 4 2 0 0
REF: B O N N N E
HYP: B N N N N N
Eval: S S

For this example, there are no deletions (D = 0) or insertions (I = 0) and only two substitutions (N = >O) and (O = >N) are happened so the number of correct words C is 4. Using the WER equation, the SCLITE program calculated the WER value for the given example as 16.67%.

Table 7. The Top 5 confusion pairs of sent cCNN+wLSTM+CRF model tested on sent+para test data (87.65% accuracy)

Freq	Confusion Pair (REF ==>HYP)
7078	N ==> O
1951	O ==> N
1229	E ==> O
1224	B ==> O
48	B ==> N

After analysis of confusion pairs, we found out that some of the confusion pairs are related to "O" Tags. Here, from Tables 7 and 8, the confusion pairs of "N == > O", "B == > O" and "O == > N" happened because of false recognition. According to Table 2, 64.97% of the tags in the mySentence corpus are "O" tags. Therefore, the models might predict most of the uncommon words as "O".

Table 8. The Top 5 confusion pairs of sent+para cCNN+wLSTM+Softmax model tested on sent test data (74.81% accuracy)

Freq	Confusion Pair (REF==>HYP)
9790	N ==> O
4193	B ==> O
541	N ==> E
321	E ==> O
64	B ==> N

8 Conclusion

According to the comparison of twelve NCRF^{++} architectures trained and tested on both sentence and sent+para data, we can see that the word LSTM with softmax inference layer and no character representation layer had the best accuracy with sent-level (**99.95%**) as well as sent+para-level (**97.40%**) data. According to the error analysis, most of the errors occurred because the models falsely recognized "O" tags, which have the highest proportion in the dataset. Further, we believe that the neural sequence labeling based sentence segmentation methods presented in this paper are applicable to informal conversations in other languages. In the future, we will investigate the impact of pre-trained word embeddings on this sequence labeling based sentence segmentation task with different embedding settings for a low-resource language - Myanmar. To these ends, we make all our configuration files, code, data, and models publicly available (https://github.com/ye-kyaw-thu/mySentence).

References

1. Pa, W.P., Thu, Y.K., Finch, A., Sumita, E.: Word boundary identification for myanmar text using conditional random fields. In: Zin, T.T., Lin, J.C.-W., Pan, J.-S., Tin, P., Yokota, M. (eds.) GEC 2015. AISC, vol. 388, pp. 447–456. Springer, Cham (2016). https://doi.org/10.1007/978-3-319-23207-2_46
2. Thu, Y.K., Finch, A., Sagisaka, Y., Sumita, E.: A study of myanmar word segmentation schemes for statistical machine translation, In: Proceedings of the 11th International Conference on Computer Applications, pp. 167–179, Yangon, Myanmar (2013). http://onlineresource.ucsy.edu.mm/handle/123456789/2335
3. Lingua::EN::Sentence. https://metacpan.org/release/KIMRYAN/Lingua-EN-Sentence-0.29/view/lib/Lingua/EN/Sentence.pm. Accessed 30 Dec 2022
4. Tomanek, K., Wermter, J., Hahn, U.: Sentence and token splitting based on conditional random fields, In: Proceedings of the 10th Conference of the Pacific Association for Computational Linguistics, vol. 49, p. 57 (2017). https://api.semanticscholar.org/CorpusID:15539970
5. Jurish, B., Würzner, K.-M.: Word and sentence tokenization with hidden Markov models, JLCL **28**(2), 61–83 (2013). https://api.semanticscholar.org/CorpusID:6659476

6. Sadvilkar, N., Neumann, M.: PySBD: pragmatic sentence boundary disambiguation. In: Proceedings of Second Workshop for NLP Open Source Software (NLP-OSS). Association for Computational Linguistics, pp. 110–114. (2020) https://doi.org/10.18653/v1/2020.nlposs-1.15
7. Yang, J., Liang, S., Zhang, Y.: Design challenges and misconceptions in neural sequence labeling. In: Proceedings of the 27th International Conference on Computational Linguistics, pp. 3879–3889. Association for Computational Linguistics, Santa Fe, New Mexico, USA (2018). https://aclanthology.org/C18-1327
8. Thu, Y.K., Chea, V., Finch, A., Utiyama, M., Sumita, E.: A large-scale study of statistical machine translation methods for khmer language. In: Proceedings of 29th Pacific Asia Conference on Language, Information and Computation, Shanghai, China, pp. 259–269 (2015). https://aclanthology.org/Y15-1030.pdf
9. Hlaing, Z.Z., Thu, Y.K., Supnithi, T., Netisopakul, P.: Improving neural machine translation with POS-tag features for low-resource language pairs. Heliyon, vol. 8. (2022). https://doi.org/10.1016/j.heliyon.2022.e10375
10. Shared By Louis Augustine. www.facebook.com/sharedbylouisaugustine. Accessed 19 Sept 2022
11. Maung Zi's Tales. www.facebook.com/MaungZiTales. Accessed 19 Sept 2022
12. Thu, Y.K.: myWord: Syllable, Word and Phrase Segmenter for Burmese. GitHub Link. https://github.com/ye-kyaw-thu/myWord. Accessed 9 Sept 2021
13. NHK World-Japan, Corona Virus Questions and Answers in Burmese. https://www3.nhk.or.jp/nhkworld/my/news/qa/coronavirus/. Accessed 19 Sept 2022
14. Thu, Y.K.: myPOS: Myanmar Part-of-Speech Corpus, GitHub Link. https://github.com/ye-kyaw-thu/myPOS. Accessed 2 Aug 2021
15. Yang, J., Zhang, Y.: NCRF++: an open-source neural sequence labeling toolkit. In: Proceedings of the 56th Annual Meeting of the Association for Computational Linguistics (ACL), pp. 74–79. Association for Computational Linguistics, Melbourne, Australia (2018). https://doi.org/10.18653/v1/P18-4013
16. Santos, C.D., Zadrozny, B.: Learning character-level representations for part-of-speech tagging. In: Proceedings of the 31st International Conference on Machine Learning (ICML), vol 5. (2014). https://proceedings.mlr.press/v32/santos14.html
17. Klakow, D., Jochen, P.: Testing the correlation of word error rate and perplexity. Speech Commun. **38**(1–2), 19–28 (2002). https://doi.org/10.1016/S0167-6393(01)00041-3
18. Zipf, G.K.: Human Behavior and Principle of Least Effort: An Introduction to Human Ecology. Addison-Wesley, Cambridge (1949)
19. SCTK, the NIST Scoring Toolkit. https://github.com/usnistgov/SCTK. Accessed 19 Sept 2022

A Compact Version of EfficientNet

Hoang Van Thanh[1](✉), Tu Minh Phuong[2], and Kang-Hyun Jo[3]

[1] Quang Binh University, Dong Hoi, Quang Binh, Vietnam
thanhhv@qbu.edu.vn
[2] Posts and Telecommunications Institute of Technology, Hanoi, Vietnam
phuongtm@ptit.edu.vn
[3] University of Ulsan, Ulsan, Korea
acejo@ulsan.ac.kr

Abstract. EfficientNet is a convolutional neural network architecture that was created by doing a neural architecture search with the AutoML MNAS framework, which optimized both accuracy and efficiency. It is based on MobileNetV2's inverted bottleneck residual blocks, as well as squeeze-and-excite blocks. With far lower parameter computation burdens on the ImageNet challenge, EfficientNet may compete with the best. This paper provides a mobile version of EfficientNet that has accuracy similar to the ImageNet dataset but runs nearly twice as fast.

1 Introduction

In recent years, deep convolutional neural networks (CNNs) have shown extraordinary performance in a variety of computer vision tasks, with bigger and deeper CNNs being the key trend for tackling significant problems [2,11]. The most accurate CNNs often include hundreds of layers and thousands of channels [2,5,12,15]. However, because many real-world applications demand real-time processing and mobile devices with limited resources, a model should be small and have a low computational cost. To solve this issue, researchers are examining the trade-off between efficiency and accuracy via model compression work and the development of efficient architectures.

One of the architectures that has gained popularity is the Inception module proposed in GoogLeNet [11], which allows one to build deeper networks without increasing the size and computational cost of the model. This architecture was further improved by factorizing convolution in subsequent work [12]. The Depthwise Separable Convolution (DWConv) generalized the factorization idea and decomposed the standard Convolution into a depthwise convolution followed by a pointwise 1×1 convolution. Recently, EfficientNets [14], a family of efficiently scaled convolutional neural nets, have emerged as state-of-the-art models for image classification tasks. EfficientNets optimize accuracy and efficiency by reducing model size and floating-point operations executed while maintaining model quality.

EfficientNets is created by using a neural architecture search to design the basic network architecture of the model. This process generates a series of models, each of which is an enlarged version of the previous one. EfficientNet-B7

N. T. Nguyen et al. (Eds.): CITA 2023, LNNS 734, pp. 297–305, 2023.
https://doi.org/10.1007/978-3-031-36886-8_25

achieved the highest accuracy at the time, with a top-1 accuracy of 84. 4% and a top-5 accuracy of 97.1% in the ImageNet dataset. Compared with other models with the highest accuracy at the time, EfficientNets reduced the size by 8.4 times and increased the efficiency by 6.1 times. EfficientNets also reached the most advanced level on multiple well-known datasets through transfer learning.

This paper investigates the EfficientNet-B0, the basic network model in the EfficientNets series. The core structure of the network is the mobile inverted bottleneck convolution (MBConv) module, which also introduces the attention thought of the compression and excitation network (Squeeze-and-Excitation Network, SENet) [4]. After that, we create a lightweight variation of EfficientNet-B0 to make it more friendly for mobile devices.

2 Related Work

2.1 Efficient Architecture for CNNs

In recent times, a great deal of research has been directed towards developing an efficient architecture approach [3,5,10,16,17]. These studies have aimed at exploring the development of efficient Convolutional Neural Networks (CNNs) that can be trained end-to-end.

MobileNetV1, as proposed by Howard et al. [3], introduced a technique called *depthwise separable convolutions* to replace traditional convolution layers for improved efficiency. This approach separates the spatial filtering and feature generation mechanisms by using two distinct layers. The first layer applies a lightweight depthwise convolution operator that uses a single convolutional filter per input channel to capture spatial information. The second layer employs a heavier pointwise convolution, a 1×1 convolution, to capture cross-channel information for feature generation. This factorization of traditional convolution layers improves the overall efficiency of CNN.

MobileNetV2, as proposed in [10], incorporated the linear bottleneck and inverted residual structure to further improve the efficiency of the layer structures. This structure, illustrated in Fig. 1a, consists of an expansion convolution of 1×1, followed by depth convolutions and a projection layer of 1×1. A residual connection is established between the input and output only if they have the same number of channels and spatial sizes. This structure maintains a compact representation at both input and output while internally expanding to a higher-dimensional feature space, thereby increasing the expressiveness of nonlinear per-channel transformations.

EfficientNets [14], which are a family of efficiently scaled convolutional neural networks, have emerged as state-of-the-art models for image classification tasks. These networks aim to optimize both accuracy and efficiency by reducing the size of the model and the number of floating point operations required while maintaining the quality of the model.

EfficientNet's architecture was created using a technique called neural architecture search, which automates the design of neural networks to optimize both accuracy and efficiency measured in terms of floating-point operations per second

(FLOPS). The basic network architecture uses the mobile inverted bottleneck convolution (MBConv). The researchers then scaled up this baseline network to obtain a family of deep learning models, collectively called EfficientNets.

Inverted bottleneck blocks and depthwise separable convolutions are the core ideas for efficient mobile-size networks, and EfficientNet further developed these ideas with the introduction of the MBConv operation. This operation is based on the mobile inverted bottleneck block used in MobileNetV2 and has been shown to achieve state-of-the-art performance in the ImageNet dataset while maintaining efficiency. The combination of these techniques has enabled the creation of highly accurate models that can be deployed on mobile and edge devices with limited computational resources.

Making mobile-size convolution models efficient is critical. The MobileNet family [10,13] has extensively studied efficient operations, focusing on sparse depthwise convolution and the inverted bottleneck block. MnasNet [13] and EfficientNet [14] built on this work by developing the MBConv operation based on the mobile inverted bottleneck from [10]. EfficientNet demonstrated that models with MBConv operations can achieve state-of-the-art performance on the ImageNet challenge while still being highly efficient.

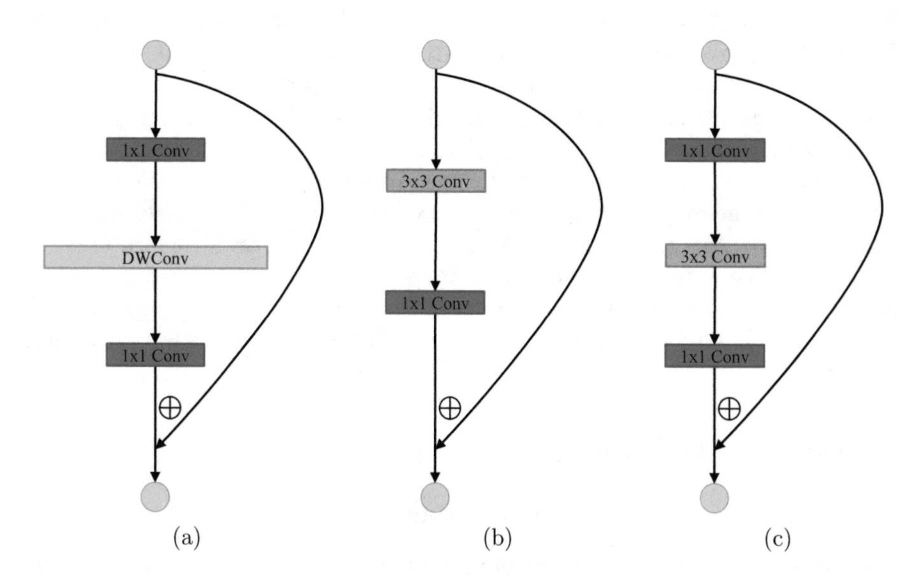

Fig. 1. Architecture of convolution blocks using in Experiments. a) The mobile inverted bottleneck convolution without SE module (MBConv) using Efficient-B0-lite. b) The convolution block with a 3×3 Conv and a 1×1 Conv (Conv31) using in proposed EfficientNet-B0-compact. c) The convolution block with a 3×3 Conv and two 1×1 Conv (Conv131). The downsampling will be in the DWConv layer or 3×3 Conv layer. The shortcut connection is disappeared if the dimensions of the input and output are different. **DWCon**: Depthwise Convolution layer. **Conv**: Convolution layer.

Table 1. EfficientNet-B0-lite Body Architecture. **MBConv**: Mobile inverted Bottleneck Convolution, the following number is the expanded factor.

Stage i	Resolution $\hat{H}_i \times \hat{W}_i$	#Channels $\hat{C}_i$	Operator $\hat{\mathcal{F}}_i$	#Blocks $\hat{L}_i$
1	224×224	32	Conv3 $\times$ 3	1
2	112×112	16	MBConv1, k3 $\times$ 3	1
3	112×112	24	MBConv6, k3 $\times$ 3	2
4	56×56	40	MBConv6, k5 $\times$ 5	2
5	28×28	80	MBConv6, k3 $\times$ 3	3
6	14×14	112	MBConv6, k5 $\times$ 5	3
7	14×14	192	MBConv6, k5 $\times$ 5	4
8	7×7	320	MBConv6, k3 $\times$ 3	1
9	7×7	1280	Conv1 $\times$ 1 & Pooling	1
10	1×1	1000	FC	1

2.2 Mobile Inverted Bottleneck Convolution Block

The mobile inverted bottleneck convolution is a module structure obtained through the search for neural network architecture, similar to the depthwise separable convolution. Figure 1a shows that it first applies a 1×1 point-by-point convolution on the input and changes the dimension of the output channel based on the expansion ratio. For example, if the expansion ratio is 3, the dimension of the channel will be increased by 3 times. However, if the expansion ratio is 1, the 1×1 point-by-point convolution and subsequent batch normalization and activation functions are omitted. Next, it performs a depthwise convolution of $k \times k$ and restores the original channel dimension at the end of the 1×1 point-by-point convolution. Finally, it performs drop-connect and skip-connection of the input. This structure results in a highly efficient and effective convolution module.

3 Network Architecture

3.1 Architecture of EfficientNet-B0-lite

EfficientNet-B0 is the simplest model in the EfficientNets series and is based on the mobile inverted bottleneck convolution module (MBConv), which also incorporates the attention mechanism of the compression and excitation module. The authors have also developed a lite version of the model that is more suitable for mobile devices and IoT devices. This version removes the squeeze-and-excite module (SE) [4] as it is not well supported by some mobile accelerators and replaces all swish activations with RELU6 to make post-quantization easier. The architecture of the model is presented in Table 1, where each layer is followed by batch normalization and ReLU non-linearity except the last fully connected layer that has no non-linearity and is followed by a softmax layer for

Table 2. Proposed EfficientNet-B0-compact Body Architecture. **Conv31**: Convolution block with a 3×3 convolution layer and a 1×1 convolution layer. **MBConv**: Mobile inverted Bottleneck Convolution, the following number is the expanded factor.

Stage i	Resolution $\hat{H}_i \times \hat{W}_i$	#Channels $\hat{C}_i$	Operator $\hat{\mathcal{F}}_i$	#Blocks $\hat{L}_i$
1	224×224	24	Conv3×3	1
2	112×112	24	MBConv3, k3×3	1
3	112×112	24	Conv31, k3×3	1
4	56×56	40	MBConv6, k5×5	2
5	28×28	80	MBConv6, k3×3	3
6	14×14	112	MBConv6, k5×5	3
7	14×14	192	MBConv6, k5×5	4
8	7×7	320	MBConv6, k3×3	1
9	7×7	1280	Conv1×1 & Pooling	1
10	1×1	1000	FC	1

classification. Down-sampling is achieved with strided convolution in the depthwise convolutions of the first block in each stage and the first layer. Finally, a global average pooling is performed to reduce the spatial resolution to 1 before the fully connected layer.

3.2 Architecture of Proposed Compact Version of EfficientNet-B0-compact

We introduced a new version called EfficientNet-B0-compact to have a more friendly model. The proposed model has some small differences in the early stages in comparison to the EfficientNet-B0-lite. They are:

- Remove the first MBConv1 block which has resolution of 112×112 (stage 2 in Table 1). We remove that MBConv1 block because, in the early stages, the operations are performed at a high resolution, so the number of FLOPS for each operation is high and also greatly affects the running speed of the model. The experiments also show that removing this block does not reduce the accuracy of the model much.
- Change the second MBConv6 block that has a resolution of 112×112 to the Conv31 block (stage 3 in Table 1 compared to stage 2 in Table 2). In the MBConv block, depthwise convolution is performed. Although this operation has a small number of parameters and FLOPS, the running speed is slower than that of traditional convolution when performed on high resolution. So we replaced it with a regular 3×3 convolution. To reduce the number of parameters, as well as to ensure that the new model runs faster than the old one, we only perform a 1×1 convolution operation afterward. Then make the residual connection as shown in Fig. 1b.

The overall architecture of EfficientNet-B0-compact is shown in detail in Table 2.

Table 3. Top-1 and Top-5 error rates (%) of EfficientNet-B0, EfficientNet-B0-lite and proposed EfficientNet-B0-compact on ImageNet datasets.

Model	#Params (10^6)	#FLOPS (10^6)	Top-1 (%)	Top-5 (%)	Speed (ms)
EfficientNet-B0	5.61	391.71	24.86	7.81	53.94
EfficientNet-B0-lite	4.97	385.69	26.75	8.72	37.69
EfficientNet-B0-compact	4.97	387.10	26.95	9.01	31.17

4 Experiments on ImageNet Dataset

This paper evaluates the EfficientNet-B0-lite variants in the ImageNet dataset [9] and compares them in terms of the accuracy metrics Top-1 and Top-5.

4.1 Dataset

The ILSVRC 2012 classification dataset, which comprises 1.2 million images for training and $50,000$ images for validation belonging to $1,000$ different classes, is used in our experiments. The same data augmentation approach used in [14] is employed to augment the training images. During the testing phase, a single crop of size 224×224 is applied. The classification accuracy of Top-1 and Top-5 in the validation set is reported, following the methodology described in [14].

4.2 Implementation Details

This study utilized Tensorflow to implement all variations of EfficientNet-B0, using the training scheme proposed in [14]. Backpropagation [6] was used for training the models using Stochastic Gradient Descent [8] with Nesterov momentum [7] (NAG) optimizer, over a period of 350 epochs. During the first five epochs, the learning rate increases linearly from 0 to 0.4, after which it follows a cosine decay (with the learning rate for epoch $t \leq 300$ set to $0.5 \times \mathrm{lr} \times (\cos(\pi \times t/300) + 1)$). Parameters were initialized using the Xavier initializer [1]. Other settings included a weight decay of 0.0001, momentum of 0.9, and a batch size of 2048.

All networks were trained in Google Colab service with a TPU environment. The speeds of all models are evaluated on a laptop with CPU, RAM 8 GB without GPU devices, when running the model with batch size, is 8.

4.3 Performance Evaluation

Table 3 shows the comparison of the top-1 and top-5 errors of EfficientNet-B0, EfficientNet-B0-lite, and the proposed EfficientNet-B0-compact on ImageNet datasets.

Table 4. Top-1 and Top-5 error rates (%) of variants of EfficientNet-B0-lite when adopting other convolution blocks to stage 2 on ImageNet datasets.

Architecture of stage 2	#Params (10^6)	#FLOPS (10^6)	Top-1 (%)	Top-5 (%)	Speed (ms)
MBConv1	4.97	385.69	26.75	8.72	37.69
DWConv3×3	4.97	398.54	26.68	8.64	37.39
Conv3×3	4.98	433.32	26.68	8.71	36.20
Conv1×1	4.97	381.94	27.21	8.92	35.93
Remove	4.97	394.78	26.78	8.87	34.39

Table 5. Top-1 and Top-5 error rates (%) of variants of EfficientNet-B0-lite when adopting other convolution blocks to stage 3 on ImageNet datasets.

Architecture of stage 3	#Params (10^6)	#FLOPS (10^6)	Top-1 (%)	Top-5 (%)	Speed (ms)
MBConv1	4.97	394.78	26.78	8.87	34.39
Conv131	4.97	388.91	27.11	8.95	31.64
Conv31	4.97	387.10	26.95	9.01	31.17

It is easy to see that although the lite and compact versions have a slightly higher error compared to the original EfficientNet-B0 they can run 2 times faster. Moreover, the compact version can achieve performance similar to that of the lite version while running 10% faster.

4.4 Ablation Study for Compact Architecture

We also do some ablation studies on variants of EfficientNet-B0-lite to find the compact version.

Experiment on Stage 2. Table 4 shows the comparison of Top-1 and Top-5 errors of variants of EfficientNet-B0-lite when adopting other convolution blocks to stage 2 on ImageNet datasets.

As can be seen, when we replace the MBConv1 blocks of stage 2 with the DWConv 3×3, Conv 3×3, and Conv 1×1 layers, respectively, the model is faster. However, the Top-1 and Top-5 errors increased. But when we put nothing to that stage, that variant is the fastest, while the error is similar to the original.

Experiment on Stage 3. Table 5 shows the comparison of Top-1 and Top-5 errors of variants of EfficientNet-B0-lite when adopting other convolution blocks to stage 3 on ImageNet datasets.

It is easy to see that when we adopt the Conv31 architecture, the accuracy does not change much, but the speed of the model can be 10% faster.

We also conduct experiments on changing the architecture of deeper stages. However, the model is not running much faster, and the accuracy drops rapidly.

5 Conclusion

This paper has studied the architecture of the EfficientNet-B0 model. On that basis, a new EfficientNet-B0-compact model is proposed with similar accuracy but faster speed than the original EfficientNet-B0 model, and the EfficientNet-B0-lite model. The new model is based on EfficientNet-B0-lite with early-stage changes. These are: 1) Remove the first MBCon1 block with resolution 112 × 112; 2) Change the second MBConv block on the 112 × 112 resolution layer to a Conv31 block. Experimental results in the ImageNet dataset show that the proposed EfficientNet-B0-compact model has a similar accuracy but is 2 times faster than the original EfficientNet-B0-model, and is 15% faster compared to the EfficientNet-B0-lite model.

References

1. Glorot, X., Bengio, Y.: Understanding the difficulty of training deep feedforward neural networks. In: Proceedings of the International Conference on Artificial Intelligence and Statistics, pp. 249–256 (2010)
2. He, K., Zhang, X., Ren, S., Sun, J.: Deep residual learning for image recognition. In: Proceedings of the IEEE Conference on Computer Vision and Pattern Recognition, pp. 770–778 (2016)
3. Howard, A.G., et al.: MobileNets: efficient convolutional neural networks for mobile vision applications. arXiv preprint arXiv:1704.04861 (2017)
4. Hu, J., Shen, L., Sun, G.: Squeeze-and-excitation networks. In: Proceedings of the IEEE Conference on Computer Vision and Pattern Recognition, pp. 7132–7141 (2018)
5. Huang, G., Liu, Z., Maaten, L.V.D., Weinberger, K.Q.: Densely connected convolutional networks. In: Proceedings of the IEEE Conference on Computer Vision and Pattern Recognition, pp. 4700–4708 (2017)
6. LeCun, Y., et al.: Backpropagation applied to handwritten zip code recognition. Neural Comput. **1**(4), 541–551 (1989)
7. Nesterov, Y.E.: A method for solving the convex programming problem with convergence rate o $(1/k^2)$. Dokl. Akad. Nauk SSSR. **269**, 543–547 (1983)
8. Robbins, H., Monro, S.: A stochastic approximation method. Ann. Math. stat. **22**(3), 400–407 (1951)
9. Russakovsky, O., et al.: ImageNet large scale visual recognition challenge. Int. J. Comput. Vis. **115**(3), 211–252 (2015). https://doi.org/10.1007/s11263-015-0816-y
10. Sandler, M., Howard, A., Zhu, M., Zhmoginov, A., Chen, L.C.: MobilenetV2: inverted residuals and linear bottlenecks. In: Proceedings of the IEEE Conference on Computer Vision and Pattern Recognition, pp. 4510–4520 (2018)
11. Szegedy, C., et al,: Going deeper with convolutions. In: Proceedings of the IEEE Conference on Computer Vision and Pattern Recognition, pp. 1–9 (2015)
12. Szegedy, C., Vanhoucke, V., Ioffe, S., Shlens, J., Wojna, Z.: Rethinking the inception architecture for computer vision. In: Proceedings of the IEEE Conference on Computer Vision and Pattern Recognition, pp. 2818–2826 (2016)
13. Tan, M., et al.: MnasNet: platform-aware neural architecture search for mobile. In: Proceedings of the IEEE Conference on Computer Vision and Pattern Recognition, pp. 2820–2828 (2019)

14. Tan, M., Le, Q.: EfficientNet: rethinking model scaling for convolutional neural networks. In: Proceedings of the International Conference on Machine Learning, pp. 6105–6114 (2019)
15. Zagoruyko, S., Komodakis, N.: Wide residual networks. In: Proceedings of the British Machine Vision Conference, pp. 1–12 (2016)
16. Zhang, X., Zhou, X., Lin, M., Sun, J.: ShuffleNet: an extremely efficient convolutional neural network for mobile devices. In: Proceedings of the IEEE Conference on Computer Vision and Pattern Recognition, pp. 6848–6856 (2018)
17. Zoph, B., Vasudevan, V., Shlens, J., Le, Q.V.: Learning transferable architectures for scalable image recognition. In: Proceedings of the IEEE Conference on Computer Vision and Pattern Recognition, pp. 8697–8710 (2018)

Using Deep Learning for Obscene Language Detection in Vietnamese Social Media

Dai Tho Dang[1](✉), Xuan Thang Tran[2], Cong Phap Huynh[1], and Ngoc Thanh Nguyen[3]

[1] Vietnam - Korea University of Information and Communication, The University of Danang, Danang, Vietnam
{ddtho,hcphap}@vku.udn.vn

[2] Information Technology Department, Tay Nguyen University, Buon Ma Thuot, Vietnam
txthang@ttn.edu.vn

[3] Department of Applied Informatics, Wroclaw University of Science and Technology, Wroclaw, Poland
Ngoc-Thanh.Nguyen@pwr.edu.pl

Abstract. Nowadays, a vast volume of text data is generated by Vietnamese people daily on social media platforms. Besides the enormous benefits, this situation creates many challenges. One of them concerns the fact that a tremendous amount of text contains obscene language. This kind of data negatively affects readers, especially young people. Detecting this kind of text is an important problem. In this paper, we investigate this problem using Deep Learning (DL) models such as Convolutional Neural Networks (CNN), Long-Short Term Memory (LSTM), and Bidirectional Long-Short Term Memory (BiLSTM). Besides, we combine LSTM and CNN in both sequence (sequential LSTM-CNN) and parallel (parallel LSTM-CNN) forms and sequential BiLSTM-CNN to solve this task. For word embedding phrase, we use Word2vec and PhoBERT. Experiment results show that the BiLSTM model with PhoBERT gains the best results for the obscene discrimination task, with 81.4% and 81.5% for accuracy and F1-score, respectively.

Keywords: Obscene language · Deep Learning · Vietnamese Social Media

1 Introduction

Nowadays, social media platforms like Facebook, Twitter, and Instagram have revolutionized communication between individuals, groups, and communities [1]. In Vietnam, there were 76.95 million social media users in January 2022. It means that social media users in Vietnam increased by 5.0 million from 2021 to 2022 [2]. The daily time spent using the Internet is 6 h 38 m. The average daily time spent on social media was 2 h and 28 min. Additionally, 85,2% of users spent time on social media platforms [2].

Facebook has been the most popular social media in Vietnam in recent years. The following is information about users of some popular social media networks in Vietnam in early 2022. Facebook, Instagram, TikTok, LinkedIn, and Twitter users were 70.40 million, 11.65 million, 39.91 million, 4.20 million, and 2.85 million, respectively [2]. As

N. T. Nguyen et al. (Eds.): CITA 2023, LNNS 734, pp. 306–317, 2023.
https://doi.org/10.1007/978-3-031-36886-8_26

a result, a vast volume and variety of texts are generated by people daily on social media platforms. Unfortunately, a huge number of texts contain obscene or profanity language. In general, the obscene language includes vulgar and pornographic [3]. The obscene language is defined as that which tends to morally corrupt the young or those otherwise "susceptible" to being corrupted. These texts negatively affect readers, especially young people.

The discovery of information with obscene language has been studied for English and other languages, such as Chinese, Turkish, and Korean [4–6]. However, to our best knowledge, this issue has not been deeply studied for Vietnamese texts. This is the motivation of this work, in which we focus on using DL for obscene language detection in Vietnamese texts.

The main contributions of this study are as follows:

- We create a new dataset for obscene language detection in Vietnamese social media.
- We propose an approach to investigate the problem of obscene language detection in Vietnamese social media using CNN, LSTM, Bi-LSTM, sequential LSTM-CNN, parallel LSTM-CNN, and sequential BiLSTM-CNN to solve this task.

The remainder of this paper is organized as follows. The related works are presented in Sect. 2. Section 3 describes our proposal for obscene language detection in Vietnamese social media. Experiments and evaluation are shown in Sect. 4. Finally, conclusions and future directions are provided in Sect. 5.

2 Related Works

The obscene language detection has been investigated for English and other languages, such as Chinese and Korean. There have been three main approaches to solving this problem: Rule-Based Linguistic, ML, and DL approaches [4–8].

For English, the authors of work [6] have studied the method for the detection of cyberbullying in YouTube comments. This method is a combination of content-based, cyberbullying-specific, and user-based features. Incorporating context in the form of users' activity histories improves cyberbullying detection accuracy. This approach measured the number of profane words in a post using a dictionary of obscene. In [9], the authors proposed learning semantic subspace into obscene model language at the word and sentence level. Advantageous of this approach is that it can deal with sparse obscene-related datasets restricted to a few languages (German, English, French).

In [7], Turkish obscene detection of search engine entries is investigated. This study applied several classical ML algorithms (Logistic Regression, SGD Classifier, Linear SVC, Multinomial NB, KNeighbors Classifier, Random Forest Classifier, and XGB Classifier) and DL models (LSTM, BERT, Electra, and T5). They compared each approach's performance from both accuracy and speed aspects. BERT and Electra models give the best results.

In [5], an obscene detection method using word embedding and the LSTM model are presented for Korean. The proposed method divides the text for training into the onset, nucleus, and coda. Further, it considers the semantics of the words and the morphology information using the FastText model. Moreover, training on the flow of context using the

LSTM model can detect obscene that cannot be detected by the methodologies proposed in previous studies. In [10], the authors focus on the problem of the filtering system mis-detecting normal phrases with Obscene phrases. They proposed a deep learning-based framework including grapheme and syllable separation-based word embedding and appropriate CNN structure. The proposed model was evaluated on the chatting contents from one of the famous online games in South Korea and generated 90.4% accuracy.

For Vietnamese language, the problem of hate speech detection has been considered to study. Bi-GRU-LSTM-CNN mode is introduced in [11]. The basic architecture of this model is CNN with 1D convolutions. Besides, it also combines two other DL models that are LSTM and Gated Recurrent Unit (GRU). Experimental results show that this proposal achieved 70.57% of the F_1-score. A model is a combination of the pre-trained PhoBERT model and the Text-CNN model presented in [12]. The combined PhoBERT-CNN model significantly improves the classification performance thanks to the resonance mechanism of the two single models, reducing errors between the predicted labels and the actual labels. The experiment results show that the proposed PhoBERT-CNN model outperforms SOTA methods and achieves an F1-score of 67,46% and 98,45% on ViHSD and HSD-VLSP, respectively. In [11], the Bidirectional LSTM is used for hate speech detection on Vietnamese social media text. Experimental results show that this proposal achieved 71.43% of the F_1-score. In [13], the authors used CNN, LSTM, BiLSTM, and B-GRU models with Phow2vec and fastText. The CNN combined with fastText embedding gives the highest accuracy value is 87.08%.

For obscene language detection in Vietnamese social media text, to our knowledge, this issue has yet to be studied.

3 Proposed Approach

Our proposed approach for obscene language detection in Vietnamese social media is shown in Fig. 1. It includes three main steps: Pre-processing, Embeddings, and Deep Learning.

3.1 Pre-processing

Several preprocessing steps are employed to process the Vietnamese written language for text normalization and input preparation for DL models. These include tokenizing compound words, converting all text to lowercase, removing HTML tags, dropping emojis, and deleting all punctuations.

The first step in this process involves identifying compound words within the text, which consist of several meaningful combinations of two or more words. This step is important in Vietnamese because compound words can lose meaning when separated. To address this, compound words are tokenized by combining the words with underscores. For example, the sentence "Hà Nội là thủ đô của Việt Nam" would become "Hà_Nội là thủ_đô của Việt_Nam".

Next, all text in the dataset is converted to lowercase to normalize the representation of words. The embedding model can misunderstand uppercase and lowercase versions

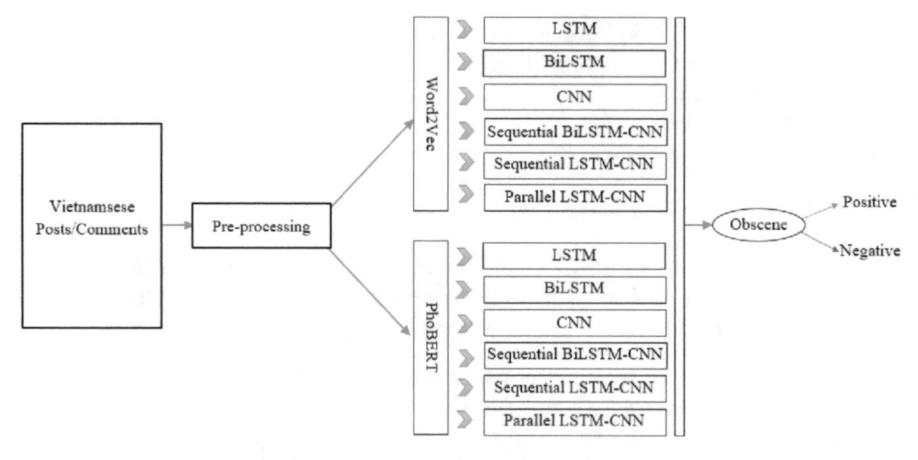

Fig. 1. The schema of the proposed approach.

of the same word as separate, which could negatively impact its accuracy. HTML tags, emojis, and punctuations are then removed from the text, as they typically do not contribute much to the meaning of a sentence. Additionally, removing these elements can help reduce the size of the word dictionary and computational resources required for the task.

These preprocessing steps are crucial for effectively processing Vietnamese written comments and posts and preparing input for models to analyze and understand their meaning accurately.

3.2 Word Embeddings

Word embeddings are a technique where individual words are transformed into a numerical representation of the word (a vector). Where each word is mapped to one vector, this vector is then learned in a way that resembles a neural network. This study employs two well-known pre-trained embedding models for the Vietnamese language, including Word2vec[1] trained on baomoi (400 dimensions) and PhoBERT[2] (768 dimensions). These embedding models outperform previous monolingual and multilingual approaches, obtaining new state-of-the-art performances on several Vietnamese NLP tasks, including Part-of-speech tagging, Dependency parsing, Named-entity recognition and Natural language inference [14].

3.3 Deep Learning (DL)

This study applies the following DL models and their combinations for detecting obscene language in Vietnamese social media.

- Long-Short Term Memory (LSTM)

---

1 https://github.com/sonvx/word2vecVN.

2 https://github.com/VinAIResearch/PhoBERT.

- Bidirectional Long-Short Term Memory (BiLSTM)
- Convolutional Neural Networks (CNN)
- Sequential LSTM-CNN
- Parallel LSTM-CNN
- Sequential BiLSTM-CNN

- Long-Short Term Memory (LSTM)

The LSTM is a kind of recurrent neural network (RNN). Compared with traditional RNNs, this model can better contact the context and process data serially. A simple LSTM network includes components: Input gate, Forget gate, and Output gate.

- Forget gate determines which information is kept and which can be discarded.
- Input gate determines which information is essential or not.
- Forget gate is used to disc Bidirectional and the information.

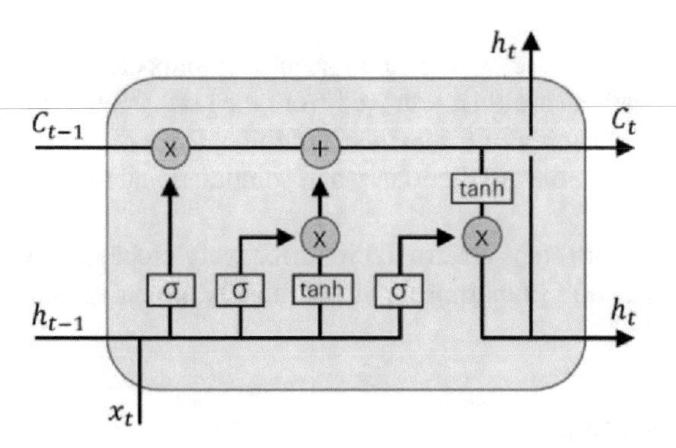

Fig. 2. The architecture of LSTM [15].

In Fig. 2, C_{t-1} and h_{t-1} are the previous neuron's state and output, respectively. Besides, x_t is the current input, and σ is the Sigmoid function. They together determine the output's current neuron h_t. The gate is adjusted using the Sigmoid function so that the gate's output value is between 0 and 1. The forgetting gate disregards all previous memories when it is 0, and the output gate disregards the newly computed states when it is 0 [15].

- Bidirectional Long-Short Term Memory (BiLSTM):

The BiLSTM is a sequence processing model that consists of two LSTMs. One takes the input in a forward direction and the other in a backward direction. BiLSTMs effectively increase the information available to the network, improving the context available to the algorithm.

This type of architecture has many advantages in real-world problems, especially in NLP. The main reason is that every component of an input sequence has information from both the past and present.

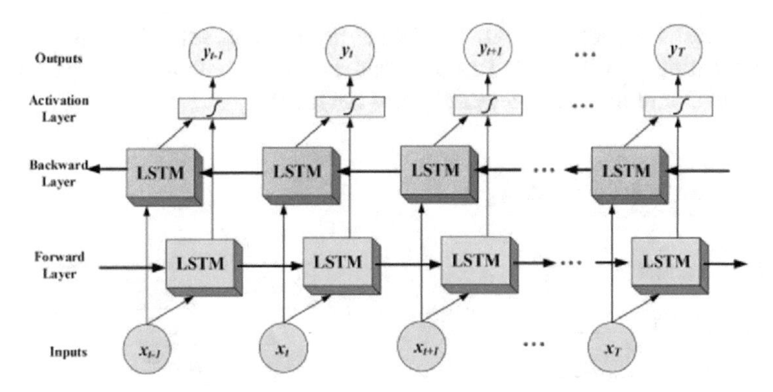

Fig. 3. The architecture of BiLSTM [16].

- Convolutional Neural Networks (CNN):

CNN was originally designed to perform DL for computer vision tasks. This model has also proven highly helpful for NLP tasks.

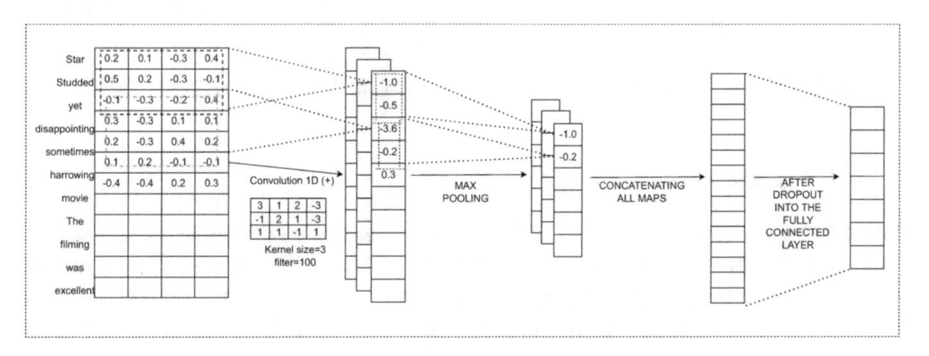

Fig. 4. Example: Text classification using CNN [17].

As shown in Fig. 4, the input of the model is a sentence represented as a matrix. In the matrix, each row is one vector that expresses a word. 1D convolution is performed onto the matrix with the kernel size being 3, along with 4 and 5. Max-Pooling is performed upon the filter maps, which are further concatenated and sent to the last fully connected layer for classification purposes [17].

- Sequential LSTM–CNN:

For the sequential LSTM–CNN model (see Fig. 5), the input data, which includes Vietnamese statuses, comments, and reviews, will be mapped into vector representations using embedding models such as Word2vec and PhoBERT. The output vectors will be fed into the LSTM model, which utilizes a method for extracting valuable features by analyzing the contextual information of sentences. Subsequently, the CNN model will sequentially operate on the LSTM's outputs to capture basic patterns by examining the holistic feature of the information. The output layer comprises a dropout layer, which aids in preventing overfitting, and a fully-connected layer with a single neuron using a sigmoid activation function to detect obscene language.

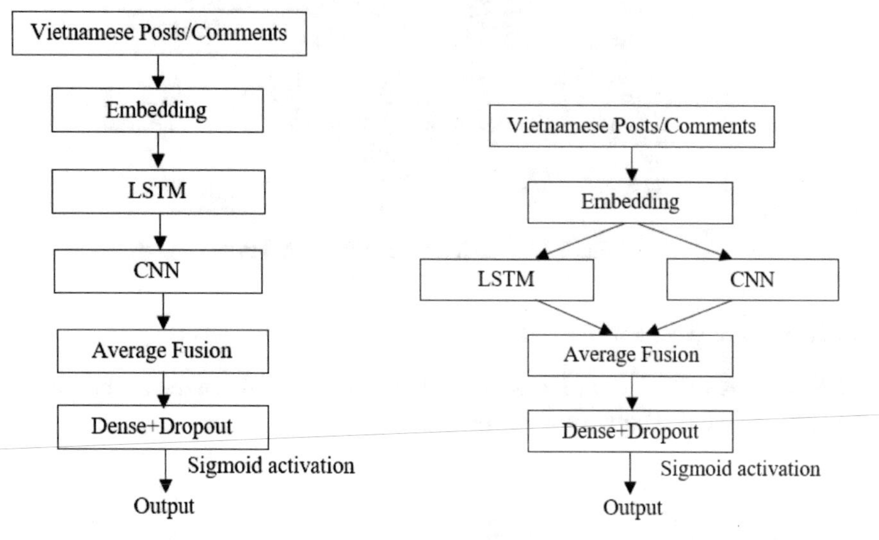

Fig. 5. The sequential LSTM–CNN.

Fig. 6. The parallel LSTM–CNN.

- Parallel LSTM–CNN:

Figure 6 shows the parallel CNN-LSTN. Given a Vietnamese dataset and its embedded vector representations, the LSTM and CNN simultaneously operate on two parallel branches. A fusion method using the average formula will merge the outputs from the two models to yield an averaged vector by assuming that LSTM and CNN contribute equally to detect obscene language.

- Sequential BiLSTM-CNN:

A sequential BiLSTM-CNN model is a combination of Bidirectional LSTM and CNN models. However, instead of utilizing LSTM, we employ BiLSTM to construct a new BiLSTM-CNN model. In the original formulation applied to named entity recognition, it learns both character-level and word-level features. The CNN component is used to induce the character-level features. For each word, the model employs a convolution and a max pooling layer to extract a new feature vector from the per-character feature vectors, such as character embeddings and (optionally) character type.

4 Experiments and Evaluation

This section performs experiments and measures the efficiency of the proposed approaches. The dataset for this task has yet to be created in Vietnamese. Thus, we first create the dataset. Next, Word2Vec and PhoBERT are applied for word embeddings. Then, we run models LSTM, BiLSTM, CNN, Sequential LSTM-CNN, Parallel LSTM-CNN, and BiLSTM-CNN on the dataset. Finally, the results are evaluated and compared, including these approaches' accuracy performance, precision, recall, and F1-score.

4.1 Creating the Dataset

The Vietnamese dataset for this problem is formed by crawling from social networks, including Facebook, Tiktok, and Twitter. It consists of Vietnamese posts and comments on various social aspects, such as culture, politics, education, and entertainment. The data is not automatically labeled, and we label it manually. Posts and comments are labeled with 1 for obscene or 0 for non-obscene.

Table 1. The examples for labeling.

No.	Comments	Labels
1	Nhị Thắng cực đ*t l*l m* mày	1
2	Sản phẩm của 1 nền giáo dục th*i n*t mà	1
3	Mỹ Duyên t kể chuyện gì dạ ☺	0
4	Jack hát nhạc vàng thì hợp,luyến láy như c*t	1

Table 1 shows examples (we replace the obscene words as *). The sentences No. 1 and No. 2 explicitly contain obscene words, while the user cleverly types deliberate typos, shorten words "c" in sample No. 4 (a Vietnamese word that sounds similar and is a abbreviation of swear words).

The dataset consists of 4,000 records, including 2,000 labeled as obscene and 2,000 non-obscene. It was then randomly split into a training set (80%) and a testing set (20%).

4.2 Experiment Setup

We employ two different embedding models: Word2vec and PhoBERT. The embedding size, 400 and 786, respectively, is the output dimension of Word2vec and PhoBERT embedding model. The number of hidden states in the LSTM layer is set to 128, and the 2-dimensional convolution layer (CNN) employs filter windows (l) of sizes 3, 4, and 5 with 100 feature maps each. The average formula for the ensemble the LSTM and the CNN model will be applied to generate the feature fusion. The output layer consists of one neuron, classifying the two classes, and utilizes a sigmoid activation function to perform binary classification. The batch size is set to 128. The loss function used is the binary cross-entropy loss, and the optimizer is the Adam algorithm with a learning rate of 0.001 and a dropout rate of 0.5

4.3 Experimental Results

The experimental results are represented in Table 2. The LSTM, CNN, BiLSTM, parallel LSTM-CNN, sequential LSTM-CNN, and sequential BiLSTM-CNN trained over Word2vec embedding method get 68.2% 76.3%, 77.1%, 77.4%, 77.5%, and 78.6% accuracies, respectively. These models trained on the PhoBERT embedding get 80.8%, 79.8%, 81.4%, 81%, 80.7%, and 81.4% accuracies, respectively. Thus, each approach with the PhoBERT embedding gave higher experimental results than that approach with Word2vec embedding.

The LSTM with Word2vec embedding archived the worst results, with 68.2% accuracy and 55.6% F1-score. This unsatisfactory result can be explained by the limitation of Vietnamese Word2vec pre-trained models, which trained on the limited corpus with only 7.1GB text (1,675,819 unique words from 97,440 documents). The parallel BiLSTM model with the PhoBERT gains the best results for the obscene discrimination task, with 81.4% and 81.5% for accuracy and F1-score, respectively. In fact, the PhoBERT is the state-of-the-art pre-trained language model for Vietnamese natural language processing tasks, which outperforms previous monolingual and multilingual approaches.

To evaluate the performance of DL models for the obscene detection task, we plot the training accuracy and loss values on the training phrase of six different deep learning models with 50 epochs each (including LSTM, CNN, BiLSTM, sequential LSTM-CNN, parallel LSTM-CNN, and sequential BiLSTM-CNN) to make a comparison. As in Fig. 7, the simple LSTM, CNN, and parallel LSTM-CNN models share a similar trend where training accuracy curves are under 90% and training loss curves are nearly 0.3. Specifically, the averaged training accuracy of LSTM, CNN, and parallel LSTM-CNN are 86.1%, 85.14%, and 83.88%, and their averaged training loss are 0.318, 0.343, and 0.356, respectively.

In contrast, the charts of accuracy and loss of BiLSTM, sequential LSTM-CNN and sequential BiLSTM-CNN model during the training phases showed a different tendency. Although the averaged accuracy performances in the training phase of these models are

Table 2. LSTM, CNN, sequential LSTM–CNN, parallel LSTN–CNN, and sequence BiLSTM-CNN.

	Word2vec				PhoBERT			
	Accuracy	Precision	Recall	F_1-score	Accuracy	Precision	Recall	F_1-score
LSTM	68.2%	84.2%	50.2%	55.6%	80.8%	81.5%	79.7%	80.6%
CNN	76.3%	75.7%	77.9%	76.6%	79.8%	80.3%	79.5%	79.7%
BiLSTM	77.1%	80.3%	72.0%	75.9%	81.4%	81.0%	82.0%	81.5%
Parallel LSTM-CNN	77.4%	76.8%	78.7%	77.7%	81.0%	81.6%	80.0%	80.8%
Sequential LSTM-CNN	77.5%	78.5%	76.0%	77.1%	80.7%	81.7%	79.3%	80.4%
Sequential BiLSTM-CNN	78.6%	79.9%	76.4%	78.1%	81.4%	82.3%	80.0%	81.1%

94.68%, 94.04%, and 92.84%, respectively, the averaged accuracy results over the testing phase of these deep learning models are only 84.1%, 80.70%, and 82.3%. This means these DL models give accurate predictions for training data rather than new data, or overfitting has occurred. This overfitting issue can be explained by the limitation of training data size, which does not contain enough data samples to represent all possible input data values accurately. Moreover, the complexity of these models is high, so it learns the noise within the training data. Thus, diversifying and scaling the data set, together with utilizing some other data science strategies such as pruning, regularization, and early stopping methods, will be taken into consideration in further research.

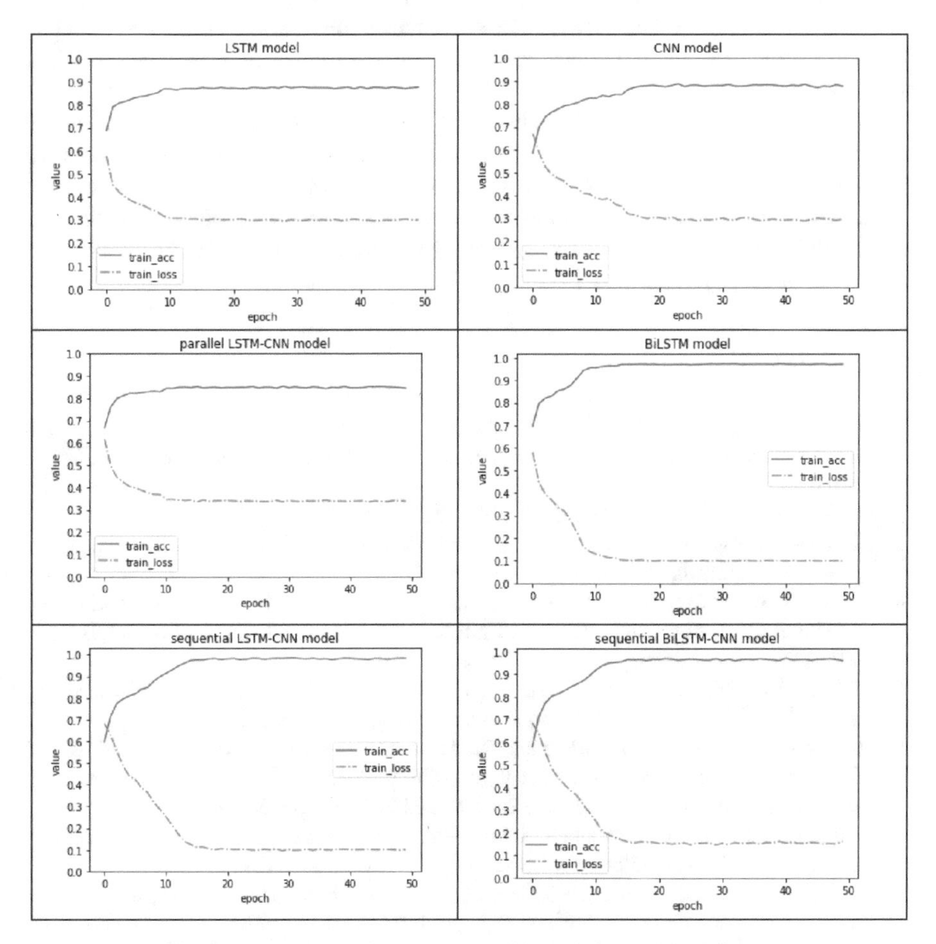

Fig. 7. Plot the models' accuracy and loss during training phrase.

5 Conclusions

This study investigated the problem of DL models for obscene language detection in Vietnamese social media. We created a Vietnamese dataset for this task. Models CNN, LSTM, Bi-LSTM, sequential LSTM-CNN, parallel LSTM-CNN, and sequential BiLSTM-CNN were used for this task. Word embedding phrase applies Word2vec and PhoBERT. The experimental results showed that the BiLSTM model with the PhoBERT embedding gained the best results, with 81.4% and 81.5% for accuracy and F_1-score, respectively.

For the future work, we will improve these approaches and look for new approaches to improve the effectiveness of this identification and detect to prevent other types of malicious language on social networks in Vietnamese. Besides, diversifying and scaling the data set, together with utilizing some other data science strategies, will be taken into consideration.

We also are planning to build a model of multi-agent systems for simulating the process of obscene language detection and to use consensus-based methods for making decisions regarding assigning obscene status for a sentence with using opinions from social media [18, 19]. For determining user opinions, consensus methods will be used to unify different user opinions for determining a common opinion that best represents a given set of user opinions [20, 21].

References

1. Dang, D.T., Nguyen, T.N., Hwang, D.: An effective method for determining consensus in large collectives. Comput. Sci. Inf. Syst. **19**(1), 435–453 (2022). https://doi.org/10.2298/CSIS210314062D
2. DataReportal: DIGITAL 2022: Viet Nam (2022). Accessed 10 Mar 2023
3. Alshehri, A., Nagoudi, E.M.B., Abdul-Mageed, M.: Understanding and detecting dangerous speech in social media. In: LREC, pp. 40–47 (2020)
4. Song, Y., Kwon, K.H., Xu, J., Huang, X., Li, S.: Curbing profanity online: a network-based diffusion analysis of profane speech on Chinese social media. New Media Soc. **23**(5), 982–1003 (2021). https://doi.org/10.1177/1461444820905068
5. Yi, M.H., Lim, M.J., Ko, H., Shin, J.H.: Method of profanity detection using word embedding and LSTM. Mob. Inf. Syst. **2021** (2021). https://doi.org/10.1155/2021/6654029
6. Dadvar, M., Trieschnigg, D., Ordelman, R., de Jong, F.: Improving cyberbullying detection with user context. In: Serdyukov, P., et al. (eds.) ECIR 2013. LNCS, vol. 7814, pp. 693–696. Springer, Heidelberg (2013). https://doi.org/10.1007/978-3-642-36973-5_62
7. Soykan, L., Karsak, C., Elkahlout, I.D., Aytan, B.: A comparison of machine learning techniques for Turkish profanity detection. In: LREC, pp. 16–24 (2022)
8. Phan, H.T., Dang, D.T., Nguyen, N.T., Hwang, D.: A new approach for predicting an important user on a topic on Twitter. In: INISTA 2020, pp. 1–6 (2020). https://doi.org/10.1109/INISTA49547.2020.9194658
9. Hahn, V., Ruiter, D., Kleinbauer, T., Klakow, D.: Modeling profanity and hate speech in social media with semantic subspaces. In: WOAH 2021 - Proceedings of the 5th Workshop on Online Abuse and Harms, pp. 6–16 (2021). https://doi.org/10.18653/v1/2021.woah-1.2
10. Woo, J., Park, S.H., Kim, H.K.: Profane or not: improving Korean profane detection using deep learning. KSII Trans. Internet Inf. Syst. **16**(1), 305–318 (2022)

11. Do, H. T.-T., Huynh, H.D., Van Nguyen, K., Nguyen, N.L.-T., Nguyen, A.G.-T.: Hate speech detection on vietnamese social media text using the bidirectional-LSTM model, pp. 4–7 (2019). http://arxiv.org/abs/1911.03648
12. Quoc Tran, K., Trong Nguyen, A., Hoang, P.G., Luu, C.D., Do, T.H., Van Nguyen, K.: Vietnamese hate and offensive detection using PhoBERT-CNN and social media streaming data. Neural Comput. Appl. **35**(1), 573–594 (2023)
13. Thị, N., Kim, T., Siểu, H.K., Phúc, P.H., Lượng, P.L.: Hate speech detection using distributed parallel training on deep learning models. In: CITA 2022, pp. 1–13 (2022)
14. Loc, C.V., Viet, T.X., Viet, T.H., Thao, L.H., Viet, N.H.: A text classification for Vietnamese feedback via PhoBERT-based deep learning. In: Yang, XS., Sherratt, S., Dey, N., Joshi, A. (eds.) Proceedings of Seventh International Congress on Information and Communication Technology. LNNS, vol. 464, pp. 259–272. Springer, Singapore (2023). https://doi.org/10.1007/978-981-19-2394-4_24
15. Wang, H., Li, F.: A text classification method based on LSTM and graph attention network. Connect. Sci. **34**(1), 2466–2480 (2022)
16. Zhang, F.: A hybrid structured deep neural network with Word2Vec for construction accident causes classification. Int. J. Constr. Manag. **22**(6), 1120–1140 (2022)
17. Soni, S., Chouhan, S.S., Rathore, S.S.: TextConvoNet: a convolutional neural network based architecture for text classification. Appl. Intell. (2022)
18. Duong, T.H., Nguyen, N.T., Jo, G.S.: A hybrid method for integrating multiple ontologies. Cybern. Syst. Int. J. **40**(2), 123–145 (2009)
19. Sliwko, L., Nguyen, N.T.: Using multi-agent systems and consensus methods for information retrieval in internet. Int. J. Intell. Inf. Database Syst. **1**(2), 181–198 (2007)
20. Katarzyniak, R., Nguyen, N.T.: Reconciling inconsistent profiles of agents' knowledge states in distributed multiagent systems using consensus methods. Syst. Sci. **26**(4), 93–119 (2000)
21. Nguyen, N.T.: Conflicts of ontologies – classification and consensus-based methods for resolving. In: Gabrys, B., Howlett, R.J., Jain, L.C. (eds.) KES 2006. LNCS, vol. 4252, Part II, pp. 267–274. Springer, Heidelberg (2006). https://doi.org/10.1007/11893004_34

Network and Communications

A New Approach for Enhancing MVDR Beamformer's Performance

Nguyen Ba Huy[1], Pham Tuan Anh[2], Quan Trong The[1](✉), and Dai Tho Dang[2]

[1] Falculty of Control Systems and Robotics, University ITMO, St. Petersburg, Russia
quantrongthe1984@gmail.com

[2] Vietnam - Korea University of Information and Communication Technology, The University of Danang, Da Nang, Vietnam
{ptanh,ddtho}@vku.udn.vn

Abstract. Minimum Variance Distortionless Response (MVDR) has many applications in the microphone array processing approach due to its advantages of suppressing background noise, interference, or third-party speaker. However, in real-life recording situations, the performance of the MVDR beamformer is often degraded by microphone mismatches, different microphone sensitivities, or errors in microphone distribution. Therefore, in this paper, the authors suggested a new approach, which uses an additive matrix for modifying the covariance matrix of observed microphone array data. The illustrated experiments were evaluated in a living room with a dual-microphone system (DMA2) and a stand speaker in the presence of noise. The theoretical results are further confirmed in terms of the signal-to-noise ratio (SNR) and the speech intelligibility of the output signal. The proposed method can be integrated into systems with multiple microphones and in annoying complex environments.

Keywords: Microphone array · Beamforming · Minimum variance distortionless response · Signal-to-noise ratio · Speech quality · Covariance matrix

1 Introduction

In numerous speech applications, such as voice-controlled systems, hearing aids, mobile phones, hands-free acoustic equipment, and surveillance equipment, the recorded signal and transmitted audio data are always degraded by considerable interference, third-party talker, and coherent and incoherent noise. The speaker is located at a certain distance from the recording microphones in complex situations with many sound sources. Generally, background noise can be considered broadband and non-stationary, and the signal-to-noise ratio (SNR) can be pretty low. Unwanted noise causes a decreased performance of preferred digital signal

Supported by Digital Agriculture Cooperative.

N. T. Nguyen et al. (Eds.): CITA 2023, LNNS 734, pp. 321–331, 2023.
https://doi.org/10.1007/978-3-031-36886-8_27

processing and substantially corrupts speech intelligibility and speech quality. Noise suppression has become an attractive approach to research the direction of microphone array beamforming, which is applied in much acoustic equipment.

Unfortunately, most of single-channel approach speech enhancement techniques use spectral subtraction in the time-frequency domain to extract the desired target speech component. However, this filtering operation can cause speech distortion and a corrupted signal in a non-stationary environment. Therefore, multiple microphones are nowadays available for capturing, saving, and enhancing the noisy mixture of useful signal, noise, and interference. Microphone arrays (MA) [12–20] have become widely used in modern speech communication systems because of their convenience, spatial diversity, and potential. Indeed, [1] established that the problem of localizing acoustic sources is very important, and MA allows for extracting desired speech sources from mixtures of sounds and noise by taking advantage of a priori spatial dimension [2,3]. When multiple observed microphone array signals are available, both information of time - domain, frequency domain, the direction of arrival (DOA) of interest signal, and the properties of recording scenarios are used for pre-precessing, post-filtering and spatial filtering to obtain the only speech components of target talker (Fig. 1).

Fig. 1. The annoying recording environment.

The single microphone-based digital signal processing causes the price of significant speech distortion [4]. The advantage of utilizing MA [11–20] has a theoretical improvement the noise suppression while minimizing the amount of speech degradation [2,3,5] and increasing the speech quality in terms of the signal-to-noise ratio (SNR). On the other hand, MA is applied for dereverberation, denoising, and enhancing speech signals [6–8] (Fig. 2).

In either frequency - or time-domain, the spatial diversity, which is characteristic of MA, takes advantage of dealing with noise reduction while saving the target speech component. Therefore, the beamforming technique is widely commonly known due to its capability of using a steerable beampattern toward the talker source and its efficiency perform removing background noise, interference, and third-party speaker. [9] presented a method that recovers the desired speech source while attenuating other competing noise sources. The MVDR beamformer is the most promising technique for speech enhancement and noise reduction in acoustic recording environments in the knowledge of the direction of arrival of the useful signal [5].

On the assumption of the available direction of arrival (DOA) of the desired speaker, the MVDR beamformer estimates the original signal and minimizes the total variance of the noise power of the resulting signal. In practical speech application, unfortunately, the DOA of the interest talker is not preferred correctly, which can lead to deteriorate the MVDR beamformer's evaluation [6]. Therefore, many research articles have been presented, evaluated, and proposed to enhance the robustness of the MVDR beamformer [7,8] that extend the susceptible direction of the original source. Nevertheless, even though the estimation of sound source localization was perfect, the microphone mismatches, the difference

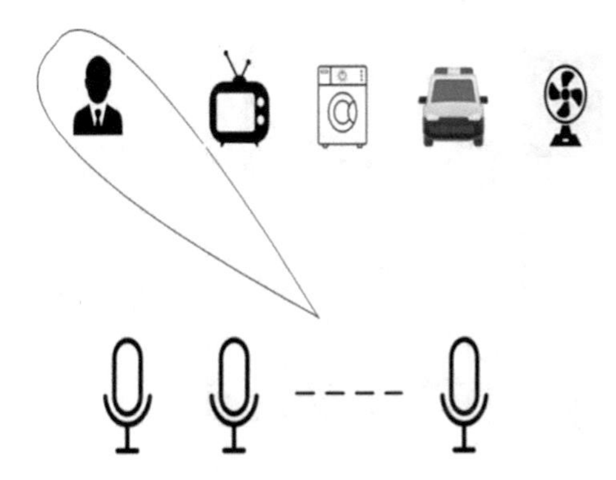

Fig. 2. The implementation of microphone array for extracting the desired target speaker.

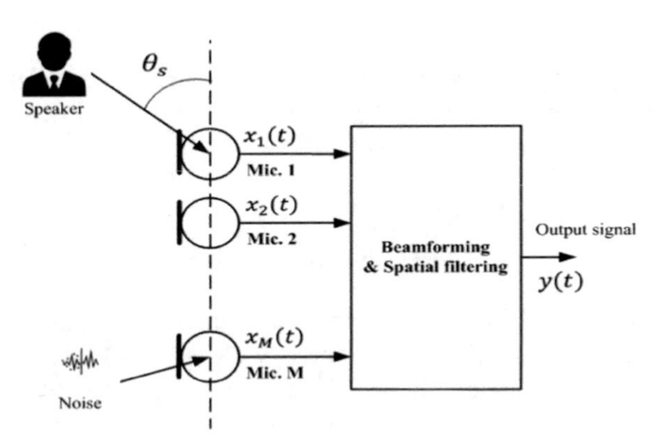

Fig. 3. The theory of microphone array's performance.

in microphone array sensitivities, the error of the distribution, or sensors may have the distinct MVDR beamformer is not able to handle well. Experiments concluded that MVDR and other existing signal processing methods could be outperformed well in the presence of an omnidirectional microphone (Fig. 3).

In this paper, the authors suggested a new technique, which uses the a priori information of coherence, the temporal signal-to-noise ratio, and the diffuse noise field to achieve an effective matrix to estimate the covariance matrix of captured signals against the speech distortion in most of the realistic acoustic environments. This contribution is organized in the following way. The following section describes the principal working of the MVDR beamformer. Section 3 presents the suggested technique to enhance the MVDR beamformer's evaluation. Sections 4 and 5 illustrated the experiments and concluded.

2 MVDR Beamformer

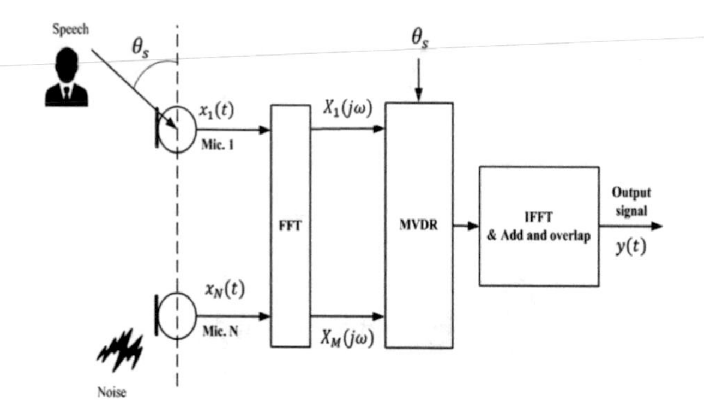

Fig. 4. The digital signal processing of MVDR beamformer.

We denote $s(t)$ a speech signal impinging on an array of N microphones with an arbitrary distribution and the distance between microphones d. The observed microphone array signals can be described as:

$$y_n(t) = g_n * s(t) + v_n(t) \tag{1}$$
$$= x_n(t) + v_n(t), n = 1, 2, ..., N \tag{2}$$

where g_n is the channel impulse response encountered by the characteristic of certain microphone n with the sound source, $x_n = g_n * s(t)$ is the speech component, which captured in microphone n in the criteria of noise-free reverberant, $*$ is the convolution operator. Indeed, numerous competing noise point sources can be expressed as noise $v_n(t)$ (Fig. 4).

In this article, we assume that it is uncorrelated with the original speech source $s(t)$ and all its associated components. The above equations of received signals can be described in the frequency domain as:

$$Y_n(j\omega) = G_n(j\omega)S(j\omega) + V_n(j\omega) \tag{3}$$
$$= X_n(j\omega) + V_n(j\omega), n = 1, 2, ..., N \tag{4}$$

where $Y_n(j\omega)$, $G_n(j\omega)$, $S(j\omega)$ and $V_n(j\omega)$ are the short-time Fourier transform of $y_n(t)$, g_n, $s(t)$ and $v_n(t)$ respectively.

All digital signal processing algorithms aim to suppress the total noisy component and preserve the signal components, $\boldsymbol{Y}(j\omega) = [Y_1(j\omega) \quad Y_2(j\omega) \quad ... \quad Y_N(j\omega)))^T$ can be applied an appropriate coefficients $\boldsymbol{W}(j\omega) = [W_1(j\omega) \quad W_2(j\omega) \quad ... \quad W_N(j\omega))]^T$ to find the desired signal, where the superscript {T} expresses the transpose of a vector in the presence of $\boldsymbol{V}(j\omega) = [V_1(j\omega) \quad V_2(j\omega) \quad ... \quad V_N(j\omega))]^T$.

The final output signal is given by:

$$Z(j\omega) = \boldsymbol{W}^H(j\omega)\boldsymbol{Y}(j\omega) \tag{5}$$
$$= \boldsymbol{W}^H(j\omega)\boldsymbol{X}(j\omega) + \boldsymbol{W}^H(j\omega)\boldsymbol{V}(j\omega) \tag{6}$$
$$= \boldsymbol{W}^H(j\omega)\boldsymbol{G}_n(j\omega)S(j\omega) + \boldsymbol{W}^H(j\omega)\boldsymbol{V}(j\omega) \tag{7}$$

where H is the transpose conjugate operator. The MVDR beamformer is based on the constrained problem of saving the desired speech component while minimizing the total output noise power. Therefore, the MVDR beamformer can be represented as the following way:

$$\begin{aligned} \min_{\boldsymbol{W}(j\omega)} \quad & \boldsymbol{W}(j\omega)^H \boldsymbol{\Phi}_{VV}(j\omega)\boldsymbol{W}(j\omega) \\ \text{s.t.} \quad & \boldsymbol{W}(j\omega)^H \boldsymbol{G}_n(j\omega) = 1 \end{aligned} \tag{8}$$

where $\boldsymbol{\Phi}_{VV}(j\omega) = E\{\boldsymbol{V}(j\omega)\boldsymbol{V}^*(j\omega)\}$ is a cross spectral matrix of noise signals. The optimum of coefficient of MVDR beamformer is computed as:

$$\boldsymbol{W}(j\omega) = \frac{\boldsymbol{\Phi}_{VV}^{-1}(j\omega)\boldsymbol{G}_n(j\omega)}{\boldsymbol{G}_n^H(j\omega)\boldsymbol{\Phi}_{VV}^{-1}(j\omega)\boldsymbol{G}_n(j\omega)} \tag{9}$$

However, in almost realistic scenarios of speech application, the information of noise always is not available, we use the covariance matrix of observed microphone arrays signals instead off $\boldsymbol{\Phi}_{XX}(j\omega) = E\{\boldsymbol{X}^H(j\omega)\boldsymbol{X}(j\omega)\}$. And the final MVDR beamformer as the following equation:

$$\boldsymbol{W}(j\omega) = \frac{\boldsymbol{\Phi}_{XX}^{-1}(j\omega)\boldsymbol{G}_n(j\omega)}{\boldsymbol{G}_n^H(j\omega)\boldsymbol{\Phi}_{XX}^{-1}(j\omega)\boldsymbol{G}_n(j\omega)} \tag{10}$$

Unfortunately, the MVDR beamformer's performance is often degraded due to many reasons, such as microphone mismatch and the different microphone sensitivities that seriously affect the final output signal. In the next section, the authors suggested using a combination of speech presence probability and matrix exchange to improve the robust covariance matrix of received microphone array signals.

3 The Proposed Method

From (10), it can be realized that the accuracy of $\Phi_{XX}(j\omega)$ get a major importance role in finding the optimal solution. The imprecise calculation of $\Phi_{XX}(j\omega)$ can lead to the speech distortion or degraded performance of MVDR beamformer in presence of incorrect the direction - of - arrival (DOA) of the interest of useful signal, the different of microphone array sensitivities or the microphone mismatches. In this paper, the author suggests using an effective matrix for dealing the speech distortion through estimating the covariance matrix of microphone array signals in diffuse noise field in the case of with dual-microphone system as:

$$\hat{\boldsymbol{\Phi}}_{XX}(j\omega) = \boldsymbol{\Phi}_{XX}(j\omega) * \boldsymbol{A} \tag{11}$$

Which the additive matrix A is can be derived from the following mathematical approach. As we know that, the coherence matrix of observed microphone array signals can be expressed as:

$$\boldsymbol{\Gamma}_{XX}(j\omega) \approx \frac{SNR(j\omega)}{1 + SNR(j\omega)}\boldsymbol{\Gamma}_S(j\omega) + \frac{1}{1 + SNR(j\omega)}\boldsymbol{\Gamma}_V(j\omega); \tag{12}$$

where $\boldsymbol{\Gamma}_{XX}(j\omega) = \begin{Bmatrix} 1 & \Gamma_{X_1X_2}(j\omega) \\ \Gamma_{X_2X_1}(j\omega) & 1 \end{Bmatrix}$, $\Gamma_{X_1X_2}(j\omega) = \frac{P_{X_1X_2}(j\omega)}{\sqrt{P_{X_1X_1}(j\omega)*P_{X_2X_2}(j\omega)}}$, $\Gamma_{X_2X_1}(j\omega) = \frac{P_{X_2X_1}(j\omega)}{\sqrt{P_{X_1X_1}(j\omega)*P_{X_2X_2}(j\omega)}}$. $P_{X_iX_i}(f,k), P_{X_iX_j}(f,k),\ i,j \in \{1,2\}$ calculated as:

$$P_{X_iX_j}(f,k) = (1-\alpha)P_{X_iX_j}(f,k-1) + \alpha X_i^*(f,k)X_j(f,k) \tag{13}$$

where α is the smoothing parameter, which in the range $\{0...1\}$.

$\boldsymbol{\Gamma}_S(j\omega) = E\{\boldsymbol{G}_n^*(j\omega)\boldsymbol{G}_n(j\omega)\}$.

And in diffuse noise field, the $\boldsymbol{\Gamma}_V(j\omega) = \begin{Bmatrix} 1 & \frac{\sin(\omega\tau_0)}{\omega\tau_0} \\ \frac{\sin(\omega\tau_0)}{\omega\tau_0} & 1 \end{Bmatrix}$. The distance between microphones is d, the sound speed is c ($343\,m/s$), $\tau_0 = d/c$ is the sound delay.

With the defined parameter $\rho = \frac{SNR(j\omega)}{1+SNR(j\omega)}$, (12) can be rewritten as:

$$\boldsymbol{\Gamma}_{XX}(j\omega) \approx \rho\boldsymbol{\Gamma}_S(j\omega) + (1-\rho)\boldsymbol{\Gamma}_V(j\omega); \tag{14}$$

The Eq. (14) is an approximately equation, and the ideal of the authors is the necessary to obtain a imprecise estimation through an additive matrix $\boldsymbol{A}$, which can be added as:

$$\boldsymbol{\Gamma}_{XX}(j\omega) = \rho\boldsymbol{\Gamma}_S(j\omega)\boldsymbol{A} + (1-\rho)\boldsymbol{\Gamma}_V(j\omega); \tag{15}$$

Therefore, the additive matrix $\boldsymbol{A}$ is derive as:

$$\boldsymbol{A} = \boldsymbol{\Gamma}_S^{-1}(j\omega) \times \frac{\boldsymbol{\Gamma}_{XX}(j\omega) - (1-\rho)\boldsymbol{\Gamma}_V(j\omega)}{\rho} \tag{16}$$

In the diffuse noise field, the temporal $SNR(j\omega)$ can be easily calculated as: The noise power $\sigma_n^2(j\omega)$:

$$\sigma_n^2(j\omega) = P_{X_1X_2}(j\omega) - \sqrt{P_{X_1X_1}(j\omega) * P_{X_2X_2}(j\omega)} \tag{17}$$

and the priori speech power $\sigma_s^2(j\omega)$

$$\sigma_s^2(j\omega) = \frac{1}{\boldsymbol{G}_n^H(j\omega)\boldsymbol{\Phi}_{XX}^{-1}(j\omega)\boldsymbol{G}_n(j\omega)} \tag{18}$$

The temporal $SNR(j\omega)$:

$$SNR(j\omega) = \frac{\sigma_s^2(j\omega)}{\sigma_n^2(j\omega)} \tag{19}$$

Consequently, the defined value ρ also be computed through the available spatial information.

In the next section, the author will evaluate the advantage of the suggested method (MVDR-st-ex) with the conventional MVDR beamformer (MVDR) in comparison in terms of the signal-to-noise ratio (SNR) and reducing speech distortion.

4 Experiments

The dual-microphone system (DMA2) is the most comfort configuration of MA, which has been widely used in almost acoustic equipments. DMA2 owns the capacity of low computation, compactness and easily installed into devices. In these experiments, (DMA2) is used, the distance between two microphones $d = 5(cm)$, the DOA of desired talker $\theta_s = 90(deg)$.

The recording experiment was conducted in a living room, in the presence of other annoying noises. For further digital signal processing, the necessaries parameters are used, the sampling frequency $Fs = 16\,\text{kHz}$, $NFFT = 512$, overlap 50%, smoothing value $\alpha = 0.1$. The observed microphone array signals are processed by MVDR, MVDR-st-ex. An objective measurement SNR is used for computing the speech quality of the resulting signals. The scheme of evaluated experiment is shown in Fig. 5. The size of the living room is $3.2 \times 4.0 \times 5.6$ (m). The distance of talker to DMA2 is $L = 2(m)$. An objective measurement [10] is used to compare the speech quality of processed signals by MVDR, MVDR-st-ex. The performance of MVDR and MVDR-st-ex are compared.

The observed microphone array signal is shown in Fig. 6.

Figures 7 and 8 depict the output signals by MVDR, MVDR-st-ex.

Figure 9 shows the energy of the original microphone, the processed signal MVDR, and MVDR-st-ex. The proposed method increases the speech quality in terms of SNR from 2.5 (dB) to 4.2 (dB) and reduces the speech distortion to 4.1 (dB). Table 1 expresses the measurement of SNR.

From these numerical results, the effectiveness of MVDR-st-ex can be proven. The combination of speech presence probability and exchange matrix improves

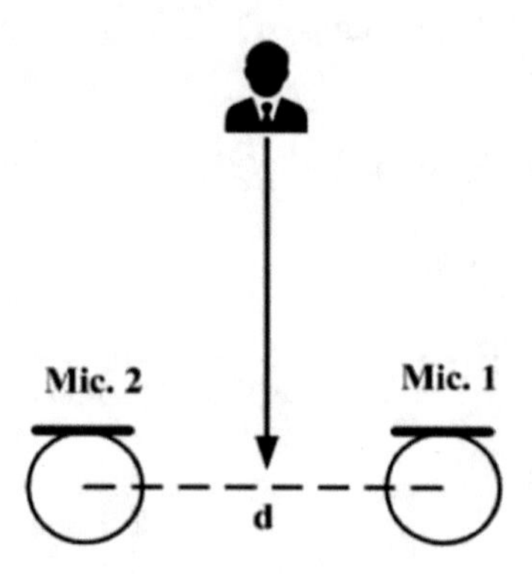

Fig. 5. The experiments with DMA2.

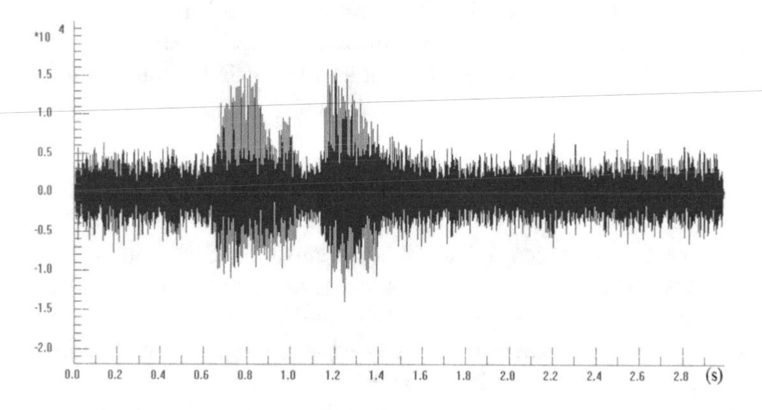

Fig. 6. The waveform of original microphone array signal.

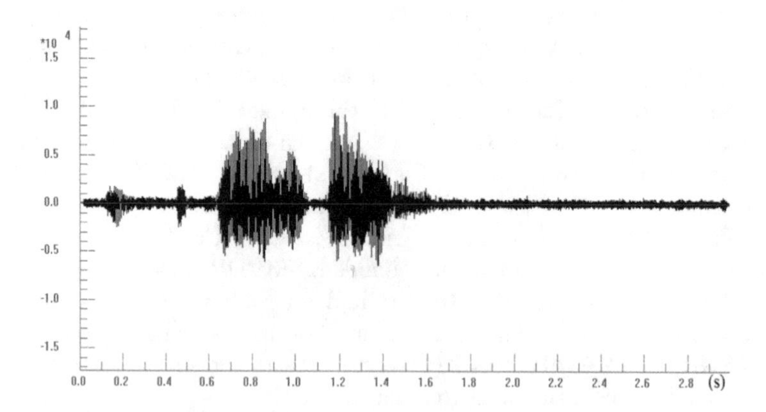

Fig. 7. The waveform of processed signal by MVDR.

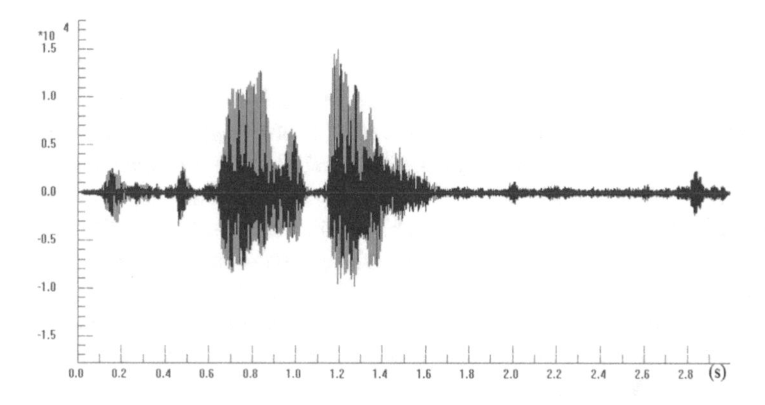

Fig. 8. The waveform of processed signal by MVDR-st-ex.

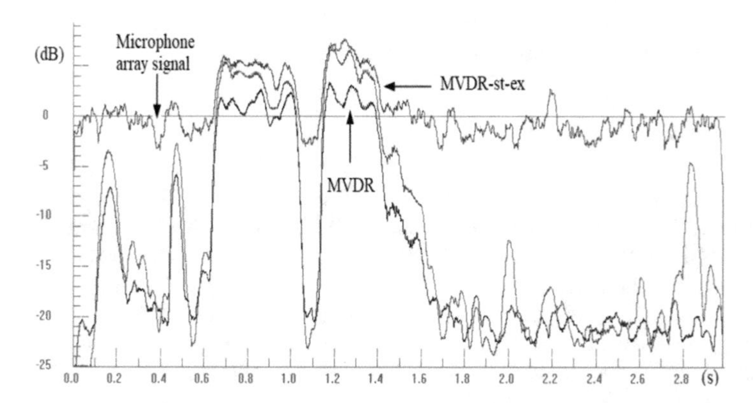

Fig. 9. The illustrated energy of microphone array signal, MVDR and MVDR-st-ex.

Table 1. The signal-to-noise ratio (dB)

Method Estimation	Microphone array signal	MVDR	MVDR-st-ex
NIST STNR	6.2	23.5	26.0
WADA SNR	4.1	24.9	29.1

the accuracy of the covariance matrix, which reduces speech distortion and enhances the SNR. The improvement of MVDR-st-ex has confirmed that the suggested technique can be installed into the system with multiple microphones.

In a noisy recording environment, MVDR-st-ex has confirmed its ability to save the target speech component while suppressing all background noise. The results have proven that the proposed method can be integrated into multiple microphone systems for better performance. In the future, the authors will investigate the combination with characteristics of recording scenarios to improve this article's success.

5 Conclusion

One of the drawbacks of the MVDR beamformer is the imprecise estimation of the covariance matrix of observed data and, accordingly, gives a lower evaluation in speech enhancement. This drawback exists for many reasons, such as the different microphone sensitivities, the error of the direction of arrival of the valuable signal, microphone mismatches, or the imprecise MA's distribution. Consequently, the corrupted evaluation of the MVDR beamformer leads to decreasing speech quality and intelligibility. This contribution presents a new technique for enhancing the performance of the MVDR beamformer. The numerical results have proven the desired advantage of reducing speech distortion and improving the speech quality of the target speaker.

Acknowledgements. This research was supported by Digital Agriculture Cooperative. The author thank our colleagues from Digital Agriculture Cooperative, who provided insight and expertise that greatly assisted the research.

References

1. Dmochowski, J., Benesty, J., Affes, S.: Direction of arrival estimation using the parameterized spatial correlation matrix. IEEE Trans. Audio Speech Lang. Process. **15**(4), 1327–1339 (2007). https://doi.org/10.1109/TASL.2006.889795
2. Benesty, J., Chen, J., Huang, Y.: Microphone Array Signal Processing. Springer, Heidelberg (2008). https://doi.org/10.1007/978-3-540-78612-2
3. Huang, Y., Benesty, J., Chen, J.: Acoustic MIMO Signal Processing. Springer, Heidelberg (2006). https://doi.org/10.1007/978-3-540-37631-6
4. Chen, J., Benesty, J., Huang, Y., Doclo, S.: New insights into the noise reduction Wiener filter. IEEE Trans. Audio Speech Lang. Process. **14**, 1218–1234 (2006). https://doi.org/10.1109/TSA.2005.860851
5. Benesty, J., Chen, J., Huang, Y., Dmochowski, J.: On microphone-array beamforming from a MIMO acoustic signal processing perspective. IEEE Trans. Audio Speech Lang. Process. **15**(3), 1053–1065 (2007). https://doi.org/10.1109/TASL.2006.885251
6. Delcroix, M., Hikichi, T., Miyoshi, M.: Dereverberation and denoising using multichannel linear prediction. IEEE Trans. Audio Speech Lang. Process. **15**(6), 1791–1801 (2007). https://doi.org/10.1109/TASL.2007.899286
7. Affes, A., Grenier, Y.: A signal subspace tracking algorithm for microphone array processing of speech. IEEE Trans. Speech Audio Process. **5**, 425–437 (1997). https://doi.org/10.1109/89.622565
8. Huang, Y., Benesty, J., Chen, J.: A blind channel identification-based two-stage approach to separation and dereverberation of speech signals in a reverberant environment. IEEE Trans. Speech Audio Process. **13**(5), 882–895 (2005). https://doi.org/10.1109/TSA.2005.851941
9. Van Veen, B.D., Buckley, K.M.: Beamforming: a versatile approach to spatial filtering. IEEE Audio Speech Signal Process. Mag. **5**(2), 4–24 (1988). https://doi.org/10.1109/53.665
10. https://labrosa.ee.columbia.edu/projects/snreval/

11. Carlson, B.D.: Covariance matrix estimation errors and diagonal loading in adaptive arrays. IEEE Trans. Aerospace Elect. Syst. **24** (1988). https://doi.org/10.1109/7.7181
12. Buck, M.: Aspects of first-order differential microphone arrays in the presence of sensor imperfections. Eur. Trans. Telecommun. **13**(2), 115–122 (2002). https://doi.org/10.1002/ett.4460130206
13. Buck, M., Wolff, T., Haulick, T., Schmidt, G.: A compact microphone array system with spatial post-filtering for automotive applications. In: Proceedings of IEEE International Conference on Acoustics, Speech, and Signal Processing (ICASSP 2009), pp. 221–224 (2009). https://doi.org/10.1109/ICASSP.2009.4959560
14. Zhang, H., Fu, Q., Yan, Y.: A frequency domain approach for speech enhancement with directionality using compact microphone array. In: Proceedings of 9th Annual Conference of the International Speech Communication Association (INTERSPEECH 2008), pp. 447–450 (2008)
15. Fischer, S., Simmer, K.: An adaptive microphone array for hands-free communication. In: Proceedings of IWAENC-1995, pp. 1–4 (1995)
16. Bitzer, J., Simmer, K.U., Kammeyer, K.-D.: Multi-microphone noise reduction techniques for hands-free speech recognition - a comparative study. In: Proceedings of Workshop on Robust Methods for Speech Recognition in Adverse Conditions (ROBUST 1999), pp. 171–174 (1999)
17. Bitzer, J., Kammeyer, K., Simmer, K.: An alternative implementation of the superdirective beamformer. In: Proceedings of IEEE Workshop on Applications of Signal Processing to Audio and Acoustics (WASPLAA 1999), pp. 7–9 (1999). https://doi.org/10.1109/ASPAA.1999.810836
18. Lotter, T., Vary, P.: Dual-channel speech enhancement by superdirective beamforming. EURASIP J. Adv. Signal Process. **2006**(1), 1–14 (2006). https://doi.org/10.1155/ASP/2006/63297
19. Buck, M., Rößler, M.: First order differential microphone arrays for automotive applications. In: Proceedings of 7th International Workshop on Acoustic Echo and Noise Control, IWAENC, pp. 19–22 (2001)
20. Elko, G.W., Pong, A.-T.N.: A steerable and variable first-order differential microphone array. In: Proceedings of IEEE International Conference on Acoustics, Speech, and Signal Processing (ICASSP), pp. 223–226 (1997). https://doi.org/10.1109/ICASSP.1997.599609

A Spectral Mask - Based on Method for Applying into Generalized Sidelobe Canceller Beamformer

Nguyen Ba Huy[1], Pham Tuan Anh[2], and Quan Trong The[1(✉)]

[1] Faculty of Control Systems and Robotics, University ITMO, St.Petersburg, Russia
quantrongthe1984@gmail.com

[2] Faculty of Software Engineering and Computer Systems, University ITMO, St.Petersburg, Russia
ptanh@vku.udn.vn

Abstract. The problem of signal processing is extracting the desired target speaker while suppressing the annoying background noise in complex recording situation, where exists third-party talker, transport vehicle or interference, noise. Speech quality or speech intelligibility are the major characteristics, which confirm the effectiveness of processed method. In speech applications, such as hearing-aids, hands-free telephony, teleconferencing, acoustic instruments, microphone array (MA) is applied for increasing the speech quality. MA uses the spatial response, the priori information about direction - of - arrival (DOA) of interest signal, the properties of recording environment to obtain speech components and remove noise. Generalized Sidelobe Canceller (GSC) is one of the most widely exploited in microphone array beamforming, due to its convenience and effective noise reduction. However, in several situations, the performance of GSC beamformer is often degraded, because of the imprecise DOA. In this paper, the authors suggested an additive spectral mask for improving the evaluation of GSC beamformer. The resulting experiment proves the received advantage of the author's proposed algorithm to increase in the term of the signal-to-noise ratio (SNR).

Keywords: Generalized Sidelobe Canceller · microphone array · noise reduction · the speech quality · the signal-to-noise ratio · beamforming technique

1 Introduction

The digital signal processing of noisy recorded signals aim at improving its perception by humans or better speech quality, speech intelligibility, which is preferred as speech enhancement. Generally speaking, a formulation of speech enhancement technique is to enhance performance of noise reduction while keeping the desired target speech component of useful speaker in annoying complex recording scenarios. It is always to retain speech undistorted with the constraint

Supported by organization Digital Agriculture Cooperative.

N. T. Nguyen et al. (Eds.): CITA 2023, LNNS 734, pp. 332–341, 2023.
https://doi.org/10.1007/978-3-031-36886-8_28

of reducing noise level and thus, the effective system's performance between noise suppression and remaining speech component.

The single-channel approach often uses the noise subtraction, which has high capacity in stationary situation, and causes speech distortion in non-stationary noise environment. The most common reason, that leads the corruption of speech quality and speech intelligibility is surrounding noise, which can't be determined by a certain formulation. A classification of speech enhancement algorithm is called spectral processing. This approach own the drawback is degradation of desired speech component of desired talker. Therefore, the scholar have studied array signal processing [1–20] and has numerous perspective developments.

Fig. 1. The complex environment around human-life.

The primary purpose of designed MA beamformer to obtain a beampattern, which towards a certain sound source and suppresses all incoming signal from other directions. MA beamformer exploits a priori information of MA's distribution, the properties of environment, the direction - of - arrival (DOA) of interest signal to extract the only useful signal while eliminating interference, noise and third-party speaker. Spatial filtering is the most importance part of MA beamformer.

Fig. 2. The convenience of microphone array for saving the only speech components.

Generalized Sidelobe Canceller (GSC) Beamformer is the most useful technique to obtain speech signal from noisy mixture, due to its convenient, lows computation and easy installed in all acoustic instrument, equipments and recorded devices. However, in realistic scenario, the performance of GSC Beamformer often degraded, due to the existing the retained speech component in the reference signal. In this article, the authors proposed using an additive spectral mask to block this component and enhance the overall performance.

The rest of this contribution is presented as following way. The second section describes the principal working of GSC beamformer. In the next section, the author analyzed the proposed spectral mask. Section IV, V are the experiments and Conclusion.

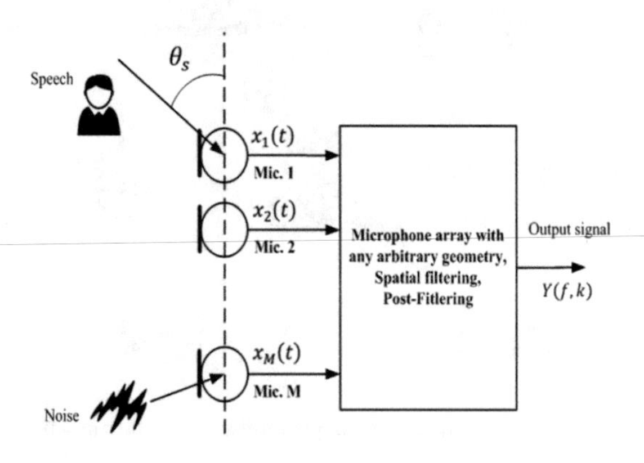

Fig. 3. The scheme of signal processing by microphone array.

2 Generalized Sidelobe Canceller Beamformer

The scheme of Generalized Sidelobe Canceller Beamformer [21–28] is shown in Fig. 4. GSC beamformer has two branches, the upper branch (FBF) often uses the delay - and - sum algorithm to form a beampattern toward the desired target speaker; the lower branch (BM) uses subtraction operation to get the only noise component. And the most importance block is adaptive filter, which removes noise from FBF signal. The common widely method used is Wiener filter.

In the STFT-domain, the equations of two recorded microphone array signals can be expressed as:

$$X_1(f,k) = S(f,k)e^{j\Phi_s} + N_1(f,k) \tag{1}$$

$$X_2(f,k) = S(f,k)e^{-j\Phi_s} + N_2(f,k) \tag{2}$$

where $\Phi_s = \pi f \tau_0 cos\theta_s$, $\tau_0 = d/c$, d is the distance between two microphones, $c = 343(m/s)$ is the sound propagation in the fresh air; f, k are the index of frequency and frame, respectively.

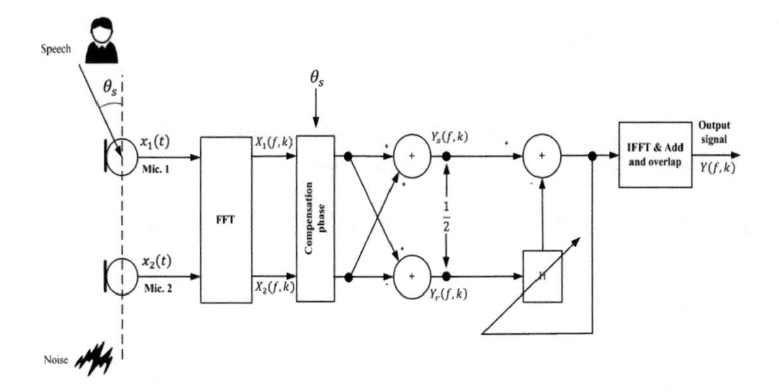

Fig. 4. The principal working of GSC Beamformer.

The output signal of FBF block: $Y_s(f,k)$ and the output of BM block: $Y_r(f,k)$ are computed as:

$$Y_s(f,k) = \frac{X_1(f,k)e^{-j\Phi_s} + X_2(f,k)e^{j\Phi_s}}{2} \tag{3}$$

$$Y_r(f,k) = \frac{X_1(f,k)e^{-j\Phi_s} - X_2(f,k)e^{j\Phi_s}}{2} \tag{4}$$

The necessary component, which used in Wiener filter, are the auto power spectral densities (PSD), and cross-PSD are calculated as the following way:

$$P_{YsYr}(f,k) = (1-\alpha)P_{YsYr}(f,k-1) + \alpha Y_s(f,k)Y_r^*(f,k) \tag{5}$$

$$P_{YrYr}(f,k) = (1-\alpha)P_{YrYr}(f,k-1) + \alpha Y_r(f,k)Y_r^*(f,k) \tag{6}$$

where α is a smoothing parameter, which in range of $\{0..1\}$. The optimum Wiener filter are determined:

$$H(f,k) = \frac{P_{YsYr}(f,k)}{P_{YrYr}(f,k)} \tag{7}$$

The final output signal is derived by:

$$Y(f,k) = Y_s(f,k) - Y_r(f,k) * H(f,k) \tag{8}$$

In fact, due to many undetermined conditions, the performance of GSC beamformer of degraded. The remaining speech component in the reference signal can lead to decrease speech quality at GSC beamformer's output. In the next section, the authors will present a spectral mask, which blocks all desired speech component to enhance the speech enhancement.

3 The Suggested Spectral Mask

The suggested spectral mask (SM) based on the estimation of desired target speech power $\sigma_s^2(f,k)$ and noise power $\sigma_n^2(f,k)$. In diffuse noise field, the covariance noise power $\sigma_n^2(f,k)$ can be calculated as:

$$\sigma_n^2(f,k) = E\{X_1(f,k)X_2^*(f,k)\} - \sqrt{E\{|X_1(f,k)|^2\} \times E\{|X_2(f,k)|^2\}} \tag{9}$$

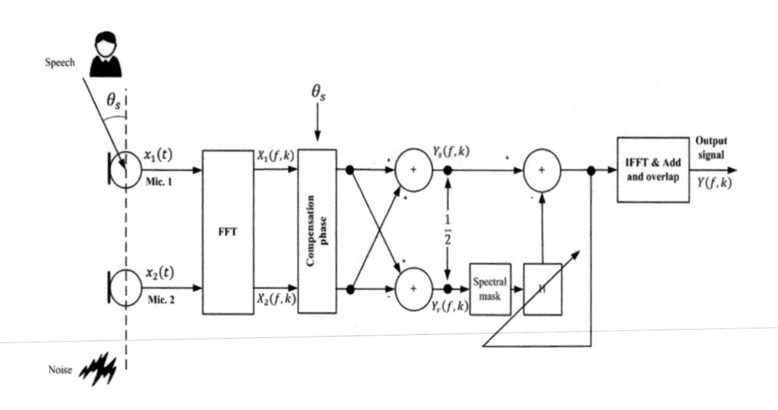

Fig. 5. The author's suggested spectral mask for the reference signal.

According to the preferred direction of arrival θ_s, the covariance speech power $\sigma_s^2(f,k)$ is computed by the following equation:

$$\sigma_s^2(f,k) = \frac{1}{\boldsymbol{D}_s^H(f,\theta_s)\Phi_{XX}^{-1}(f,k)\boldsymbol{D}_s(f,\theta_s)} \tag{10}$$

where the steering vector $\boldsymbol{D}_s(f,\theta_s) = [e^{j\Phi_s} \quad e^{-j\Phi_s}]^T$, and $\boldsymbol{\Phi}_{XX}(f,k) = E\{\boldsymbol{X}^H(f,k)\boldsymbol{X}(f,k)\}$, $\boldsymbol{X}(f,k) = [X_1(f,k) \quad X_2(f,k)]^T$. Consequently, the priori signal-to-noise ratio $SNR_prio(j\omega)$ can be derived as the following way:

$$SNR_{prio}(j\omega) = \frac{\sigma_s^2(f,k)}{\sigma_n^2(f,k)} \tag{11}$$

In the reverberation and very complex noisy recording situations, due to the similarity of diffuse noise field's properties and the speech leakage. When the power of diffuse noise is high, the relevant considered frame is determined as target speaker, there the performance of GSC beamformer is corrupted and the canceled the speech quality.

As we know that, the existing of speech component in the reference signal of GSC beamformer indicates that in the low frequency, the speech component is not perfect suppressed. Therefore, the filter own the following equation: $1 - \frac{\sin(\omega\tau_0)}{\omega\tau_0)}$ can block and filtered out speech leakage.

The authors proposed an combination between the convention Wiener filter and the above filter. This combination is the author's proposed spectral mask, and can be written as:

$$SM(f,k) = (1 - \frac{\sin(\omega\tau_0)}{\omega\tau_0)}) \times \frac{1}{1 + SNR_{prio}(j\omega)} \tag{12}$$

This spectral mask ensure to eliminate the amplitude and the low frequency component of speech leakage in the reference signal. And the reference signal was subtracted the remained speech component by the operator:

$$\hat{Y_r}(f,k) = Y_r(f,k) * SM(f,k) \tag{13}$$

In the next section, an illustrated experiments with the real-life recording environment to verify and confirm the effectiveness of the author's technique.

4 Experiments

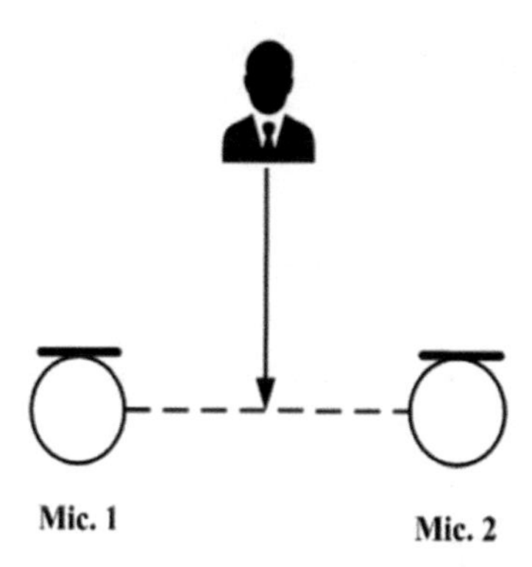

Fig. 6. The scheme of illustrated experiments.

The purpose of this section is confirmation of the suggested method to reduce speech distortion and increase the speech quality. An objective measurement [29] is used for computing the obtained SNR by the conventional GSC beamformer (GSC-con-beam) and the proposed method.

The author exploited the using of dual - microphone systems (DMA2), due to low computation, compactness and easy installed microphone array beamforming algorithm. The distance between two microphones is $d = 5(cm)$, and the speaker stand at $2(m)$ from DMA2 at $\theta_s = 90(deg)$. The diffuse noise field existed.

For further signal processing, these parameters: $NFFT = 512$, overlap 50%, the smoothing parameter $\alpha = 0.1$, the sampling frequency $Fs = 16kHz$. An objective measurement [29] is used for comparing the signal-to-noise ratio between GSC-con-beam and the proposed method.

Figure 8 shows the obtained signal by using GSC-con-beam.

And the resulting signal, which received by the proposed method.

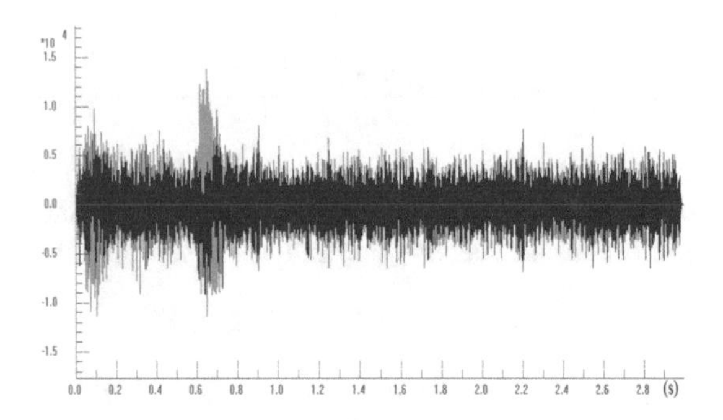

Fig. 7. The waveform of the recored microphone array signal.

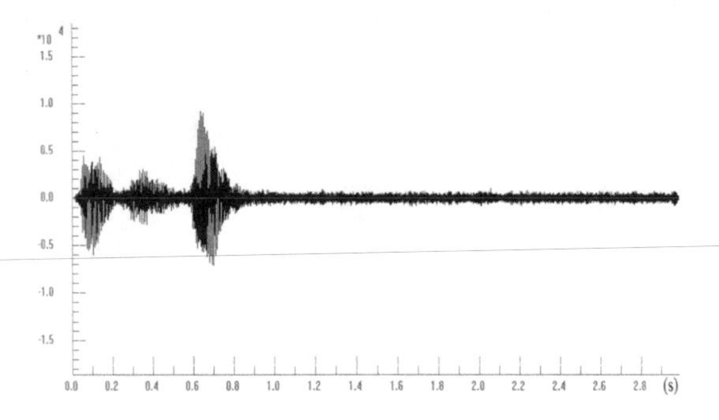

Fig. 8. The waveform of processed signal by GSC-con-beam.

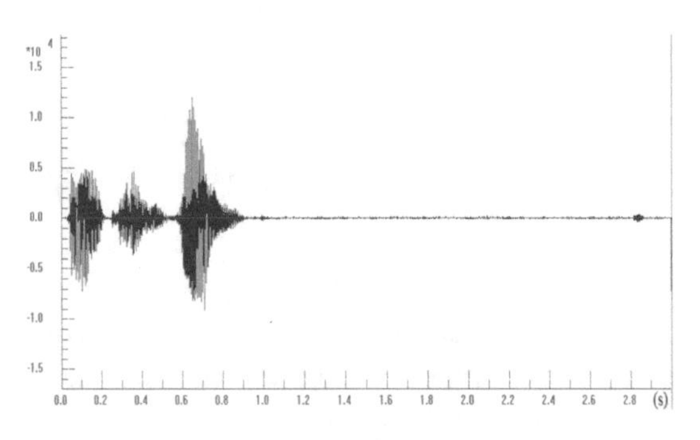

Fig. 9. The waveform of resulting signal by the proposed method.

A comparison of energy between the original microphone array signal, GSC-con-beam and the proposed method is shown in Fig. 10.

From Fig. 10, we can see that, the proposed technique reduce the speech distortion while suppressing the background noise in comparison with GSC-con-beam. Therefore, the speech quality in term of SNR is increased from 9.9(dB) to 10.1(dB). The measured SNR is depicted in Table 1.

The demonstrated waveform of microphone array signal, processed signal by GSC-con-beam and the proposed method has show us about the advantage

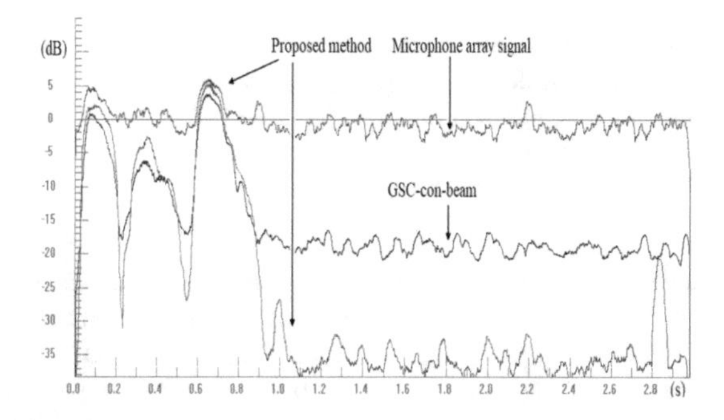

Fig. 10. The energy of microphone array signal, GSC-con-beam and the proposed method.

Table 1. The signal-to-noise ratio (dB)

Method Estimation	Microphone array signal	Conventional GSC	The proposed method
NIST STNR	4.8	18.5	28.6
WADA SNR	2.0	16.4	26.3

of suggested technique. The proposed method has increased noise reduction in the noisy frame and decreased speech distortion at the frame with presence of speech.

The numerical results proven the effectiveness of the suggested technique in reducing the speech distortion and removing annoying noise. The obtained advantage shows that, this approach can be integrated into multiple microphones system to solve the existing drawback of MA beamforming.

Saving the target speaker while removing background noise is the goal of almost signal processing algorithms. In this contribution, the proposed is suitable these requirements. In the future, the author will improve the robust of blocking matrix in non-stationary environment to achieve better performance.

5 Conclusion

In recent years, with the development of acoustic devices, the importance of automatic speech recognition, speech acquisition, speech enhancement has considerably attracted the scholars. MA own the common implementation of small devices with built-in DMA2, such as mobile phone, voice recorders; consequently, efficient digital signal processing are particular requisite. MA beamforming's evaluation often corrupted due to numerous undetermined conditions. In this paper, the author have presented a spectral mask to block the speech component, which remained in the reference signal, and increase the GSC beamformer's performance. The illustrated result has confirmed that, the proposed method allow achieving satisfactory speech enhancement in noisy complex scenario and open a new research approach.

Acknowledgements. This research was supported supported by Digital Agriculture Cooperative. The author thank our colleagues from Digital Agriculture Cooperative, who provided insight and expertise that greatly assisted the research.

References

1. Yuen, N., Friedlander, B.: Asymptotic performance analysis of ESPRIT higher order ESPRIT and virtual ESPRIT algorithms. IEEE Trans. Signal Process. **44**(10), 2537–2550 (1996). https://doi.org/10.1109/78.539037
2. Zhang, W., Wu, S., Wang, J.: Robust Capon beamforming in presence of large DOA mismatch. Electron. Lett. **49**(1), 75–76 (2013). https://doi.org/10.1049/el.2012.3182
3. Steinwandt, J., Roemer, F., Haardt, M.: Performance analysis of ESPRIT-type algorithms for non-circular sources. Proc. IEEE Int. Conf. Acoust. Speech Signal Process. (ICASSP), 3986–3990 (2013). https://doi.org/10.1109/ICASSP.2013.6638407
4. Lee, J., Hudson, R.E., Yao, K.: Acoustic DOA estimation: an approximate maximum likelihood approach. IEEE Syst. J. **8**(1), 131–141 (2014). https://doi.org/10.1109/JSYST.2013.2260630
5. Wen, J., Zhou, X., Zhang, W., Liao, B.: Robust adaptive beamforming against significant angle mismatch. Proc. IEEE Radar Conf. (Radar-Conf), 713–716 (2017). https://doi.org/10.1109/RADAR.2017.7944295
6. Gannot, S., Burshtein, D., Weinstein, E.: Signal enhancement using beamforming and nonstationarity with applications to speech. IEEE Trans. Signal Process. **49**(8), 1614–1626 (2001). https://doi.org/10.1109/78.934132
7. Wan, H., Huang, H., Liao, B., Quan, Z.: Robust beamforming against direction-of-arrival mismatch via signal-to-interference-plus-noise ratio maximization. In: 2017 9th International Conference on Wireless Communications and Signal Processing (WCSP), pp. 1–5 (2017). https://doi.org/10.1109/WCSP.2017.8171102
8. Rahmani, M., Bastani, M.H., Shahraini, S.: Two layers beamforming robust against direction-of-arrival mismatch. IET Signal Process. **8**(1), 49–58 (2014). https://doi.org/10.1049/iet-spr.2013.0031
9. Zohourian, M., Enzner, G., Martin, R.: Binaural speaker localization integrated into an adaptive beamformer for hearing aids. IEEE/ACM Trans. Audio Speech Lang. Process. **26**(3), 515–528 (2018). https://doi.org/10.1109/TASLP.2017.2782491
10. Massoud, A., Noureldin, A.: Angle of arrival estimation based on warped delay-and-sum (WDAS) beamforming technique. Proc. OCEANS (2011). https://doi.org/10.23919/OCEANS.2011.6107316
11. Ahmed, M.F.A., Vorobyov, S.A.: Collaborative beamforming for wireless sensor networks with Gaussian distributed sensor nodes. IEEE Trans. Wireless Commun. **8**(2), 638–643 (2009). https://doi.org/10.1109/TWC.2009.071339
12. Zhao, L., Benesty, J., Chen, J.: Design of robust differential microphone arrays. IEEE/ACM Trans. Audio Speech Lang. Process. **22**(10), 1455–1466 (2014). https://doi.org/10.1016/j.apacoust.2016.03.015
13. Frost, O.L.: An algorithm for linearly constrained adaptive array processing. Proc. IEEE **60**(8), 926–935 (1972). https://doi.org/10.1109/PROC.1972.8817
14. Liu, L., Li, Y., Kuo, S.M.: Feed-forward active noise control system using microphone array. IEEE/CAA J. Automat. Sinica **5**(5), 946–952 (2018). https://doi.org/10.1109/JAS.2018.7511171

15. Gogineni, V.C., Mula, S.: Logarithmic cost based constrained adaptive filtering algorithms for sensor array beamforming. IEEE Sensors J. **18**(24), 5897–5905 (2018). https://doi.org/10.1109/JSEN.2018.2841430
16. Comminiello, D., Scarpiniti, M., Parisi, R., Uncini, A.: Combined adaptive beamforming techniques for noise reduction in changing environments. In: Proceedings of the 36th International Conference on Telecommunications and Signal Processing (TSP), pp. 690–694, Jul. (2013). https://doi.org/10.1109/TSP.2013.6614025
17. Khan, Z., Kamal, M.M.D., Hamzah, N., Othman, K., Khan, N.I.: Analysis of performance for multiple signal classification (MUSIC) in estimating direction of arrival. In: Proceedings of the IEEE International RF and Microwave Conference, pp. 524–529 (2008). https://doi.org/10.1109/RFM.2008.4897465
18. Ma, X., Dong, X., Xie, Y.: An improved spatial differencing method for DOA estimation with the coexistence of uncorrelated and coherent signals. IEEE Sens. J. **16**(20), 3719–3723 (2016). https://doi.org/10.1109/JSEN.2016.2532929
19. Zhong, X., Premkumar, A.B., Madhukumar, A.S.: Particle filtering for acoustic source tracking in impulsive noise with alpha-stable process. IEEE Sens. J. **13**(2), 589–600 (2013). https://doi.org/10.1109/JSEN.2012.2223209
20. Yan, Y., Jin, M., Qiao, X.: Low-complexity DOA estimation based on compressed MUSIC and its performance analysis. IEEE Trans. Signal Process. **61**(8), 1915–1930 (2013). https://doi.org/10.1109/TSP.2013.2243442
21. Kong, Y., Zeng, C., Liao, G., Tao, H.: A new reduced-dimension GSC for target tracking and interference suppression. In: Proceedings of 2011 IEEE CIE International Conference on Radar, pp. 785–788 (2011). https://doi.org/10.1109/CIE-Radar.2011.6159658
22. Kim, S.M., Kim, H.K.: Multi-microphone target signal enhancement using generalized sidelobe canceller controlled by phase error filter. IEEE Sens. J. **16**(21), 7566–7567 (2016). https://doi.org/10.1109/JSEN.2016.2602375
23. Buckley, K.: Broad-band beamforming and the generalized sidelobe canceller. IEEE Trans. Acoust. Speech Signal Process. ASSP-34(5), 1322–1323 (1986). https://doi.org/10.1109/TASSP.1986.1164927
24. Tanan, S.C., Nathwani, K., Jain, A., Hegde, R.M., Rani, R., Tripathy, A.: Acoustic echo and noise cancellation using Kalman filter in a modified GSC framework. In: Proceedings of the 48th Asilomar Conference on Signals, Systems and Computers, pp. 477–481 (2014). https://doi.org/10.1109/ACSSC.2014.7094489
25. Liu, Z., Zhao, S., Zhang, G., Jiao, B.: Robust adaptive beamforming for sidelobe canceller with null widening. IEEE Sens. J. **19**(23), 11213–11220 (2019). https://doi.org/10.1109/JSEN.2019.2936681
26. Yu, C., Su, L.: Speech enhancement based on the generalized sidelobe cancellation and spectral subtraction for a microphone array. In: Proceedings of the 8th International Congress on Image and Signal Processing (CISP), pp. 1318–1322 (2015). https://doi.org/10.1109/CISP.2015.7408086
27. Triki, M.: Performance issues in recursive least-squares adaptive GSC for speech enhancement. In: Proceedings of the IEEE International Conference on Acoustics, Speech and Signal Processing, pp. 225–228 (2009). https://doi.org/10.1109/ICASSP.2009.4959561
28. Gomes, F., Neto, A., De Araujo, A.: Kalman filter applied to GSC in adaptive antennas array. In: Proceedings of the IEEE Radio and Wireless Conference (RAWCON), pp. 125–130 (2000). https://doi.org/10.1109/RAWCON.2000.881871
29. https://labrosa.ee.columbia.edu/projects/snreval/

An Integrated Model of Placement Optimization and Redundancy Elimination in RFID Network Planning

Van Hoa Le[1], Thanh Chuong Dang[2], Hong Quoc Nguyen[3], and Viet Minh Nhat Vo[4](✉)

[1] School of Hospitality and Tourism, Hue University, Hue, Vietnam
levanhoa@hueuni.edu.vn
[2] University of Sciences, Hue University, Hue, Vietnam
[3] University of Education, Hue University, Hue, Vietnam
[4] Hue University, Hue, Vietnam
vvmnhat@hueuni.edu.vn

Abstract. Radio frequency identification network planning (RNP) is the issue of placing readers in a work area so that the readers cover most of the tags, while satisfying some constraints such as the minimum number of readers used, the minimum interference, the minimum energy consumption, etc. RNP is assessed as an NP-hard problem and natural evolution-based approaches are often used to solve them. The paper proposes an integrated model of placement optimization and redundancy elimination, in which the optimal position of readers is found by a genetic-based method and the elimination of redundant readers is accomplished by a filtering policy. Gridding the work area is also analyzed, in which the finer the cell size is, the more efficient the reader placement is. The simulation results show that the cell whose size equal to the radius of the reader's interrogation area gives the best efficiency in terms of coverage, amount of used readers and fast convergence, but it also suffers a little extra interference.

Keywords: RFID network planning · gridding · cell size · GA-based placement optimization · redundany elimination

1 Introduction

Radio frequency identification (RFID) technology has demonstrated many outstanding advantages such as no physical contact, fast transmission, high security and high storage capacity. Unlike traditional barcode recognition technology, RFID is more widely used to tag of physical objects for monitoring in many different fields such as healthcare, supply chain management, logistics, transportation and agriculture. RFID technology is seen as the foundation for the Internet of Things (IoT), where RFID systems provide the information infrastructure for large-scale IoT applications [1, 2].

An RFID network consists of a collection of RFID tags, one or more RFID readers connected, and a central server to store and process collected data. An RFID tag can be

N. T. Nguyen et al. (Eds.): CITA 2023, LNNS 734, pp. 342–354, 2023.
https://doi.org/10.1007/978-3-031-36886-8_29

active, when it has its power supply, or it can be passive, when it has no power source. In order to respond to an interrogation or transmit its data, a passive RFID tag must draw power from readers. Passive tags are quite cheap, so they are widely used in reality, such as public security, traffic control, warehouse monitoring, etc. [3, 4].

Each RFID reader has a limited coverage range, so it is necessary to place them properly so that the network of RFID readers can cover almost all tags. In addition, the minimum number of used readers, the minimum interference, the minimum energy consumption, etc. are other requirements of an RFID system. Finding an appropriate placement of RFID readers in a work area is known as RFID network planning (RNP) [5] and solving the issue is often hard work because it is rated as NP-Hard [6]. Therefore, the methods inspired by natural evolution, such as genetic algorithms, particle swarm optimization, cuckoo search, etc. are often exploited to solve this problem [7].

The paper proposes an integrated model of placement optimization and redundancy elimination. A genetic algorithm (GA) is used to find the optimal position of readers in terms of maximum tag coverage, minimum number of readers used and minimum interference. For the redundant reader elimination, a policy is proposed to eliminate the redundant readers without or with little impact on the tag interrogation efficiency. To reduce the computational complexity, the number of candidate readers to place is to be limited. The workspace is thus gridded, where each cell is a candidate position to place a reader. The finer the gridding, the smoother the placement of readers will be, but the higher the number of candidate positions and the more complicated the calculation. The paper investigates and analyses some cases of cell size to determine the best gridding.

The main contributions of the article include the following:

(1) proposing an integrated model of GA-based placement optimization and policy-based redundancy elimination; and
(2) investigating and analyzing the impact of the cell size on the reader placement efficiency.

The centents of the next sections are as follows. Section 2 presents works related to the use of GA in solving RNP. Section 3 introduces the proposed model, which describes in detail finding the optimal reader placement based on GA and removing redundant readers based on the proposed policies. The implementation and analysis of simulation results are shown in Sect. 4 and the conclusion is in Sect. 5.

2 Related Works

Since RNP is rated as NP-hard [6], the commonly used methods to solve RNP are based on natural evolution, in which the genetic algorithm (GA) has attracted much attention in literature because of its robust and global search. The following are the reviews of using genetic algorithms in solving RNP.

The first application of GA in solving RNP was proposed by Guan et al. [8], where the goals are to minimize the number of placed readers, maximize the coverage, minimize the interference, and guarantee uplink/downlink signals. These goals are formulated into component objective functions and a fitness function is built as the weighted sum of the component objective functions. To implement GA, Guan et al. proposed a multilevel encoding model for chromosomes, where level 1 identifies the readers' position, level 2 identifies the antenna type, and level 3 contains other parameters (such as signal attenuation). Experimental results on a rectangular area of 120 m^2 show only six readers are needed to cover 92% of a work area, while that proposed by Vasquez et al. [9] requires up to 7 readers, but covers only 90%.

Similarly, Yang et al. [10] proposed a GA-based solution to the multi-objective RNP optimization by mapping the planning into the structure of genes, the chromosome and the operations such as selection, crossover and mutation. This proposal not only eliminates the search errors of the traditional multi-objective optimization method, but also provides an effective solution to RNP.

With Botero & Chaouchi [11], RNP is considered with six objectives, including the minimum overlap area, the minimum number of used readers, the maximum number of covered tags, the minimum number of readers placed out of the work area, the minimum number of redundant readers and the minimum number of tags located in the overlap range. These objectives are formulated as component objective functions and weighted sum to form the fitness function. The chromosome is encoded as a 21-bit sequence that carries information about the energy level and readers' position. Experiments performed on two propagation models of Friis and ITU in a square area of 20×20 m^2 show that ITU model has a smaller coverage, fewer number processing loops and faster processing time than the Friis model.

Xiong et al. [12] use a genetic algorithm to determine the minimum number of readers and their optimal reader placement in a work area of 30×30 m^2 with 99 randomly distributed tags. In the case of using only ten readers, the interrogation area of readers only covered 76 tags, however, it also outperformed those of the previous studies that covered only 72 tags. To cover all 99 tags, Xiong's proposal required 21 readers, but these was still much less than the previous studies that required 30 readers.

Different from the above studies, Tang et al. [13] consider the case of the heterogeneous coverage range. Component objectives including the minimum overlap, reader collision, and interference are formulated into a multi-objective function. The proposed algorithm is an improvement of GA by integrating with a divide-and-conquer greedy heuristic algorithm. The results show that the multi-objective GA achieves better results than some other recently developed evolution-based methods.

Although the genetic algorithm was successfully applied to RNP, the fixed-length encoding scheme limits the adjustment of the number of encoded readers in each chromosome. Therefore, Zhang et al. [14] developed a flexible genetic algorithm in which chromosomes are variable-length, crossover is performed by sub-region swapping and mutation is based on Gaussian-distribution. The experimental results show that the flexible genetic algorithm achieves higher efficiency, in terms of coverage, interference and convergence, compared to traditional genetic approaches.

The proposals for applying GA to solve RNP mainly use fixed-length encoding for chromosomes, but this approach is inefficient in terms of resource usage. The variable-length encoding approach depending on the number of readers in each candidate solution presents more advantages, but the sub-region swapping-based crossover approach does not prove its efficiency in terms of convergence speed. This paper proposes a 2-phase model, in which Phase 1 uses a GA to find the optimal position of readers according to different cell sizes and Phase 2 uses a set of policies of redundancy elimination to minimize the number of used readers.

3 Integrated Model of Placement Optimization and Redundancy Elimination

3.1 Description of RFID Network Planning

We consider a work area of size X × Y m^2 (e.g. 50 × 50), in which the tags are distributed evenly and randomly. To limit the number of candidate positions to place readers, the work area is gridded into cells. Coarse gridding will make the algorithm converge quickly, but it won't find the best placement for readers; while finer gridding increases the chances of finding the optimal placement position, it also increases the computational volume. The problem is how to determine the best cell size without increasing computational complexity.

Assuming that the readers are isomorphic with their transmission frequency of 915 MHz, their transmission power of 2 watts (W), their receive power threshold of 0.1 milliwatts (mW), and their equipped scalar antenna with circular coverage, the antenna coverage radius (r) is determined by $r = \frac{\lambda}{4\pi}\sqrt{\frac{P_t G_t G_r}{P_r}}$, where P_t is the power transmitted by a reader (2 W), P_r is the power transmitted by a tag (0.1 mW or −10 dBm), Gt and Gr, are the gain of the reader and the tag (assumed to be 1), and λ is the signal wavelength (0.3278 m).

By using the above values, the coverage radius (r) of each reader is determined to be 3.69 m. The hexagonal packing-based approach in [15] is an example of the reader placement that covers the entire work area with minimal interference. With a radius of 3.69 m, the distance of two consecutive readers is 2 × 3.69 × cos(30°) ≈ 6.4 m, as shown in Fig. 1. So to cover the entire work area of 50 × 50 m^2 we need (50/6.4)^2 ≈ 64 readers. This is also the maximum number of readers to use (n_{max}). However, depending on the distribution of tags, the number of used readers (n) may be less.

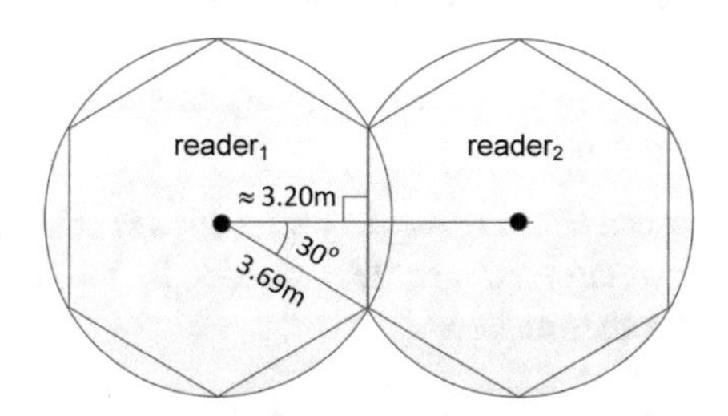

Fig. 1. The circular coverage area is "organised" into a hexagonal cell.

We can choose a smaller cell size to increase the smoothness and flexibility of placement. However, this increases the number of candidate positions. As shown in Fig. 2, by reducing the cell size by half (3.2 m), the number of candidate locations increases to 16 × 16 = 256. This also increases the number of candidate solutions when n readers are placed on these 256 locations and increases computational complexity.

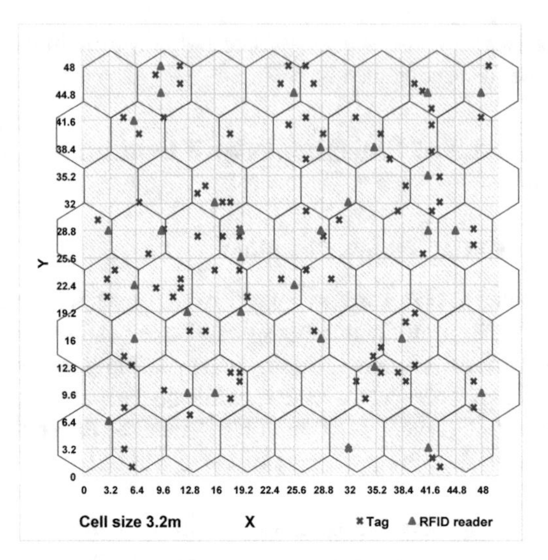

Fig. 2. Example of a gridded work area.

The objective of RNP is to find the position of readers so that the minimum number of used readers and the maximum number of covered tags. Let x_i be indexed to determine the state of reader i, $x_i = 1$ if reader i is used and $x_i = 0$ if reader i is free. The objective of RNP is to minimize the number of used readers:

$$f_1 = \sum\nolimits_{i=1}^{n_{\max}} x_i \tag{1}$$

subject to cover up to the tags in the work area:

$$f_2 = \sum\nolimits_{i=1}^{n_{\max}} T_i x_i \tag{2}$$

where T_i is the number of tags covered by reader i and T is the total number of tags in the work area.

In addition, the signal-to-interference ratio (SIR) should also be kept to a minimum. There are two types of interference:

(1) the downlink interference (from reader to tag), which is calculated as the sum of the peak signals received at tag j from reader i divided by the sum of the above signals and interference from other readers:

$$f_3 = \frac{\sum_{j=1}^{n} \sum_{i \in R_j} \max(D_{i,j} x_i)}{\sum_{j=1}^{n} \sum_{i \in R_j} \max(D_{i,j} x_i) + \sum_{j=1}^{n} \sum_{i \in R_j} \sum_{k \in S_j} \max(D_{i,j} x_i)} \tag{3}$$

and the uplink interference (from tag to reader) which is calculated similarly to downlink interference but in the opposite direction:

$$f_4 = \frac{\sum_{j=1}^{n} \sum_{i \in N_j} \max(U_{i,j} x_i)}{\sum_{j=1}^{n} \sum_{i \in N_j} \max(U_{i,j} x_i) + \sum_{j=1}^{n} \sum_{i \in N_j} \sum_{k \in S_j} \max(U_{i,j} x_i)} \tag{4}$$

where R_j is the set of readers i covering tag j (whose distance to tag j is less than the reader coverage radius) and S_j is the set of readers i generating interference at tag j (assumed to be readers have the same distance to tag j). $D_{i,j}$ and $U_{i,j}$ are downlink and uplink signals received/sent at tag j from reader i. For simplicity, $D_{i,j}$ and $U_{i,j}$ are considered as Euclidean distances in coordinates between tag j and reader i.

3.2 Formulation

To solve RNP, an 2-phase model (Fig. 3) is proposed, in which, Phase 1 (the placement optimization) exploits a genetic algorithm to find the optimal position of n readers and, Phase 2 (the redundancy elimination) uses some proposed policies to eliminate redundant readers without or with little effect on tag coverage. Detailed descriptions of Phase 1 and Phase 2 are presented in the next subsections. In fact, eliminating redundant readers can reduce tag coverage, so Phase 1 can be repeated several times to fine-tune the position of used readers. However, before implementing the 2-phase model, the determination of the best gridding should be made to limit the number of candidate positions when readers are trying to be placed.

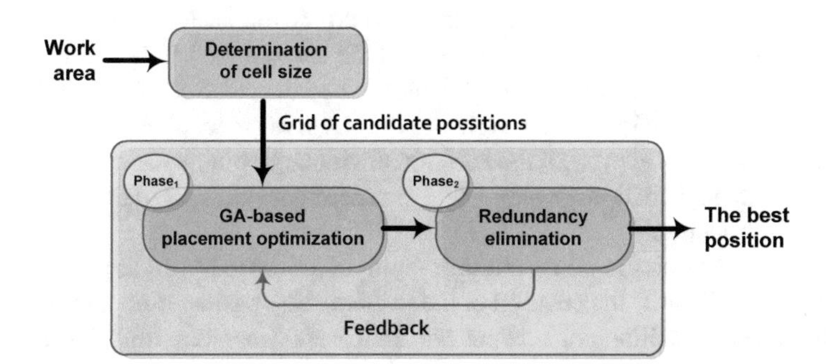

Fig. 3. The 2-phase model of placement optimization and redundant reader elimination

Phase 1: GA-Based Placement Optimization

A genetic algorithm is a search method inspired by natural evolution, in which good parental features are passed to offspring. To be able to apply a GA to solve a real problem, the candidate solutions need to be encoded into chromosomes. Then, evolutionary operations such as selection, crossover (recombination), and mutation are performed to produce offspring. The individuals, including parents and offspring, are then selected to form a new population. The evolution process is repeated until the best individual is found or the convergence condition is satisfied [16]. The operations are described in detail as follows.

To formulate RNP into GA, candidate solutions in the phenotype space need to be mapped to chromosomes in the genotype space. With the goal of determining the optimal position of readers in a work area, the coordinates of readers on the grid are of primary interest. In our formulation, each candidate solution is encoded as a sequence of genes, where 2 consecutive genes represent the coordinate of a reader. As a result, the chromosome is represented as a sequence of $2 \times n$ genes, where n is the number of used readers.

The population needs to be well initiated before performing the algorithm. Initiating the population involves determining the population size and the initial position of individuals. This is important because they affect the speed of convergence and the found optimal solution. For RNP, the initial population is randomly initialized and the number of individuals per population is 5% of the total of candidate solutions. The initial position of individuals has a Gaussian distribution.

At each iteration, the selection of parents for crossover, as well as the selection of individuals for the next generation population, is done based on their fitness. The goal of RNP is to minimize the number of used readers and is subject to the constraints of maximum coverage and minimum interference. Therefore, the fitness function is the weighted sum of the component functions from Eq. (1) to Eq. (4):

$$fitness = f_1 w_1 + f_2 w_2 + (f_3 + f_4) w_3 \tag{5}$$

where w_k is the weight of the k-th objective function and $\sum_{k=1}^{3} w_k = 1$. The best individual has the greatest fitness.

There are different ways to select parents to crossover, such as: randomly, roulette wheel-based, tournament-based, etc., among which the roulette wheel-based selection is the fairest because all individuals have a chance to be selected. Specifically, the selection probability of each individual is converted to the size of a wheel pie corresponding to its fitness value. With this approach, the individual with large fitness has a high probability to be selected (thanks to its large pie), while the individual with small fitness has the low probability of selection.

Crossover can be single-point, multiple-point or uniform. Multiple-point or uniform crossover helps “disturb” the position of the readers at each generation, thus the chances of finding a good position for readers are increased. However, this slows down the algorithm convergence. Therefore, for RNP, the two-point crossover is chosen for the first few generations and a single-point crossover for next generations to increase the chance of global optimization and speed up the convergence.

The mutation is an operation that creates diversity for a population and escapes local optimal positions. However, to ensure that the mutant is not too different from its parent, the mutation probability is often chosen to be very small. With RNP, the mutation is in only one gene with a probability of 0.05. The mutation value must belong to the gridded work area.

The selection of individuals for a new population is made based on the fitness values of offspring and their parents, where 30% of elite parents (with the best fitness) are transferred directly to the new population; the remaining 70% are individuals from the offspring with the best fitness.

The algorithm stops when the fitness does not change anymore over a few generations (e.g. 5 generations), or a given threshold for the maximum number of generations is reached (e.g. 50). The individual with the best fitness in the current population is selected as the encoding of the best-found candidate solution.

Phase 2: Redundant Reader Elimination

The output of Phase 1 is the best position for readers. However, there may be some redundant readers whose elimination does not or significantly affects tag coverage. Therefore, in Phase 2, a set of policies is proposed to eliminate redundant readers and thus reduce the number of used readers. A reader that is considered to be eliminated or disabled must meet the following three criteria:

- Eliminating does not or only reduces less than 1% tag coverage but the total coverage does not reduce below 90%;
- Eliminating can reduce more than 10% of interference; and
- Eliminating does not or only reduce less than 1% of the fitness value.

With the set of redundancy elimination policies, the overall system efficiency is not significantly affected.

4 Simulation and Analysis

The simulation is performed in a work area of 50 × 50 m^2, in which 99 tags were randomly distributed (Fig. 2). The number of readers used is n = 0.5 × n_{max} = 32. The algorithm stops after 50 generations or the fitness does not change after 5 generations. The fitness function is chosen in Eq. (5). Table 1 describes other simulation parameters.

Table 1. Simulation parameters

Parameters	Value
Experimental grid sizes	3.2, 1.6 and 0.8
Individuals/population	8
Selection method	Roulette wheel
Crossover	Single-point and two-point
Mutation (with probability)	Only one gen (0.05)
New population	30% elite parents, 70% best offspring
Weights of the component objective functions	$w_1 = 0.2$; $w_2 = 0.6$ and $w_3 = 0.2$

Simulation objectives include:

- Determining the optimal cell size of the work area at which the optimal position of readers is achieved; and
- Reducing the number of used readers by eliminating redundant readers.

4.1 Determining the Optimal Cell Size

Three surveyed cell sizes include 3.2, 1.6 and 0.8 m respectively, representing 1, 1/2 and 1/4 of the reader coverage radius. The placement efficiency, in terms of fitness, convergence and coverage ratio, is shown in Table 2.

Table 2. Comparison of placement efficiency with different cell sizes

No.	Grid size (m)	Candidate solutions	Best fitness	Convergence (loops)	Coverage ratio
1	3.2	$16^2 = 256$	0.904147	16	94%
2	1.6	$32^2 = 1024$	0.932430	22	98%
3	0.8	$64^2 = 4096$	0.936176	30	98%

From Table 2, we see that the smaller the cell size is, the higher the number of candidate solutions is. That increases the computational complexity and the convergence time. Specifically, with the grid size of 3.2 m, the number of candidate solutions is only 16^2, so only 16 iterations are needed to converge. However, when the grid decreases to 0.8 m, the number of candidate solutions increases to 64^2, so it takes 30 iterations to converge.

Large cell gridding does not help achieve the best fitness because there are few candidate solutions that can choose from. But as the cell size gets smaller, the reader placement becomes smooth and thus an optimal solution can be found where the fitness gets the best value. However, too small cell gridding explodes the computational complexity and prolongs the convergence time, however it improves little fitness and tag coverage. As described in Table 2, the cell size 1.6 m gives good results in terms of fitness, coverage ratio and convergence as that of the cell size 0.8 m. Figure 4 shows that after 30 generations, the fitness value corresponding to the cell size 1.6 m is 3% better than that of the cell size 3.2 m and its fitness value is approximately equal to that of the cell size 0.8 m.

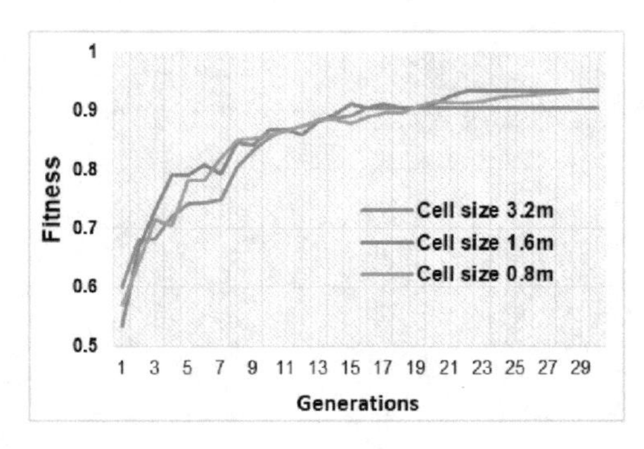

Fig. 4. Comparison of the fitness corresponding to different cell sizes

4.2 Reducing the Number of Used Readers

After finding the optimal position of readers, reducing the number of used readers by the policies of redundancy elimination is considered. As shown in Fig. 5, 4 readers (dotted circles) are eliminated for all 3 cell sizes of 3.2 m, 1.6 m and 0.8 m. However, the reduction of these readers also has an impact on the tag coverage as shown in Table 3, where the fitness corresponding to 2 cell sizes of 1.6 m and 0.8 m is slightly reduced, but the fitness corresponding to the cell size of 3.2 m is slightly increased. It is clear that eliminating some readers has increased the number of uncovered tags. The coverage corresponding to the cell size of 3.2 m is reduced by 2%, but the interference is also reduced by about 7%. In the case of cell sizes of 1.6 m and 0.8 m, the coverage is reduced by 4% and the interference is also reduced by about 7%.

Table 3. Comparison of the coverage and the interference before and after eliminating redundant readers

No.	Grid size (m)	Fitness value		Readers		Coverage (%)		Interference (%)	
		Before	After	Before	After	Before	After	Before	After
1	3.2	0.904147	0.909925	32	28	94	92	17.8	11.4
2	1.6	0.932430	0.924471	32	28	98	94	17.8	11.0
3	0.8	0.936176	0.925513	32	28	98	94	16.6	10.7

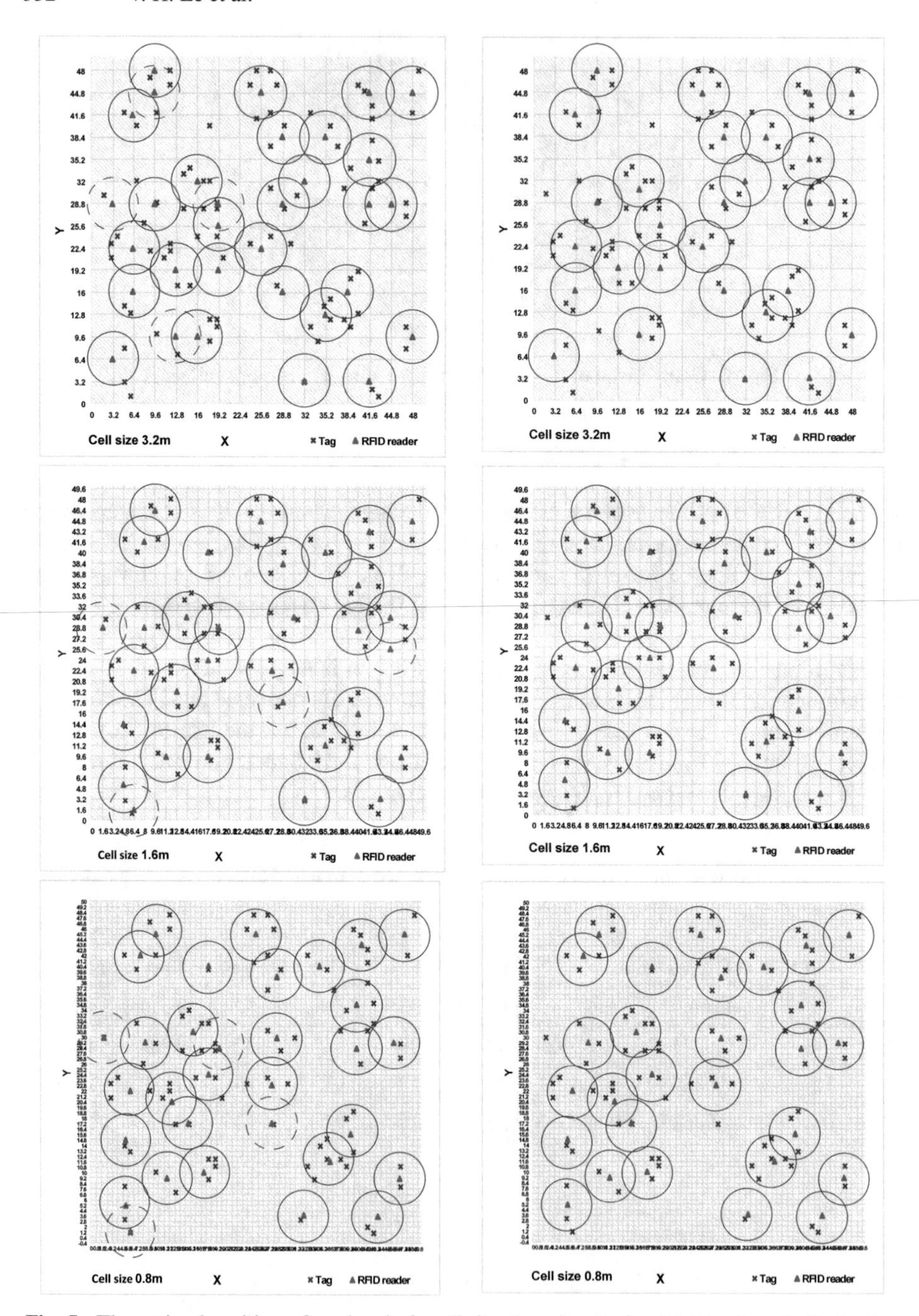

Fig. 5. The optimal position of readers before (left column) and after (right column) eliminating redundant readers with 3 cell sizes of 3.2, 1.6 and 0.8 m.

5 Conclusion

The article has successfully applied the genetic algorithm to find the optimal position of readers with the constraints of maximum coverage and minimum interference. In addition, the article also proposes a set of policies to eliminate redundant readers, so that the redundancy elimination does not or significantly affect tag coverage, interference and fitness. A 2-phase model is proposed, in which Phase 1 optimizes the position of readers based on a genetic algorithm and Phase 2 reduces the number of used readers by eliminating redundant readers. The article investigated and analyzed the impact of different cell sizes on the efficiency of RFID network planning. Simulation results show that our proposal has determined the best cell size (1.6 m) where the minimum number of readers is needed but it still ensures the overall tag reading efficiency for the whole system.

Acknowledgements. This work was supported/partially supported by Hue University under the Core Research Program, Grant No. NCM.DHH.2019.05.

References

1. Elbasani, E., Siriporn, P., Choi, J.S.: A survey on RFID in industry 4.0. In: Kanagachidambaresan, G.R., Anand, R., Balasubramanian, E., Mahima, V. (eds.) Internet of Things for Industry 4.0. EICC, pp. 1–16. Springer, Cham (2020). https://doi.org/10.1007/978-3-030-32530-5_1
2. Costa, F., Genovesi, S., Borgese, M., Michel, A., Dicandia, F.A., Manara, G.: A review of RFID sensors, the new frontier of internet of things. Sensors **21**(9) (2021)
3. Ibrahim, A.A.A., Kisar, Y., Hzou, K., Welch, I.: Review and analyzing RFID technology tags and applications. In: 2019 IEEE 13th International Conference on Application of Information and Communication Technologies (AICT), pp. 1–4, Baku, Azerbaijan (2019)
4. Gee, M., Anandarajah, P., Collins, D.: A review of chipless remote sensing solutions based on RFID technology. Sensors **19**(22), 4829 (2019)
5. Azizi, A.: Applications of Artificial Intelligence Techniques in Industry 4.0. Springer, Singapore (2019). https://doi.org/10.1007/978-981-13-2640-0
6. Suriya, A., David Porter, J.: Genetic algorithm based approach for RFID network planning. In: TENCON 2014 - 2014 IEEE Region 10 Conference, pp. 1–5, Bangkok, Thailand (2014)
7. Chen, H., Zhu, Y.: RFID networks planning using evolutionary algorithms and swarm intelligence. In: 2008 4th International Conference on Wireless Communications, Networking and Mobile Computing, pp. 1–4, Dalian, China (2008)
8. Guan, Q., Liu, Y., Yang, Y., Yu, W.: Genetic approach for network planning in the RFID systems. In: Sixth International Conference on Intelligent Systems Design and Applications, vol. 2, pp. 567–572, Jian, China (2006)
9. Vasquez, M., Hao, J.K.: A heuristic approach for antenna positioning in cellular networks. J. Heurist. **7**, 443–472 (2001)
10. Yang, Y., Wu, Y., Xia, M., Qin, Z.: A RFID network planning method based on genetic algorithm. In: 2009 International Conference on Networks Security, Wireless Communications and Trusted Computing, vol. 1, pp. 534–537, Wuhan, China (2009)
11. Botero, O., Chaouchi, H.: RFID network topology design based on Genetic Algorithms: In: 2011 IEEE International Conference on RFID-Technologies and Applications, pp. 300–305, Sitges, Spain (2011)

12. Xiong, Y., Giuseppe, V., Occhiuzzi, C., Marrocco, G., Caizzone, S., Quijano, J.A.: Optimization of multichip RFID tag antenna with genetic algorithm and method of moments. In: 2013 IEEE International Symposium on Antennas and Propagation and USNC/URSI National Radio Science Meeting, pp. 416–417 (2013)
13. Tang, L., Zheng, L., Cao, H., Huang, N.: An improved multi-objective genetic algorithm for heterogeneous coverage RFID network planning. Int. J. Prod. Res. **54**(8), 2227–2240 (2016)
14. Zhang, Y.H., Gong, Y.J., Gu, T.L., Li, Y., Zhang, J.: Flexible genetic algorithm: a simple and generic approach to node placement problems. Appl. Soft Comput. **52**, 457–470 (2017)
15. Huang, H.P., Chang, Y.T.: Optimal layout and deployment for RFID systems. Adv. Eng. Inform. **25**(1), 4–10 (2011)
16. Kumar, M., Husain, M., Upreti, N., Gupta, D.: Genetic algorithm: review and application. SSRN Electron. J. **2**(2), 451–454 (2010)

Software Engineering and Information System

Conjunctive Query Based Constraint Solving for Feature Model Configuration

Alexander Felfernig, Viet-Man Le(✉), and Sebastian Lubos

Institute of Software Technology, Graz University of Technology, Graz, Austria
{alexander.felfernig,vietman.le,slubos}@ist.tugraz.at

Abstract. Feature model configuration can be supported on the basis of various types of reasoning approaches. Examples thereof are SAT solving, constraint solving, and answer set programming (ASP). Using these approaches requires technical expertise of how to define and solve the underlying configuration problem. In this paper, we show how to apply conjunctive queries typically supported by today's relational database systems to solve constraint satisfaction problems (CSP) and – more specifically – feature model configuration tasks. This approach allows the application of a wide-spread database technology to solve configuration tasks and also allows for new algorithmic approaches when it comes to the identification and resolution of inconsistencies.

Keywords: Constraint Solving · Knowledge-based Configuration · Feature Model · Conjunctive Query · Relational Database

1 Introduction

Feature models (FM) can be used to represent commonality and variability aspects of highly-variant software and hardware systems [1,2,5,11,17]. These models can be specified on a graphical level and then translated into a corresponding formal representation that allows for feature model configuration including associated reasoning tasks such as conflict detection (e.g., induced by user requirements [16]) and conflict resolution (e.g., resolving the existing conflicts in a void feature model [3,15,18,24,26]).

There are different reasoning approaches supporting feature model configuration [9,23]. First, SAT solving [14] allows for a direct translation of individual features into a corresponding set of Boolean variables specifying feature inclusion or exclusion. SAT solving has shown to be an efficient reasoning approach supporting feature model analysis and configuration [19]. Second, constraint satisfaction problems (CSPs) can be used to represent feature model configuration tasks [6]. Compared to SAT solving, CSPs allow more "direct" constraint representations expressing typical constraint types such as logical equivalence and implications. Finally, answer set programming (ASP) supports the definition of object-centered configuration problems [22]. On the reasoning level, these

N. T. Nguyen et al. (Eds.): CITA 2023, LNNS 734, pp. 357–367, 2023.
https://doi.org/10.1007/978-3-031-36886-8_30

problems are translated into a corresponding Boolean Satisfiability (SAT) based representation which can then be used by a SAT solver.

All of the afore mentioned knowledge representations require additional technical expertise in representing feature model configuration knowledge and identifying corresponding configurations (on the reasoning level either in the form of SAT solving, constraint solving, or answer set programming). Furthermore, the resulting knowledge bases have to be included using the corresponding application programming interfaces. An alternative to SAT solving, constraint satisfaction problems, and ASP is to represent feature model configuration problems in terms of *conjunctive queries*.

The major contributions of this paper are the following: (a) we show how conjunctive queries can be used to represent and solve configuration tasks, (b) we report initial results of a corresponding performance evaluation, and (c) we sketch how such approaches can be exploited to make related consistency management tasks (e.g., feature model diagnosis) more efficient.

The remainder of this paper is organized as follows. In Sect. 2, we introduce an example feature model that is used as working example throughout this paper. In Sect. 3, we show how configuration tasks can be supported on the basis of conjunctive queries and introduce corresponding definitions of a configuration task and a feature model configuration. Alternative approaches to conflict resolution are discussed in Sect. 4. Thereafter, we present initial results of a performance analysis that compares the efficiency of conjunctive query based configuration with constraint solving (see Sect. 5). Section 6 includes a discussion of potential threats to validity. Finally, in Sect. 7 we conclude the paper with a discussion of open issues for future work.

2 Example Feature Model

For demonstration purposes, we introduce a simplified *survey software* feature model (see Fig. 1). Typically, features in such models are organized in a hierarchical fashion [5] where relationships between features specify the overall model structure: (a) *mandatory* relationships indicate that specific features have to be included in a configuration, for example, each survey software must include a corresponding payment, (b) *optional* relationships indicate that specific features can be included in a configuration, for example, the AB testing feature is regarded as optional in a survey software configuration, (c) *alternative* relationships indicate that exactly one of a given set of subfeatures must be included in a configuration (if the parent feature is included), for example, a payment is either a license or a no license, (d) *or* relationships indicate the optional inclusion of features from a given set of subfeatures given that the parent feature is included in the configuration, for example, questions (feature QA) can additionally include (beside single-choice) multiple-choice and multimedia-style questions.

Furthermore, so-called cross-tree constraints specify additional properties which are orthogonal to the discussed relationships: (a) *excludes* relationships between two features specify that the two features must not be included at the same time, for example, the *no license* feature does not allow the inclusion of AB testing and vice-versa, (b) *requires* relationships between two features specify the fact that if a specific feature f_1 is included, then another feature f_2 must be included, i.e., f_1 requires f_2. For example, the inclusion of AB testing requires the inclusion of statistics.

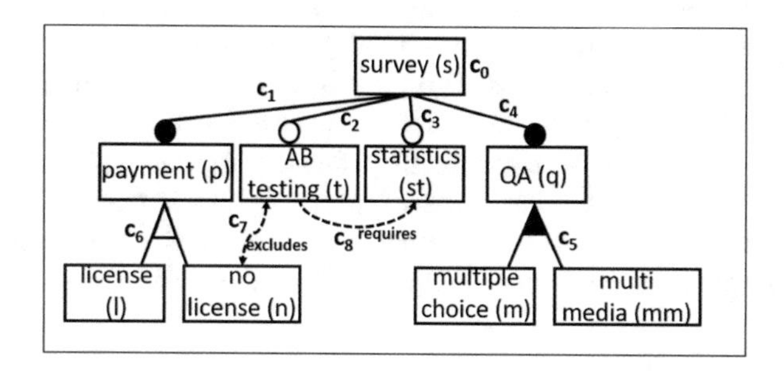

Fig. 1. An example feature model (*survey software*).

To support reasoning operations on feature model properties, these models have to be translated into a formal representation. Examples of such representations are SAT problems [13,19], constraint satisfaction problems [7,12], and answer set programs [22]. In the following, we show how a feature model configuration task can be represented and solved on the basis of *conjunctive queries.*

3 Conjunctive Query Based Configuration

Representing a feature model configuration task in terms of a conjunctive query allows the application of relational database technologies to support configuration tasks. In the following, we introduce the concepts of a feature model configuration task (see Definition 1) and a feature model configuration (see Definition 2) on the basis of a conjunctive query $F_{[c]}S$ where F can be (a) a table representing all possible feature model configurations (i.e., an enumeration of all possible feature model configurations), (b) a set of individual tables where each table represents an individual feature from the feature model, and (c) a set of individual tables representing tuples consistent with individual constraints of the feature model. Furthermore, $[c]$ is a set of criteria that have to be fulfilled – this includes feature model constraints as well as requirements defined by a configurator user (customer). Finally, S represents a so-called projection, i.e., those attributes (and values) that should be included in a query result.

Definition 1 (FM Configuration Task). A feature model (FM) *configuration task* can be defined as conjunctive query $F_{[c]}S$ where F represents the available features in tabular form (explicitly or implicitly), $[c]$ is the selection part of a conjunctive query representing criteria to be taken into account by the feature model configuration, and S is a set of projection attributes (features), i.e., features whose settings should be included as a part of the shown configuration result. Furthermore, $c = cr \cup cf$ where cr represents a set of customer requirements (i.e., feature inclusions and exclusions) and cf represents a set of constraints derived from the feature model.

In our working example, $cf = \{c_0..c_8\}$ (c_0 is the root constraint assuming that we are not interested in empty feature models). Furthermore, if we assume that the user of the feature model configurator is not interested in paying licenses and not interested in a multimedia based question support, $cr = [s = 1, p = 1, n = 1, mm = 0]$. The translation of a feature model representation into a corresponding conjunctive query based representation is shown in Table 1. For simplicity, in this formalization we assume a single-table representation which includes a complete enumeration of the whole configuration space.

Table 1. Translating feature model constraints into queries ($cf = \{c_0..c_8\}$).

constraint	conjunctive query based representation
c_0	$F_{[s=1]}\{s..mm\}$
c_1	$F_{[s=1 \wedge p=1 \vee s=0 \wedge p=0]}\{s..mm\}$
c_2	$F_{[\neg t=1 \vee s=1]}\{s..mm\}$
c_3	$F_{[\neg st=1 \vee s=1]}\{s..mm\}$
c_4	$F_{[s=1 \wedge q=1 \vee s=0 \wedge q=0]}\{s..mm\}$
c_5	$F_{[q=1 \wedge (m=1 \vee mm=1) \vee q=0 \wedge m=0 \wedge mm=0]}\{s..mm\}$
c_6	$F_{[((\neg l=1 \vee n=0 \wedge p=1) \wedge (\neg(n=0 \wedge p=1) \vee l=1)) \wedge ((\neg n=1 \vee l=0 \wedge p=1) \wedge (\neg(l=0 \wedge p=1) \vee n=1))]}\{s..mm\}$
c_7	$F_{[\neg t=1 \vee \neg n=1]}\{s..mm\}$
c_8	$F_{[\neg t=1 \vee st=1]}\{s..mm\}$

Given the definition of a feature model configuration task $F_{[c]}S$, we are able to introduce the definition of a corresponding feature model configuration (see Definition 2).

Definition 2 (FM Configuration). A feature model (FM) *configuration* for a FM configuration task $F_{[c]}S$ is a tuple representing a result of executing the corresponding conjunctive query.

Following the "single table" approach which follows the idea of explicitly representing the whole configuration space, an example of a configuration result is the tuple $\{s = 1, p = 1, l = 0, n = 1, t = 0, st = 1, q = 1, m = 1, mm = 0\}$.

(a) All Possible Feature Model Configurations. One possibility to represent the feature configuration space is to explicitly enumerate all possible configurations[1] and regard feature model configuration as basic query on a single table F (see Table 2). Such a representation can make sense only if the configuration space is small and corresponding conjunctive queries on F can be executed in an

[1] This enumeration can also be performed on the basis of a conjunctive query where [c] represents the set of feature model constraints.

efficient fashion. In this context, a conjunctive query returning a configuration could be the following: $F_{[s=1,p=1,n=1,mm=0]}\{s,p,l,n,t,st,q,m,mm\}$ resulting, for example, in a feature model configuration $\{s = 1, p = 1, l = 0, n = 1, t = 0, st = 1, q = 1, m = 1, mm = 0\}$. On the level of the structured query language (SQL), a related query is: SELECT *s,p,l,n,t,st,q,m,mm* FROM *F* WHERE *s=1 and p=1 and n=1 and mm=0.*[2]

Example Query Optimizations. A possibility of improving the performance of queries on F is to reduce the number of tuples in F by assuming that the root feature in F (in our case, the feature *survey* (s)) is always *true* (1).

Table 2. Explicit configuration space description in one table F including the attributes s .. mm.

s	p	l	n	t	st	q	m	mm
1	1	1	0	0	0	1	1	0
1	1	1	0	0	1	1	1	0
..	..	..	..	..	..	..	..	..

(b) "One Table per Feature" Representation. With an increasing configuration space size, an implicit representation is needed since an enumeration of all possible feature combinations is impossible or would lead to serious performance issues. An alternative approach to represent the configuration space of a feature model is depicted in Table 3 where each feature of the feature model in Fig. 1 is represented by a separate table, for example, feature p (payment) is represented by the table p including the attribute *val* and corresponding table entries 0 and 1 expressing feature exclusion or inclusion. In this context, a conjunctive query returning a feature model configuration could be formulated as follows: $F_{[s.val=1,p.val=1,n.val=1,mm.val=0]}\{s.val..mm.val\}$ where F is regarded as implicit representation of the Cartesian product $s \times p \times l \times n \times t \times st \times q \times m \times mm$. On the level of SQL, a corresponding query is: SELECT *s.val.. mm.val* FROM *s .. mm* WHERE *s.val=1 and p.val=1 and n.val=1 and mm.val=0.*

Example Query Optimizations. Possibilities of improving the performance of such queries are (a) to reduce the domain of the root feature to *true* (1), (b)

Table 3. Implicit configuration space description where each feature is represented by a single table with one attribute *val* and the table entries 0 and 1.

Table	s	p	l	n	t	st	q	m	mm
Attribute	val	val	val	val	val	val	val	val	val
Domain	0,1	0,1	0,1	0,1	0,1	0,1	0,1	0,1	0,1

[2] To assure that only one tuple is returned as configuration, we assume a query setting such as *LIMIT=1* (this is database-specific).

to reduce the domain of dead features to *false*, and (c) to reduce the domain of false optional features to *true* – for related details on feature model analysis, we refer to Benavides et al. [5].

(c) "Local Consistency" based Representation. An alternative to the "One Table per Feature" representation is to use tables for expressing local consistency properties. For example, the constraint *AB testing* (t) requires *statistics* (st) can be represented as corresponding consistency table *t-st* (see Table 4) expressing all possible combinations of t and st taking into account solely the *requires* relation between the two features.

Table 4. Implicit configuration space representing locally consistent feature assignments (a_i), e.g., in table *t-st*.

Feature	t	st
a_1	1	1
a_2	0	1
a_3	0	0

Table 5. Locally consistent feature assignments a_i of *excludes* relationship between feature n and t represented by table *n-t*.

Feature	n	t
a_1	0	1
a_2	1	0
a_3	0	0

Since Tables 4 and 5 include the feature t, we need to assure that in the final configuration, t has exactly one value – this can be achieved be integrating join conditions into the conjunctive query. In SQL, a corresponding query is: SELECT *n-t.n,n-t.t,t-st.st* FROM *n-t,t-st* WHERE *n-t.t=t-st.t*. Using this representation, feature names are regarded as table attributes, for example, feature t (*AB testing*) is regarded as an attribute of table *n-t* as well as *t-st*.

Example Query Optimizations. Following the concept of *k-consistency* in constraint solving [25], the amount of tuples in a constraint table (e.g., *n-t*) can be reduced by taking into account further feature model constraints. In the extreme case, this would lead to global consistency [25] which assures that each constraint table contains only tuples (representing feature value combinations) that can be part of at least one configuration. For complex feature models (FM), global consistency is impossible since this would basically require the enumeration of all possible configurations.

Summarizing, there are three basic options for representing FM configuration tasks as conjunctive queries: (a) enumerating all possible configurations (see Table 2), (b) implicit representation with one table per feature (see Table 3), or (c) using tables to represent locally consistent features settings.

4 New Approaches to Conflict Resolution

In configuration scenarios, the conjunctive query $F_{[c]}S$ could result in the empty set. Such situations can occur, for example, if the feature model is void, i.e., no solution can be identified or the given set of customer requirements induces an inconsistency with the feature model constraints [8,10]. In such situations, we need to identify minimal sets of constraints (in the feature model or the customer requirements) that have to be adapted in order to restore consistency. In the context of feature model configuration, there exist a couple of approaches supporting corresponding feature model diagnosis operations [4,15]. An alternative to the application of diagnosis algorithms is to directly exploit information included in table-based representations. An example thereof is depicted in Table 6. Assuming a set of customer requirements cr, we are able to directly compare these requirements with a set of potential configurations. If cr is *inconsistent* with a configuration $conf_i$, those feature inclusions (exclusions) of cr inconsistent with $conf_i$ represent a diagnosis Δ where $cr - \Delta$ is consistent with $conf_i$.

Table 6. Direct derivation of diagnoses.

Feature	s	p	l	n	t	st	q	m	mm
cr	1	1	0	1	1	1	1	1	0
$conf_1$	1	1	0	1	**0**	1	1	1	0
$conf_2$	1	1	**1**	**0**	1	1	1	1	0
..	..	..	..	..	..	..	..	..	..

This way, we are able to determine diagnoses without the need of applying corresponding diagnosis algorithms. In the example depicted in Table 6, we are able to identify two potential diagnoses: $\Delta_1 = \{t = 1\}$ and $\Delta_2 = \{l = 0, n = 1\}$. Applying Δ_1 basically means to inform the user about a potential exclusion of feature t which could then lead to a consistent configuration.

5 Performance Analysis

We compared the performance of three approaches for representing FM configuration tasks as conjunctive query with constraint solving on the basis of four real-world feature models selected from the S.P.L.O.T. feature model repository [20]. Table 7 provides an overview of selected feature models. Due to space complexity, not all configurations could be determined for *WebArch* within reasonable time limits.

For each feature model, we randomly synthesized[3] and collected 25,000 user requirements that cover 40–60% of the leaf features in the feature model. We

[3] To ensure the reproducibility of the results, we used the seed value of 141982L for the random number generator.

applied the systematic sampling technique [21] to select 10 *no-solution* user requirements and 10 user requirements with at least one solution. In Table 8, each setting shows the average runtime of the corresponding approach after executing the queries on the basis of these 20 user requirements. We used CHOCO SOLVER[4] as a reasoning solver and HSQLDB[5] as an in-memory relational database management system. All reported experiments were run with an Apple M1 Pro (8 cores) with 16-GB RAM, and an HSQLDB maximum cache size of 4 GB.

Table 8 shows the results of this evaluation of selected feature models represented as (a) an explicit enumeration of *all configurations*, (b) an implicit representation of the feature model configuration space (*one table per feature*), (c) an implicit representation where individual tables represent locally consistent feature assignments (*one table per constraint*), and (d) constraint satisfaction problem (CSP). This initial evaluation shows similar runtimes for small feature models and significantly longer runtimes for conjunctive query based approaches in the case of more complex models. Basically, these initial results of our performance evaluation show the applicability of conjunctive query based approaches to the task of feature model configuration.

Table 7. Properties of selected feature models (IDE = IDE product line, DVS = digital video system, DELL = DELL Laptop feature model, WebArch = web architectures).

feature model	IDE	DVS	DELL	WebArch
#features	14	26	47	77
#leaf features	9	16	38	46
#hierarchical constraints	11	25	16	65
#cross-tree constraints	2	3	105	0
#configurations	80	22,680	2,319	–

Table 8. The average runtime (*msec*) of conjunctive query and constraint-based feature model configuration.

Feature model	IDE	DVS	DELL	WebArch
All configurations	0.78	9.43	6.49	–
One table per feature	0.49	0.45	2.53	327.75
One table per constraint	0.51	0.78	3.32	294.55
CSP	0.73	0.78	1.09	1.1

[4] choco-solver.org.
[5] hsqldb.org.

6 Threats to Validity and Open Issues

In this paper, we have shown how to apply conjunctive queries to the identification of feature model configurations. We have provided results of an initial performance analysis of explicit and implicit table-based representations of feature models. We have compared the performance of conjunctive (database) queries with corresponding constraint solver based implementations. Initial results are promising and show the applicability of our approach in terms of search efficiency. We are aware of the fact that further in-depth evaluations with industrial datasets and different types of knowledge representations are needed to gain further insights into the applicability of our approach. We also sketched alternative approaches to deal with inconsistencies in feature model configuration scenarios. We are aware that related evaluations are needed to be able to estimate in detail the potential improvements in terms of runtime performance of diagnosis and conflict search.

7 Conclusions and Future Work

We have introduced a conjunctive query based approach to feature model (FM) configuration. In this context, we have compared the performance of three alternative conjunctive query based knowledge representations. With this, we provide an alternative to SAT solving, constraint solving, and ASP which can reduce overheads due to new technology integration. Initial results of our performance analysis are promising, however, we are aware of the fact that technologies such as SAT solving can outperform conjunctive queries. In addition, we have sketched how to exploit table-based representations for efficient conflict resolution.

There are a couple of open issues for future work. First, we will extend our performance analyses to comparisons with a broader range of industrial feature models. We will also include SAT solving and ASP in our evaluations. We also plan to analyze the potentials of combining different worlds, for example, we will analyze to which extent approaches from constraint solving (e.g., forward checking) can help to improve the efficiency of conjunctive queries. Vice-versa, we will analyze to which extent search techniques from relational databases (and also machine learning) can help to further advance the field of SAT solving, constraint solving, and ASP. Finally, we want to analyze the properties of phase transitions, i.e., when to move from an explicit to an implicit search space representation.

Acknowledgments. The work presented in this paper has been developed within the research project ParXCel (*Machine Learning and Parallelization for Scalable Constraint Solving*) which is funded by the Austrian Research Promotion Agency (FFG) under the project number 880657.

References

1. Acher, M., Temple, P., Jézéquel, J.M., Galindo, J., Martinez, J., Tiadi, T.: VARY-LATEX: learning paper variants that meet constraints. In: 12th International Workshop on Variability Modelling of Software-Intensive Systems, pp. 83–88. Madrid, Spain (2018)
2. Apel, S., Kästner, C.: An overview of feature-oriented software development. J. Object Technol. **8**(5), 49–84 (2009)
3. Bakker, R., Dikker, F., Tempelman, F., Wogmim, P.: Diagnosing and solving over-determined constraint satisfaction problems. In: Proceedings of IJCAI-93, pp. 276–281. Morgan Kaufmann (1993)
4. Benavides, D., Felfernig, A., Galindo, J.A., Reinfrank, F.: Automated analysis in feature modelling and product configuration. In: Favaro, J., Morisio, M. (eds.) ICSR 2013. LNCS, vol. 7925, pp. 160–175. Springer, Heidelberg (2013). https://doi.org/10.1007/978-3-642-38977-1_11
5. Benavides, D., Segura, S., Ruiz-Cortes, A.: Automated analysis of feature models 20 years later: a literature review. Inf. Syst. **35**, 615–636 (2010)
6. Benavides, D., Trinidad, P., Cortés, A.: Using constraint programming to reason on feature models. In: Chu, W.C., Juzgado, N.J., Wong, W.E. (eds.) Proceedings of the 17th International Conference on Software Engineering and Knowledge Engineering (SEKE 2005), Taipei, Taiwan, Republic of China, July 14–16, 2005, pp. 677–682 (2005)
7. Benavides, D., Trinidad, P., Ruiz-Cortés, A.: Automated reasoning on feature models. In: Pastor, O., Falcão e Cunha, J. (eds.) CAiSE 2005. LNCS, vol. 3520, pp. 491–503. Springer, Heidelberg (2005). https://doi.org/10.1007/11431855_34
8. Felfernig, A., Boratto, L., Stettinger, M., Tkalcic, M.: Group Recommender Systems. Springer (2018)
9. Felfernig, A., Hotz, L., Bagley, C., Tiihonen, J.: Knowledge-based Configuration - From Research to Business Cases. Elsevier (2014)
10. Felfernig, A., Schubert, M., Zehentner, C.: An efficient diagnosis algorithm for inconsistent constraint sets. AI for Engineering Design, Analysis, and Manufacturing (AIEDAM) **26**(1), 53–62 (2012)
11. Felfernig, A., Le, V.M., Popescu, A., Uta, M., Tran, T.N.T., Atas, M.: An overview of recommender systems and machine learning in feature modeling and configuration. VaMoS 2021, ACM, New York (2021)
12. Rossi, F.: Peter van Beek. Handbook of Constraint Programming. Elsevier, T.W. (2006)
13. Gomes, C., Kautz, H., Sabharwal, A., Selman, B.: Satisfiability Solvers. Handbook of Knowledge Representation **3**, 89–134 (2008)
14. Gu, J., Purdom, P.W., Franco, J., Wah, B.W.: Algorithms for the satisfiability (sat) problem: a survey. In: DIMACS Series in Discrete Mathematics and Theoretical Computer Science, pp. 19–152. American Mathematical Society (1996)
15. Hentze, M., Pett, T., Thüm, T., Schaefer, I.: Hyper explanations for feature-model defect analysis. In: 15th International Working Conference on Variability Modelling of Software-Intensive Systems. VaMoS'21. Association for Computing Machinery, New York (2021)
16. Junker, U.: QUICKXPLAIN: preferred explanations and relaxations for over-constrained problems. In: 19th National Conference on Artifical Intelligence, AAAI 2004, pp. 167–172. AAAI (2004)

17. Kang, K., Cohen, S., Hess, J., Novak, W., Peterson, S.: Feature-oriented Domain Analysis (FODA) – Feasibility Study. Technical Report SEI-90-TR-21 (1990)
18. Le, V.M., Felfernig, A., Uta, M., Benavides, D., Galindo, J., Tran, T.N.T.: DIRECT-DEBUG: automated testing and debugging of feature models. In: 2021 IEEE/ACM 43rd International Conference on Software Engineering: New Ideas and Emerging Results (ICSE-NIER), pp. 81–85. IEEE/ACM (2021)
19. Mendonça, M., Wasowski, A., Czarnecki, K.: SAT-based analysis of feature models is easy. In: Muthig, D., McGregor, J.D. (eds.) SPLC 2009, vol. 446, pp. 231–240. ACM (2009)
20. Mendonca, M., Branco, M., Cowan, D.: S.P.L.O.T.: software product lines online tools. In: Proceedings of the 24th ACM SIGPLAN Conference Companion on Object Oriented Programming Systems Languages and Applications, OOPSLA 2009, pp. 761–762. ACM, New York (2009)
21. Mostafa, S.A., Ahmad, I.A.: Recent developments in systematic sampling: a review. J. Stat. Theory Practice **12**(2), 290–310 (2018). https://doi.org/10.1080/15598608.2017.1353456
22. Myllärniemi, V., Tiihonen, J., Raatikainen, M., Felfernig, A.: Using answer set programming for feature model representation and configuration. In: 16th International Workshop on Configuration, pp. 1–8. CEUR, Novi Sad, Serbia (2014)
23. Popescu, A., Polat-Erdeniz, S., Felfernig, A., Uta, M., Atas, M., Le, V., Pilsl, K., Enzelsberger, M., Tran, T.: An overview of machine learning techniques in constraint solving. J. Intell. Inf. Sys. **58**(1), 91–118 (2022)
24. Reiter, R.: A theory of diagnosis from first principles. AI J. **32**(1), 57–95 (1987)
25. Rossi, F., van Beek, P., Walsh, T.: Handbook of Constraint Programming. Elsevier, Amsterdam (2006)
26. White, J., Benavides, D., Schmidt, D., Trinidad, P., Dougherty, B., Ruiz-Cortes, A.: Automated diagnosis of feature model configurations. J. Syst. Softw. **83**(7), 1094–1107 (2010)

ESpin: Analyzing Event-Driven Systems in Model Checking

Nhat-Hoa Tran(✉)

Hanoi University of Science and Technology, Hanoi, Vietnam
hoatn@soict.hust.edu.vn

Abstract. Multiple occurrences of the events make the behaviors of an event-driven system usually less logical and explicit. Thus, an event-driven system is hard to be analyzed and found errors. In this paper, we introduce an approach to analyze event-driven systems following the properties related to the occurrence of the events. Our method is based on the exploration of the states of the system using model checking techniques. To that end, we introduced a domain-specific language (DSL) to easily describe the behaviors of the system. The specification of the system in the DSL is then translated into the model of the system to explore the states of the system. The property to be checked is used to realize the model checking algorithm that is used to label the graph realized from the state space. The experiment results show that we can facilitate the specification of the system and analyze the behaviors of the system correctly.

Keywords: Event-driven systems · Model checking techniques · Domain-specific language · CTL/RTCTL properties

1 Introduction

Event-driven systems are popularly used in social life to capture, communicate, and process events continuously. However, developing an event-driven system is error-prone because the flow of the corresponding program is usually less logical and obvious. Therefore, checking the correctness of an event-driven system is necessary. Because there are various complex events and the occurrences of these events are also different, it is hard to analyze the behaviors of the systems. Taking the combination of the events manually to check the possible cases is always costly and time-consuming. Model checking [3] is a method for checking whether a finite-state model of a system meets a given specification exhaustively and automatically. Thus, model checking is appropriate to use to verify event-driven systems.

In model checking, we specify the system's behaviors in a language. Then the model is used to explore all of the states of the system by a model checker to verify the properties. Promela is such a modeling language, which is used by Spin [7] tool. These languages and the corresponding tools are successfully applied to

N. T. Nguyen et al. (Eds.): CITA 2023, LNNS 734, pp. 368–379, 2023.
https://doi.org/10.1007/978-3-031-36886-8_31

verify asynchronous systems and multi-threaded software systems. However, it is hard to manually specify the behaviors of an event-driven system in these languages because we need to model many events with their constraints and properties. To address these problems, in the previous work [10], we proposed an approach using a domain-specific language (DSL) to provide an easy and flexible way to specify event-driven systems. We also proposed a translation approach to convert the specification of the system in the DSL into a Promela program and then verify the program using a model checker. However, with this work, we only consider the events separately and only check the states of the system after the occurrence of each event. In many requirements, we need to verify whether the system satisfies the properties related to the occurrence of the events (such as the system will be suspended in three events). The properties here relate to time and can be in the forms of the Computation Tree Logic (CTL) and Real-Time Computation Tree Logic (RTCTL) formula. For example, the RTCTL formula $\mathbf{EF}^{\leq 5}\varphi$ means that property φ eventually holds within the bounded time, namely, 5-time units. Indeed, the analysis following these properties is necessary.

To address these problems, in this paper, we introduce an approach to analyze event-driven systems with the properties following the occurrence of the events. First, we extend our DSL to specify the system with the CTL/RTCTL properties related to the sequence of the events. Second, we apply the translation approach to a) generate the model of the system from its specification and b) generate the model checking algorithm following the CTL/RTCTL formula. Third, the graph is now built from the exploration using the model of the system. Fourth, the graph is then labeled by the model checking algorithm above to check the property. The labeling indicates the result of the analysis. Our method is shown in Fig. 1.

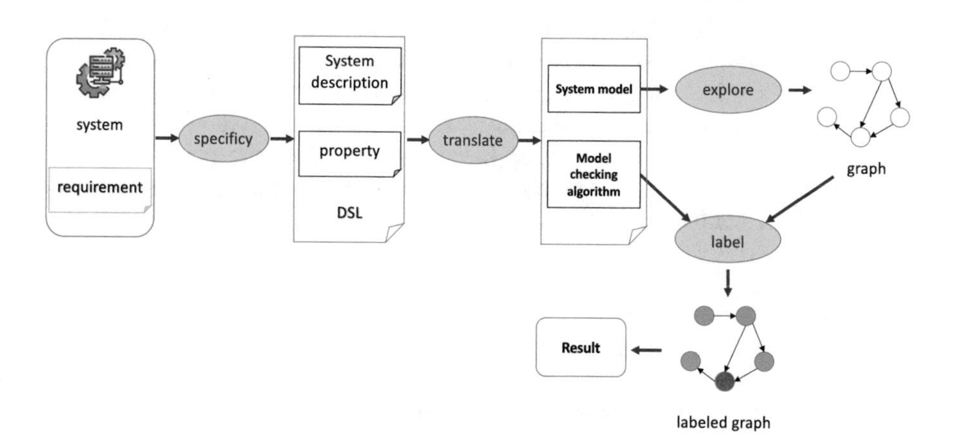

Fig. 1. The method for analyzing event-driven systems.

The rest of this paper is organized as follows: Sect. 2 introduces the DSL and the translation approach for event-driven systems. The method of analyzing the systems is described in Sect. 3. In Sect. 4, we present the implementation of our method and introduce some case studies with the experimental results. Section 5 shows the related works. Finally, the conclusion and future work are given in Sect. 6.

2 The DSL for Event-Driven Systems

2.1 Specifying Event-Driven Systems

We extend the DSL introduced in the previous work [10] to provide a high-level language and an easy way for specifying the behaviors of an event-driven system with the properties to check. The summary of the language[1] is indicated below.

1. The specification of a system contains the declarations of the variables and the definition of the events.
2. Several configurations of the system corresponding to the initial values of the variables can be defined.
3. The occurrences of the events (called scenarios) are defined as ordered lists of events or following the permutation of the events.
4. The invariant properties are defined as rules. Each rule has the form of a Hoare triple (*pre-condition, event, post-condition*) for verifying the behaviors of the system. When the corresponding *event* occurs, if the *pre-condition* holds, the rule is checked whether the *post-condition* holds or not.
5. The properties related to the occurrences of the events are specified as CTL/RTCTL formula for the analysis.

We use a demonstration of a heater system (in Fig. 2) as an example of the DSL. The specification of the system is shown in Fig. 3. Here, the system can increase or decrease (event `increase`/`decrease`) the temperature (`request_temp`) following the requests. Variable `current_temp` indicates the current temperature. The system can also randomly reset the temperature value for preventing overactive operations (event `reset`). The alarm is triggered if the temperature requested is greater than the allowed maximum temperature (`max_temp`). A configuration named `config1` is used to initialize the values of these variables.

We consider a scenario in which the events will happen randomly with the number of occurrences of these events being equal to 25. We check that there is a case that the alarm will be triggered in 25 actions of the system (i.e. $\texttt{EF} <= 25(\texttt{alarm} == 1)$). Here, we can easily see that the number of cases needed to be considered following the permutation of the three events with 25 occurrences is large. Thus, checking the system manually is time-consuming and error-prone. That means automatic verification is needed.

[1] The reader can refer to [10] for more information of the DSL.

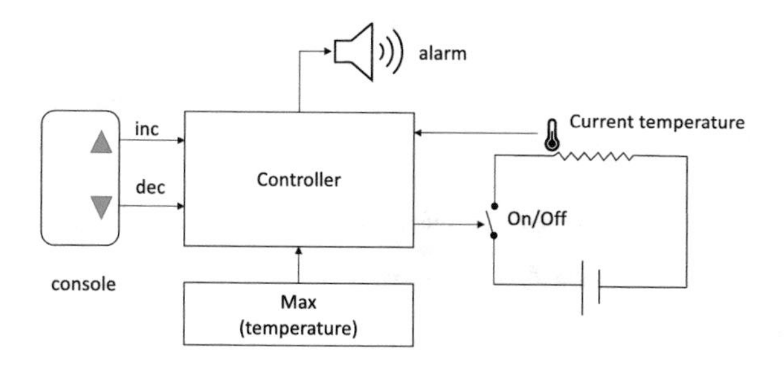

Fig. 2. A heater system.

```
system {
    specification {
        int current_temp;  int request_temp;
        int max_temp; int alarm;  int turn_on;
        event increase {
            request_temp++;
            if
                :: request_temp > current_temp -> turn_on = 1;
                    if
                        :: request_temp > max_temp -> alarm = 1;
                        :: skip;
                    fi;
                :: skip;
            fi;
        }
        event decrease {
            request_temp--;
            ...
        }
        event reset {
            turn_on = 0;
            requeset_temp = current_temp;
        }
    }
    configuration {
        config1 {
            request_temp = 9;
            current_temp = 8;
            max_temp = 29;
        }
    }
    scenario permutation with step = 25;
    verify {
        EF<=25 (alarm == 1)
    }
}
```

Fig. 3. The specification of a heater system.

2.2 The Translation Approach

In our method, we generate the corresponding *Promela program* from the specification of the event-driven system in the DSL. The translation result of the specification of the heater system (in Fig. 3) is shown in Fig. 4. Each event of the system in the DSL is translated into an independent process (with keyword `proctype`). In general, there is no interruption during the execution of the event, therefore, the behaviors of the system for each event are now specified as an atomic action using a `d_step` statement in Promela. In this example, we use only one scenario in which the events will happen randomly, thus a corresponding process named `VerificationCase0` is generated. The configuration defined in the DSL is now used to initialize the state of the system. In the code, the `run` statements are for executing the corresponding processes. A special process named `test` with the keyword `active` is generated to perform the analysis. The property indicated in the `verify` part of the specification is used to realize the model checking algorithm to label the states of the system (see Sect. 3) including the graph definition (nodes and labels) and the labeling process.

```
proctype increase() { /* for event increase */
    do
        :: d_step {
            /* action for event increase*/
            request_temp++;
            if
                :: ((request_temp)>(current_temp)) -> turn_on=1;
                if
                    :: ((request_temp)>(max_temp)) -> alarm=1;
                    :: skip ;
                fi ;
                :: skip ;
            fi ;
        }
    od
}
...
proctype VerificationCase0(){
    d_step {
        request_temp = 9 ;
        current_temp = 8 ;
        max_temp = 29 ;
        run increase() ;
        run decrease() ;
        run reset() ;
    }
}
active proctype test() {
    if
        :: run VerificationCase0()
        :: else -> skip
    fi
}
verify {
    EF <=25(((alarm)==1))
}
```

Fig. 4. The result of the generation.

3 Model Checking Event-Driven Systems

In this section, we explain the model checking algorithm to check whether formula f_0 is true or not in the model of the system, which is represented as a graph (Kripke structure) $M = (S, I, R, L)$, where S is a finite set of states, $I \subseteq S$ is a set of initial states, $R \subseteq S \times S$ is the transition relation and $L : S \rightarrow 2^{AP}$ is a labeling function. The property is written as a CTL/RTCLT formula. The algorithms include two steps: 1) building the state graph and 2) labeling the graph following the corresponding formula to check the property.

3.1 Exploring the State Space

We introduce an algorithm to explore the states of the system to construct the graph (Algorithm 1). Similar to the depth-first search (DFS), this algorithm uses a stack to record the search steps with two operations *Push* and *Pop*. The graph is built with the starting node (line 8) created from the initial state of the system (line 5). Here, the function `NODE` is to indicate and create a new node if it does not exist (lines 8, 16). All possible events which can be taken are considered (line 14). The event is processed by the function `PROCESS_EVENT` (line 15). The function `CREATE_EDGE` creates an edge to connect the current node (corresponding to the state of the system before the happening of the event) with the new node (line 16). The *depth* of the stack is checked with the allowed number of the occurrence of the events (*step*) defined in the specification (line 12).

3.2 The Properties

We can see that each occurrence of an event corresponds to a transition from one state of the system to another one. Indeed, checking the properties related to the sequence of the transitions is needed. If we consider each transition as one

Algorithm 1. Construct the state graph

```
1: Stack: S = ∅
2: State space: V = ∅
3: Start node: StartNode = null
4: procedure INIT
5:     s0 = INITIALIZE()                    ▷ initializes the state
6:     Push(s0, S)    ▷ pushes the state to the stack and increases the stack's depth
7:     Add(s0, V)                           ▷ adds the state to the state space
8:     StartNode = NODE(s0)
9:     EXPLORE
10: end procedure
11: procedure EXPLORE
12:     if (S.depth < step) then
13:         s = Top(S)
14:         for (e ∈ E) do                  ▷ E is the set of possible events
15:             s'= PROCESS_EVENT(e, s)
16:             CREATE_EDGE(NODE(s), NODE(s'))
17:             if (not Contains(V, s')) then
18:                 Push(s', S)
19:                 Add(s', V)
20:                 EXPLORE
21:             end if
22:         end for
23:     end if
24:     Pop(S)    ▷ pops the state from the stack and decreases the stack's depth
25: end procedure
```

step (one-time unit), the discrete time can be used in checking the behaviors of the system. In this research, we consider the properties expressed in the form of the Computation Tree Logic (CTL) and the Real-Time Computation Tree Logic (RTCTL) proposed by Emerson [6] to analyze the systems.

CTL is a branching-time logic, it models the time as tree-like structures. The formula contains temporal operators `AX`, `EX`, `AF`, `EF`, `AG`, `EG`, `AU`, and `EU`. The operators refer to a path as a sequence of states as follows: `A` means All, `E` means Exists, `X` means Next, `F` means Finally, `G` means Globally, and `U` means Until.

Indeed, a CTL formula can specify the order of behaviors of the system and the properties in CTL formulas can be used for verifying reachability, and safety properties. RTCTL proposed by Emerson [6] is an extension of CTL to consider the discrete time to handle quantitative assertions. Here, each action of the system consumes 1 tick (one-time unit or 1 step). For example, the RTCTL formula $\mathtt{EF}^{\leq 3}\varphi$ represents the bounded time states that property φ eventually holds within 3-time units. The meaning of the operations is as follows.

- $s_0 \models \mathtt{AF}^{\leq n}\ \psi$ iff for all paths $\sigma = s_0...$, $\exists i, 0 \leq i \leq n$, such that $s_i \models \psi$
- $s_0 \models \mathtt{EF}^{\leq n}\ \psi$ iff for some path $\sigma = s_0...$, $\exists i, 0 \leq i \leq n$, such that $s_i \models \psi$
- $s_0 \models \mathtt{AG}^{\leq n}\ \psi$ iff for all paths $\sigma = s_0...$, $\forall i, 0 \leq i \leq n$, such that $s_i \models \psi$
- $s_0 \models \mathtt{EG}^{\leq n}\ \psi$ iff for some path $\sigma = s_0...$, $\forall i, 0 \leq i \leq n$, such that $s_i \models \psi$

- $s_0 \models \mathtt{A}(\varphi\ \mathtt{U}^{\leq n}\ \psi)$ iff for all paths $\sigma = s_0...$, $\exists i, 0 \leq i \leq n$, such that $s_i \models \psi$ and $\forall j, 0 \leq j < i, s_j \models \varphi$
- $s_0 \models \mathtt{E}(\varphi\ \mathtt{U}^{\leq n}\ \psi)$ iff for some path $\sigma = s_0...$, $\exists i, 0 \leq i \leq n$, such that $s_i \models \psi$ and $\forall j, 0 \leq j < i, s_j \models \varphi$.

The superscript n of the formula specifies the maximum number of permitted transitions (steps) along a path (the time-bounded).

3.3 Model Checking Algorithm

To analyze the behaviors of the system with the corresponding property, we follow the method proposed by Clarke-Emerson-Sistla [5] and the algorithm introduced in [6] to label the state graph.

1. The checking formula is rewritten in a prefixed form and the lengths of each sub-formula are also determined (e.g. the formula $\mathtt{f} = \mathtt{A}\ (\varphi\ \mathtt{U}\ \psi)$ will be rewritten as $\mathtt{f} = \mathtt{AU}\ \varphi\ \psi$, the lengths of AU, φ, and ψ are now 3, 2, and 1, respectively).
2. The labeling process starts from the last sub-formula to the first sub-formula of f with the state s_0 (corresponding to the starting node of the graph). The states are labeled with all sub-formulas of f with length 1 in the creation of the graph. In the second step, based on the results in step 1, we label the states with sub-formula of length 2. The next steps are taken in the same way.
3. We use a DFS algorithm to label the current node of the graph when the next state (child node) is not labeled. The superscript in the formula, such as 5 in $\mathtt{AU}^{\leq 5}$, determines the depth for the DFS algorithm. This value indicates the maximum number of steps that can be taken when labeling the graph. As a result, each state will be labeled with a set of sub-formulas of the length less than or equal to i after the completion of the ith step. The algorithm terminates at the step $n = length(f)$ after completing all the sub-formulas. Now the entire formula f is checked. We use $label(s)$ to denote this set of state s. We have $M, s \models f$ iff $f \in label(s)$.

4 Experiments

4.1 Implementation

Following the method proposed, we have implemented the tool named ESpin [1] based on the SpinJa [8] model checker. The architecture of the tool is shown in Fig. 5. There are three main components in this tool: a *converter*, a *compiler*, and an *analysis*. The input of our tool is a specification of an event-driven system in the DSL. This specification is then used to generate a) the *Promela program* and b) the *model checking algorithm* by the *converter*, which is built on the XText framework [4]. The *compiler* will compile the *program* into a model in Java. The *analyzer* component will use the model to explore the state of the system

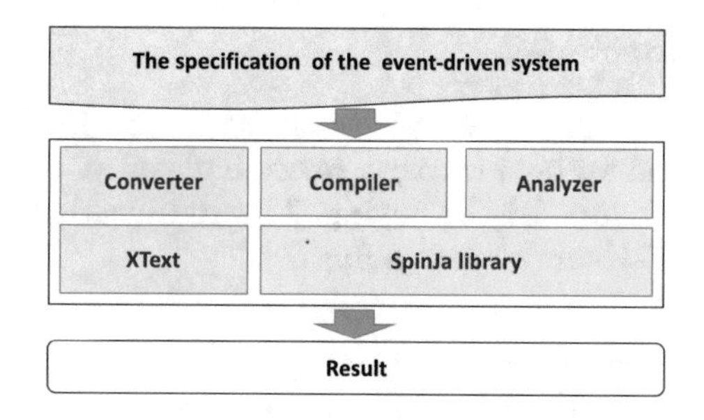

Fig. 5. Architecture of ESpin.

to build the system graph and then label this graph using the *model checking algorithm* above to produce the results of the analysis.

Below, we introduce some experiments with our tool. The experiments were conducted on a 2.4 GHz CPU Intel Core i7 with 8G RAM.

4.2 The Heater System

We use the heater system introduced in Sect. 2 for the analysis. In the first experiment, we check that after changing the temperature, the current temperature may be greater than the maximum one, and the alarm is triggered. The property is $\texttt{EF} <= \texttt{25}(\texttt{alarm} == \texttt{1})$ (Is there any case that the alarm is triggered in 25 steps?). This property holds with the configuration $config1$: $(request\_temp, current\_temp, max\_temp) = (9, 8, 29))$. It is easy to see that from the starting state if the system increases the temperature continuously in 25 steps, the current temperature will be greater than the maximum temperature. Now we keep this property but change the configuration as follows: $config2$: $(request\_temp, current\_temp, max\_temp) = (2, 2, 29)$. The system was analyzed again. As a result, the property does not hold. That is because the $request\_temp$ and the $current\_temp$ are lower than that used in the previous experiment, and the heater system can not increase the temperature to reach the max_temp value. The results of checking are shown in Table 1.

Table 1. Results of analyzing with $\texttt{EF} <= \texttt{25}(\texttt{alarm} == \texttt{1})$

	No. State	Time (s)	Memory (Mb.)	Result
config1	54	0.055	8.781	**holds**
config2	54	0.052	8.003	**not hold**

In the second experiment, we changed the property to $\texttt{AG} <= \texttt{20}(\texttt{alarm} == \texttt{0})$ to check that the alarm is not triggered in all cases within 20 steps of changing the temperate. With this experiment, the property holds with both of the two configurations `config1` and `config2` above.

4.3 A Vending Machine

We analyzed the behaviors of a simple vending machine in this case study. The system is implemented for buying cakes, water, and coffee. Users can insert coins values of 1, 2, or 5 dollars for the buying. The specification of the system and the property in the DSL are shown in Fig. 6.

```
system {
  specification {
    int water; int cake; int coffee; int input;
    int existing;
    event insert1 {
      input = input + 1;
    }
    event insert2 {
      input = input + 2;
    }
    event insert5 {
      input = input + 5;
    }
    event buy_water{
      when input >= water
      then input = input - water; existing= existing+ water;
    }
     event buy_cake{
      when input >= cake
      then input = input - cake; existing= existing+ cake;
    }
    event buy_coffee{
      when input >= coffee
      then input = input - coffee; existing= existing+ coffee;
    }
    event out {
      input = 0;
    }
  }

  configuration {
    config1 { water = 1; cake = 5; coffee = 10; existing= 0 ; }
    config2 { water = 1; cake = 6; coffee = 8; existing= 10 ; }
  }

  scenario permutation with step = 30 ;

  verify {
    AG <= 20 (existing>=0)
  }
}
```

Fig. 6. The specification of a vending machine system.

In the experiments, we use two configurations of the system with different prices of the products and different amounts of money existing in the machine. We set the limit behaviors of the systems to 30 and check that the money in the system is always greater than 0 in any case. The property used in the analysis is $\texttt{AG} <= 20(\texttt{existing} >= 0)$. The result of the analysis is shown in Table 2.

Table 2. Results of analyzing the vending machine system

	No. State	Time (s)	Memory (Mb.)	Result
config1	4820	1.151	11.983	**holds**
config2	4534	1.161	11.721	**holds**

4.4 Discussion

Analyzing event-driven systems is always hard because they usually perform the events freely. However, to make sure that the system is correct, we need to consider all of the possible occurrences of the events in the verification. Indeed, the number of cases is always large. If the system is analyzed and verified manually, it is error-prone and time-consuming. Using model checking techniques, all possible cases are taken into account in the analysis. Besides, with our approach, the verification can cover multiple configurations of the system and different strategies for the occurrences of the events. Thus, it helps the verification and analysis more accurate in comparison with other techniques, such as testing which considers only some executions of the systems.

In addition, the analysis following the properties represented in the form of the CTL/RTCTL formula considers the paths corresponding to the occurrence of the events. It is difficult to show whether the property is held or not just by considering each event individually (only a state of the system at a time). Using the labeling the graph approach following the property we can now do this easily. Besides, the conditions for the events, which are specified as the guards, will limit the behaviors of the system. In Sect. 4.3, if we do not consider the amount of money, which is stored in variable `input`, and the cost of the products, the number of cases executing the events within 30 steps will be large. However, the real number of states of the system was really small (as shown in Table 2). It means that the conditions for the occurrences of the events are important and need to be considered in the verification.

We can see that an event-driven system can have many initial states. The behaviors of the system depend on its configurations. Therefore, the results of the analysis following the properties may be different. For example, in the experiment for analyzing the heater system in Sect. 4.2, the two configurations produced different results for checking the property $\texttt{EF} <= 25(\texttt{alarm} == 1)$ due to the initial values of the variables. Therefore, to get accurate results in the analysis, different configurations also need to be taken into account. With testing, setting up several configurations needs executing the test case(s) again and again. With that point, by applying our method, we can produce several configurations efficiently with the DSL by changing the values of the variables, and all of the executions are considered fully automatic.

Furthermore, using a DSL can help facilitate the behaviors of the systems following the events. Doing this task with a programming language, such as C or Java (i.e. making a program), is always difficult. As shown in Table 3, we can see that the number of lines of the specification code (of the systems in the experiments) in the DSL is really small in comparison with the number of lines of *Java* code generated for the execution of the analysis (as indicated in column *Percentage*).

Table 3. Experiments summary

Experiment	Lines of DSL code	Lines of Java code	*Percentage*
Heater System	50	631	**7.92%**
Vending Machine	50	941	**5.31%**

5 Related Works

In this research, we consider the execution of a system following the sequence of events. From that point of view, the most used notations for the systems are Automata, Event-B, etc. The logic behind them is Higher-Order Logic, Linear Temporal Logic (LTL), Computational Tree Logic (CTL), etc. which could be used to express the properties to be verified and analyzed. Event-B [2] is a formal method for modeling and analysis at the system level. The use of set theory is a key feature of Event-B. The model of a system is refined to represent the system at different abstraction levels. Then, the mathematical proof is used to verify the consistency between refinement levels. Using Event-B, we can verify and analyze event-based systems. The expression of the systems in our DSL has some parts similar to the Event-B language, especially for describing the events. However, there are two main differences between using this method and ours. First, we apply other techniques for the verification and analysis. Event-B uses mathematical proof based on deduction methods. Thus, it may not fully automatic and may require user intervention to complete the proof. With model checking techniques as our method does, all of the states of the systems are considered and checked automatically. Second, the properties we apply in our method can express the sequence of the events with the time bounded (RTCTL formula) is not supported by Event-B.

Testing techniques are used to validate that a system satisfies the requirements using test cases. Several works used the specification of a system to check the behaviors of the system by generating test cases using model-based testing techniques such as [9]. However, these works did not deal with event-based systems. In addition, testing is usually applied after the programming phase of development, but our method aims to apply at the earlier phase of development (i.e. at the design phase). Moreover, testing is difficult to deal with the properties related to the sequence of the events.

In the previous work [10], we introduced an approach to deal with verifying the event-driven systems by applying model checking techniques. We proposed a DSL for describing the behaviors of the system corresponding to the events with configurations and validation rules. However, the properties which can be handled with this approach are only represented as invariants (i.e. Hoard trip formulas). In this paper, we extend the DSL above to deal with CTL/RTCTL formulas. We also introduce a method to check the properties by generating the algorithm to label the state space realize from the specification of the system in the DSL.

6 Conclusion

This paper introduces a method to verify and analyze event-driven systems using model checking techniques. We consider the properties related to the sequence of the events. Our main contributions include 1) the extension of the DSL and the translation approach for the DSL to deal with different properties, 2) the model checking algorithm to construct the graph from the system state space and analyze the behaviors of the system, and 3) the implementation of a tool following the approach to analyze the behaviors of the system. In the future, we will extend the DSL to describe more behaviors of the system with constraints and relations between the events, such as an event that will lead to the occurrences of other ones.

References

1. ESpin. https://github.com/nhathoatran/ESpin. Accessed 06 Feb 2023
2. Abrial, J.R.: Modeling in Event-B: System And Software Engineering. Cambridge University Press, Cambridge (2010)
3. Baier, C., Katoen, J.P., Larsen, K.G.: Principles Of Model Checking. MIT press, New York (2008)
4. Bettini, L.: Implementing Domain-Specific Languages with Xtext and Xtend. Packt Publishing Ltd. (2013)
5. Clarke, E.M., Emerson, E.A., Sistla, A.P.: Automatic verification of finite-state concurrent systems using temporal logic specifications. ACM Trans. Program. Lang. Syst. (TOPLAS) **8**(2), 244–263 (1986)
6. Emerson, E.A., Mok, A.K., Sistla, A.P., Srinivasan, J.: Quantitative temporal reasoning. Real-Time Syst. **4**(4), 331–352 (1992)
7. Holzmann, G.J.: The Spin Model Checker: Primer And Reference Manual, vol. 1003. Addison-Wesley Reading (2004)
8. de Jonge, M., Ruys, T.C.: The SpinJa model checker. In: van de Pol, J., Weber, M. (eds.) SPIN 2010. LNCS, vol. 6349, pp. 124–128. Springer, Heidelberg (2010). https://doi.org/10.1007/978-3-642-16164-3_9
9. Patel, P.E., Patil, N.N.: Testcases formation using UML activity diagram. In: 2013 International Conference on Communication Systems and Network Technologies (CSNT), pp. 884–889. IEEE (2013)
10. Tran, N.H.: A specification-based approach to model checking event-driven systems. In: Proceedings of the Tenth International Symposium on Information and Communication Technology, pp. 449–456 (2019)

Blockchain Solution for Electronic Health Records Using Hyperledger Fabric

Vu-Thu-Nguyet Pham, Quang-Chung Nguyen, Van-To-Thanh Nguyen, Thanh-Phong Ho, and Quang-Vu Nguyen(✉)

The University of Danang - Vietnam-Korea University of Information and Communication Technology, Danang, Vietnam
{pvtnguyet.19it1,nqchung.19it1,nvtthanh.19it1,htphong.19it1, nqvu}@vku.udn.vn

Abstract. Electronic health records (EHRs) are vital digital documents that contain patient medical information. The use of blockchain technology in healthcare has the potential to transform the industry's data management practices while also addressing security, privacy, and data-sharing concerns. In this paper, we propose an architecture for a blockchain-based system for storing and sharing EHRs that prioritizes privacy and improves healthcare data security. The paper introduces various Hyperledger Fabric concepts and proposes a permissioned blockchain system solution for storing and sharing healthcare records based on the Hyperledger Fabric framework and some cryptography techniques. Patients have complete control over their medical information and can authorize doctors to view it by using grant and revoke access mechanisms.

Keywords: Blockchain · Hyperledger Fabric · Electronic Health Records · Data privacy · Security · Interoperability

1 Introduction

Electronic health records (EHRs) provide the foundation for storing and managing critical patient data, playing a significant role in contemporary healthcare. However, the conventional paper-based management and sharing methods for EHRs are prone to mistakes, consistency issues, and security flaws. Due to its decentralized and immutable nature, which enables secure and open administration of patient data, the emergence of blockchain technology has shown significant promise in addressing these issues.

Although public blockchain storage solutions for EHRs have been investigated, because of their open availability, they are vulnerable to security vulnerabilities that may jeopardize patient privacy. As a result, a stronger solution that emphasizes security and privacy is needed. A good choice in this case is using permissioned blockchain technology combining with some cryptography techniques.

In this paper, we offer a blockchain-based system for the storage and exchange of electronic health records (EHRs) that makes use of cutting-edge privacy-enhancing technologies including homomorphic encryption and zero-knowledge proof. Our strategy

Q.-C. Nguyen, V.-T.-T. Nguyen and T.-P. Ho—These authors contributed equally to this work.

N. T. Nguyen et al. (Eds.): CITA 2023, LNNS 734, pp. 380–390, 2023.
https://doi.org/10.1007/978-3-031-36886-8_32

comprises creating a consortium blockchain utilizing Hyperledger Fabric v2, allowing healthcare providers to maintain control over patient data while limiting access to only those who are allowed. This guarantees the safety and confidentiality of patient information, which is crucial for maintaining the relationship of trust between patients and healthcare professionals.

2 Related Works

Several studies have been conducted to look into the use of blockchain technology in the healthcare industry. Azaria et al. [1] proposed the MedRec blockchain-based system for managing EHRs and medical research data. The system stores data on a private blockchain and access to the data is controlled by smart contracts. The authors conducted user surveys and interviews to assess the system's feasibility and potential.

Hasselgren et al. [2] proposed a blockchain-based decentralized EHR system. The system stores data on a public blockchain and access to the data is controlled by smart contracts. Various metrics were used by the authors to assess the system's performance and security.

Misbhauddin et al. [3] proposed MedAccess, a blockchain-based architecture for managing EHRs. This platform supports on-chain data storage and processing, allowing doctors, technicians, and patients to securely manage their medical records. However, the system does not fully address some issues concerning the processing and storage of patients' personal data.

A later study proposed MedChain [4]. The downside of this study is that it stores data entirely on-chain, which consumes resources and makes scaling difficult. Furthermore, this study proposes an Ethereum-based architecture that does not guarantee patient data privacy policies.

While previous research has shown the potential of blockchain technology in healthcare, there are still challenges and limitations to address. These include interoperability, scalability, governance, and regulatory compliance issues. Furthermore, the use of blockchain technology in healthcare necessitates a significant investment in terms of both technology infrastructure and human expertise.

3 Proposed Solution

In order to manage patients' medical records, we must map Hyperledger Fabric's components to EHRs system requirements. All hospitals will operate as organizations within a fabric network. Data of a patient has been classified as an asset and is recorded in the ledger. It is also possible to store the reference of the EHR data in the ledger, but since the application is not managed by real data, a separate database containing patient data will be required and integrated with production or real-world hospital EHR data, this can be a workable solution. For the time being, the patient record contains only a few fields that include personal and medical information such as age, address, allergies, symptoms, treatment, and follow-up. When a doctor administers medication to a patient, the patient's medical history is available to assist doctors in assigning appropriate treatment. It is intended to provide additional steps in application for patients to improve record privacy. A patient can choose to give a specific doctor permission to access his or her data. Doctors can only see a subset of patient data, such as all medical fields, ages, and allergies, whereas patients can see all fields but only edit personal fields.

3.1 Hyperledger Fabric

Hyperledger Fabric is an open-source enterprise-grade permissioned distributed ledger technology (DLT) platform designed to be used in enterprise contexts. It offers some key advantages over other popular distributed ledger or blockchain platforms.

Fabric's architecture is highly modular and configurable, allowing for innovation, versatility, and optimization across a wide range of industry use cases, such as banking, finance, insurance, healthcare, human resources, supply chain, and even digital music delivery.

The Fabric platform is also permissioned, which means that, unlike a public permissionless network, the participants know each other rather than being anonymous and thus completely untrustworthy. This means that, even if the participants do not fully trust one another, a network can be run under a governance model based on the trust that exists between them, such as a legal agreement or a framework for handling disputes.

One of the platform's most important differentiators is its support for pluggable consensus protocols, which allow the platform to be more effectively customized to fit specific use cases and trust models. Fabric can use consensus protocols that do not require a native cryptocurrency to incentivize expensive mining or power smart contract execution. The absence of a cryptocurrency reduces some significant risk/attack vectors, and the platform can be deployed with roughly the same operational cost as any other distributed system.

Fabric is the best choice for our system due to the combination of these distinguishing design features.

Ledger

In Hyperledger Fabric, a ledger consists of two distinct, though related, parts – a world state and a blockchain. Each of these represents a set of facts about a set of business objects.

World state is a database that holds current values of a set of ledger states. Ledger states are, by default, expressed as key-value pairs. Blockchain, on the other side, is a transaction log that records all the changes that have resulted in the current the world state.

In the application, the ideology of a ledger can be thought of as one logical database in a Hyperledger Fabric network. In reality, the network contains multiple copies of a ledger for each hospital's peers – which are kept consistent with every other copy through consensus. The ledger in our system is utilized as follow:

- The world state stores all the patient data. All patients that are queried and updated are stored in this database. All the query transactions are retrieved from the world state.

- The transaction log is a sequential structured interlinked blocks of log files, where each block contains a sequence of transactions. Each transaction represents a query or an update to the world state.

Identity and Certificate Authorities

The different actors in a blockchain network include peers, orderers, client applications, administrators and more. An actor or a node is able to participate in the blockchain network, via the means of a digital identity issued for it by an authority trusted by the system. In the most common case, digital identities (or simply identities) have the form of cryptographically validated digital certificates that comply with X.509 standard and are issued by a Certificate Authority (CA).

It is because CAs are so important that Fabric provides a built-in CA component to allow you to create CAs in the blockchain networks you form. This component—known as Fabric CA is a private root CA provider capable of managing digital identities of Fabric participants that have the form of X.509 certificates.

In our application, each hospital has its own CA server that creates the identities using client of the server to prove that the identity belongs to their hospital and can be trusted. All of the identities created by a CA run of a hospital share the same trust of the root CA. Figure 1 describes the CA system.

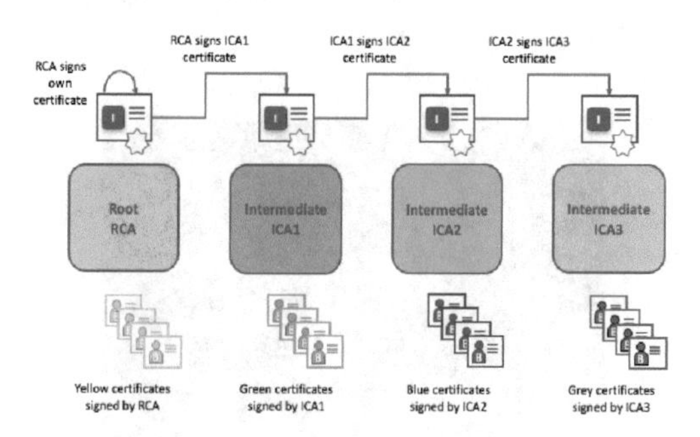

Fig. 1. Certificate Authorities [5]

Membership Service Providers

Because Fabric is a permissioned network, in order to transact on the network, blockchain participants must be able to prove their identity to the rest of the network. Certificate Authorities create identities by generating a public and private key pair that can be used to verify identity. A mechanism is required to enable that proof because a private key can never be shared publicly. An membership service provider (MSP) is a component that defines the rules that govern an organization's valid identities. In other words, MSP is used to validate the private key associated with the transaction.

The implementation of the MSP requirement, which is shown in Fig. 2, is a set of folders that are added to the network configuration and are used to define an organization both internally and externally.

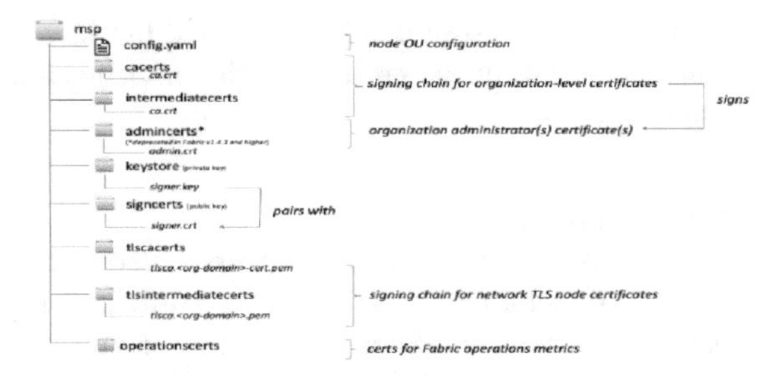

Fig. 2. MSP structure [6]

Endorsement Policies

Every smart contract in a chaincode package includes an endorsement policy that specifies the minimum number of peers from various channel members required to execute and validate a specific smart contract's transactions before the transaction is considered valid. As a result, the endorsement policies specify the entities that must "endorse" the plan's execution.

Our system's endorsement policy is specified in the configtx.yaml file. We made it clear in Fig. 3 and Fig. 4 that every peer of the parties is an endorser, and a chaincode definition is only committed to the channel after the majority of channel members have approved it.

Fig. 3. Policies defined for hospital 1

Consensus Protocol

The consensus protocol used by Hyperledger Fabric is called Raft. Based on an implementation of the Raft protocol, Raft is a crash fault tolerant (CFT) ordering service. Raft operates on the "leader and follower" principle. According to this, any one of the three states - leader, candidate, or follower, can be maintained by any node in a replicated state machine (server cluster). A node can remain in any of the three states mentioned above under normal circumstances. Any requests made to the follower node are forwarded to the leader node; only a leader can communicate with the client. A candidate for leadership may solicit votes. A follower exclusively reacts to the leader or the candidates.

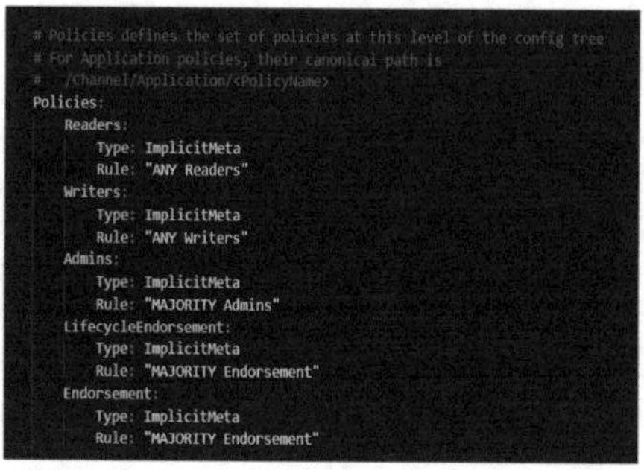

```
# Policies defines the set of policies at this level of the config tree
# For Application policies, their canonical path is
#   /Channel/Application/<PolicyName>
Policies:
    Readers:
        Type: ImplicitMeta
        Rule: "ANY Readers"
    Writers:
        Type: ImplicitMeta
        Rule: "ANY Writers"
    Admins:
        Type: ImplicitMeta
        Rule: "MAJORITY Admins"
    LifecycleEndorsement:
        Type: ImplicitMeta
        Rule: "MAJORITY Endorsement"
    Endorsement:
        Type: ImplicitMeta
        Rule: "MAJORITY Endorsement"
```

Fig. 4. Chaincode-level endorsement policy

In Raft, transactions (such as proposals or configuration modifications) are automatically forwarded to the active leader of that channel by the ordering node that receives the transaction. As a result, peers and applications are not required to be aware of the leader node's identity at any given moment. The ordering nodes alone should be aware.

The transactions are organized, bundled into blocks, approved, and delivered after the orderers validation procedures are finished.

3.2 System Architecture

Figure 4 shows a system's high-level architecture. The architecture also uses color schemes to depict the fabric network in great detail.

Every component of the application is pluggable, as stated, so the backend code and smart contract are written in JavaScript, and NodeJS serves as the REST API server. Using Angular v11, the user interface was created. Using REST calls with a JSON web token for authentication, the front and back ends communicate with one another.

In our system, patients use Zero-Knowledge proof (ZK-proof) to prove owner-ship of their EHRs. This allows them to grant specific access permissions, such as the right to their EHRs. Authorized doctors can then update these EHRs. In particular, our system is combined of the below components.

- Client application: includes a user interface and functions that allow users to interact with the system. It has three main users: patients, doctors, and administrators. Patients can manage their EHR access permissions, while doctors can update them. Admin can create new accounts and manage access permissions. All EHR-related operations are secure, and ZKP is used to improve security.
- Server: serves as a bridge between the client application, IPFS, and the Fabric network. It invokes functions to encrypt data, generate ZK-proofs, and generate smart contract transactions.
- Docker: a technology that packages applications and related components into containers for easy deployment on any Docker-supported platform. In the Hyperledger Fabric EHRs Blockchain system, Docker is used to contain application components such as peer nodes, orderer nodes, and CA nodes, ensuring consistency and flexibility across different environments. Each system component is packaged as a Docker

image, which includes all required dependencies, configuration, and executable files. These images are then deployed on a server as containers, providing isolation and security. Each container is a standalone process with its own file system, network interface, and resource constraints.
- Fabric network: The Hyperledger Fabric network consists of peer nodes, order nodes, and CA nodes that are in charge of verifying transactions, maintaining the ledger, and ensuring network security. Peer nodes maintain the blockchain's state, verify transactions, and execute chaincode to manage EHRs. Orderer nodes sort transactions and create new blockchain blocks. CA nodes issue and manage certificates for network members.
- InterPlanetary File System (IPFS): is a distributed file system that allows for storage and retrieval of files and metadata over a peer-to-peer network. IPFS is used in the Hyperledger Fabric EHRs Blockchain to store EHR metadata such as personal information, medical history, and patient treatment records.

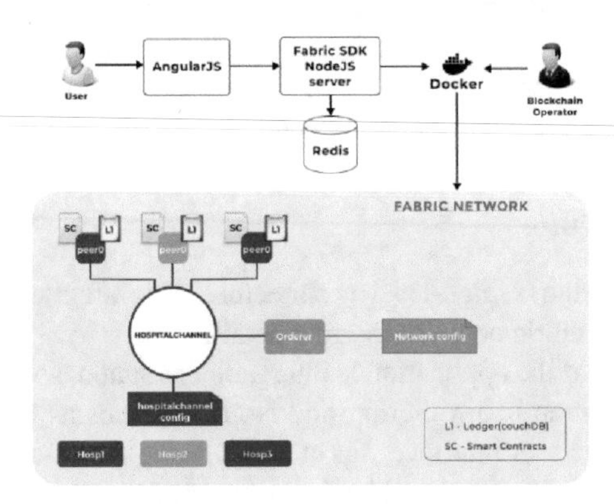

Fig. 5. System architecture

3.3 Use Case Diagram

There are three actors in the system which are admin, patient, and doctor. Each role has its own use cases which are described in Fig. 5 (Fig. 6).

4 Implementation

4.1 Fabric SDK

The Hyperledger Fabric Client SDK provides APIs for interacting with the Hyperledger Fabric blockchain, such as interacting with smart contracts, submitting transactions to a ledger, and querying the ledger.

The Fabric SDK comes with the following packages:

- fabric-ca-client: the fabric-ca APIs provide participants (admin/patient/doctor) to register and enroll in order to establish trusted identities on the blockchain network. The

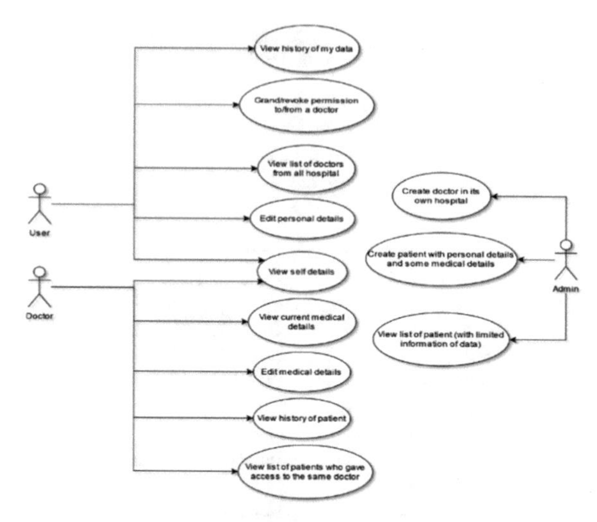

Fig. 6. Use case diagram

package generates a new CA client for interacting with the hospital's CA server to register and enroll participants.

- fabric-common: encapsulates the common code used to send transaction invocations by all fabric-sdk-node packages supporting fine-grained interactions with the Fabric network. Provides APIs for monitoring events, logging, and configuring environment variables, program arguments, and in-memory settings.
- fabric-network: This package includes the APIs needed to connect to the Fabric network, submit transactions, and query or edit the ledger. Provides APIs to manage the wallet, which is used to manage identities, and to create a connection profile based on the connection profile JSON generated when the CA is created. The Gateway class is the main class that allows the Fabric SDK to interact with the network. When the object is instantiated, it creates a gateway to a peer within the blockchain and grants access to the channels for which that peer is a member, as well as the chaincodes installed on those channels.

4.2 Smart Contracts

A smart contract defines the transaction logic that governs the lifecycle of a business object in the world state. Our system is primarily composed of three smart contracts packaged into a single chaincode, as shown in Fig. 7. Each actor executes their smart contract.

4.3 Homomorphic Encryption

Homomorphic encryption refers to a class of cryptographic techniques that allow for the execution of computations on encrypted data without requiring prior decryption. Homorphic encryption can provide a secure and privacy-preserving way to perform operations on sensitive patient information in the management of EHRs.

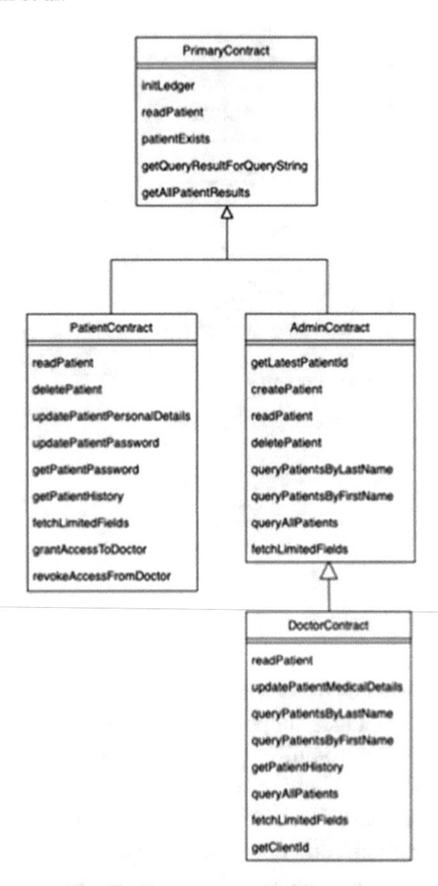

Fig. 7. Smart contracts hierarchy

The fully implemented homomorphic encryption algorithm in this study consists of the following steps:

- Initialize polynomial modulus of degree 8192.
- Initialize coefficient modulus consisting of three 60-bit prime numbers.
- Initialize plaintext modulus as a 20-bit prime number.
- Generate key pair (P_k, S_k, E_k).
- When a new patient is registered at the hospital, the administrator initializes the patient's EHR and encodes it into a vector of integers using the ASCII code. The vector's elements are then converted to binary form.
- The original EHR is transformed into a plaintext polynomial before being encrypted with the BFV algorithm.
- When updates are made, the system computes the encrypted version of the updated information as described above and then uses the EvalAdd function to combine the new encrypted version with the existing one.

4.4 Zero-Knowledge Proof

Zero-Knowledge Proof (ZKP) is a cryptographic protocol that enables one party—the prover—to demonstrate to another—the verifier—that a specific statement is true without disclosing any information beyond the statement's veracity. ZKP, or ZK-proof, is the process wherein one person does a computation without demanding that others perform the same computation and then declares with absolute certainty that the claim is true without disclosing the computation to anybody else.

ZK-proof can be applied to a Blockchain-based EHRs system to guarantee the confidentiality and security of patient data while keeping it verifiable by authorized parties. This implies that patients or healthcare service providers can demonstrate their legal authority to access certain health information without disclosing any sensitive information about the patient's medical history.

Consider a scenario where a patient wants to share their medical records with a healthcare provider but only wants to provide a limited amount of information (such prescriptions or allergies) and wants to keep the rest of their medical history private. Without actually disclosing the contents of the necessary medical information, the patient can use ZKP to demonstrate to the healthcare service provider that they have access rights to those documents.

The authors of this study raised the issue of patients having to provide proof of ownership of their electronic health records (EHRs) in order to allow access rights to a doctor updating those EHRs. This is how the algorithm operates:

- The hospital gets a new patient, creates an account for him, and starts up his EHRs.
- For the new patient, the system computes and generates a public-private key pair. The patient's EHR address is connected to the public key on the blockchain.
- The patient creates a ZKP proving ownership of the private key that corresponds to the public key linked to their EHR address in order to confirm their ownership rights to an EHR.
- The ZKP is received and verified by the system. The system accepts the request to grant access permissions to a chosen physician for updating the EHRs if the verification process is successful.

4.5 Results

This study's suggestion of a private blockchain system for handling delicate data, such medical data, is one of its most important contributions. Our method circumvents security and permission flaws, in contrast to other studies that depended on public blockchain technology, guaranteeing that data is secure and available to only authorized individuals. Additionally, we employ IPFS to store off-chain data, which lowers the cost of storage and transactions and allows the system to grow more effectively.

Modern cryptographic techniques are also used in the research to increase system security. The server uses the BFV technique and homomorphic encryption to compute and update changes to patients' electronic health records without having access to the original data. Additionally, a patient's ownership rights in any EHR are verified using a zero-knowledge proof technique. This technique makes sure the system can authenticate and provide access to the right person without jeopardizing the integrity of the record's real contents.

Considering the aforementioned, this study significantly contributes to the problem of keeping and exchanging electronic health records by addressing critical issues including security, privacy, and scalability.

5 Conclusion

Hyperledger fabric is a promising blockchain framework that has some concepts of policies, smart contracts, and provision of secure identities which make the records secure and controlled. In this paper, we presented a blockchain-based system for storing and sharing EHRs using Hyperledger Fabric v2. The proposed solution of the study has addressed the shortcomings of security and permission for sensitive data, such as personal data of patients, by using private blockchain instead of public blockchain. In addition, using IPFS to store off-chain data has minimized transaction and storage costs while also improving the system's scalability. The study also used typical cryptographic technologies to enhance the proposed system's security, such as homomorphic encryption with the BFV scheme and non-interactive zero-knowledge proofs to verify a patient's ownership of any EHR.

This study has contributed significantly to the problem of storage and sharing of electronic health records. However, to meet market and societal demands, in the future, the study may continue to be developed to improve the system's performance and scalability. The study may also be expanded to include other functions, such as processing and analyzing health data to support healthcare decisions or providing patients with more control over their health records.

References

1. Azaria, A., Ekblaw, A., Vieira, T., Lippman, A.: Medrec: using blockchain for medical data access and permission management. In: 2016 2nd International Conference on Open and Big Data (OBD) (2016)
2. Hasselgren, A., Kralevska, K., Gligoroski, D., Pedersen, S.A., Faxvaag, A.: Blockchain in healthcare and health sciences—a scoping review. Int. J. Med. Inform. **134**, 104040 (2020)
3. Misbhauddin, M., AlAbdulatheam, A., Aloufi, M., Al-Hajji, H., AlGhuwainem, A.: MedAccess: a scalable architecture for blockchain-based health record management. In: 2020 2nd International Conference on Computer and Information Sciences (ICCIS) (2020)
4. Shen, B., Guo, J., Yang, Y.: MedChain: efficient healthcare data sharing via blockchain. Appl. Sci. **9**(6), 1207 (2019)
5. Hyperledger Fabric Documentation. https://hyperledger-fabric.readthedocs.io/en/release-2.4/_images/identity.diagram.1.png
6. Hyperledger Fabric Documentation. https://hyperledger-fabric.readthedocs.io/en/release-2.4/_images/membership.diagram.6.png

Test Criteria for Context-Aware Mobile Applications

Thi Thanh Binh Le[1,2](✉), Oum -El- Kheir Aktouf[2], Ioannis Parissis[2], and Thanh Binh Nguyen[3]

[1] The University of Danang, University of Science and Education, Da Nang, Vietnam
lttbinh@ued.udn.vn, ntbinh@vku.udn.vn
[2] University Grenoble Alpes, Grenoble INP, LCIS, Valence, France
oum-el-kheir.aktouf@lcis.grenoble-inp.fr,
ioannis.parissis@grenoble-inp.fr
[3] The University of Danang, Korea University of Information and Communication Technology, Da Nang, Vietnam

Abstract. Context-aware mobile apps provide adaptive services that depend on the changing environments. The challenge in testing context-aware mobile apps holds in that these apps adapt their behavior when context conditions are changing. According to an exhaustive survey of the testing context-aware mobile apps research area, current testing approaches take into consideration the apps functions and different contexts but do not consider testing coverage criteria nor evaluate the coverage of various situations. Our work is intended to fill this gap. In our previous research work, we presented a test model for context-aware mobile apps dealing with changing location context based on the combination of a Bigraph Reaction System and a Dynamic Feature Petri Net. In this paper, we propose a new test criterion for context-aware mobile apps. This criterion results from the combination of pattern-flow-based coverage criteria and boundary-based coverage criteria. With this criterion, we select the user's location coordinates inside partition boundaries, which help narrowing the input data space and cover all test situations in test process.

Keywords: Test Criteria · Context-Aware Mobile Apps Testing · Model-Based Testing

1 Introduction

Today, mobile phones have a huge impact on people lifestyle. They allow users to run mobile applications (apps) and access mobile Internet services at anytime, anywhere. Thanks to their high portability, they are very useful to access information wherever the users go through GPS-enabled devices and mobile networks. Context-aware mobile apps provide adaptive services responding to the dynamically changing contexts in the environment [2]. As a result of the availability of mobile devices and mobile network infrastructures, context-aware mobile apps can position the mobile device on the Earth and use its current location to provide suitable services to the user.

N. T. Nguyen et al. (Eds.): CITA 2023, LNNS 734, pp. 391–403, 2023.
https://doi.org/10.1007/978-3-031-36886-8_33

Testing context-aware mobile apps is challenging due to the high interaction complexity between the applications and their environments [3]. In our previous paper [1], we proposed a test model for context-aware mobile apps dealing with changing location context. The model consists of: 1) describing the static structure of the environment by using place graphs and link graphs, called bigraphs and modeling the dynamic behaviors of the environment by reaction rules, which indicate the changes of the contexts in the environment, a collection of bigraphs enriched with reaction rules is called a Bigraph Reaction System (BRS); 2) using a Dynamic Feature Petri Net (DFPN) to model the middleware (the system is assumed to include a middleware and a set of services); 3) combining BRS and DFPN to describe the interactions between the environment and the system. Additionally, our work [1] introduced a pattern-flow testing criterion [2] to select test paths. However, while experimenting context-aware mobile apps testing with the proposed test model, we encountered some issues when selecting test data set. This paper proposes a new test criterion which eases the selection of test data sets and the coverage of test situations. It results from the combination of pattern-flow-based coverage criteria and boundary-based coverage criteria.

The remaining parts of this paper are structed as follows. Section 2 presents the background on test criteria. Section 3 introduces a case study used to illustrate the proposed approach. Section 4 briefly introduces the test model for context-aware mobile apps presented in [1]. Section 5 proposes a new test criterion for context-aware mobile apps. Section 6 provides an experimental analysis of the effectiveness of the proposed test criterion. Conclusion and future work are given in Sect. 7.

2 Background

In this section, we introduce the definitions related to location contexts. These definitions are prerequisites to develop the selection of test data sets for context-aware mobile apps.

Definition 1. A location $l_i = (long_i; lat_i)$ is a pair of real numbers which refer to the longitude longi and the latitude lati of the location on the Earth surface [9].

Definition 2. A path $path_i(l_{i1}; l_{i2}; ...; l_{ik})$ is defined as a sequence of locations [9].

Definition 3. $P_x(long_x; lat_x)$ is the location information of a service $service(P_x)$ which is known as Point of Interest (POI) [7].

Definition 4. A test case is a sequence of input stimuli to be fed into a system and the corresponding expected behavior of the system. A test case comprises abstract test cases and concrete test cases.

Definition 5. An abstract test case consists of abstract information about the sequence of inputs and outputs. The missing information is often concrete parameter values or function names. Abstract test cases are often the first step in test case creation. They are used to get an idea of the test case structure or to get information about satisfied coverage criteria.

Definition 6. A concrete test case is an abstract test case plus all the concrete information that is missing to execute the test case. Concrete test cases comprise the complete test information and can be executed on the system under test (SUT).

Definition 7. Control-flow-based coverage criteria are defined on the basis of a control flow graph representing a program.

Definition 8. Data-flow-based coverage criteria are focused on the data flow of variables. Expressions can define and use variables. A def is a location where a value for a variable is stored into memory. A use is a location where a variable's value is accessed.

Definition 9. Boundary-based coverage criteria [4] are constraints that specify value partition of objects. Objects that satisfy a value partition are said to be inside/outside the partition.

3 Case Study

The TripAdvisor application illustrated on Fig. 1 is an example of context-aware mobile app. This app provides Tourist Spots service, Hotels service, and Restaurants service, which help travelers to search suitable services based on their locations.

Fig. 1. The Tripadvisor application.

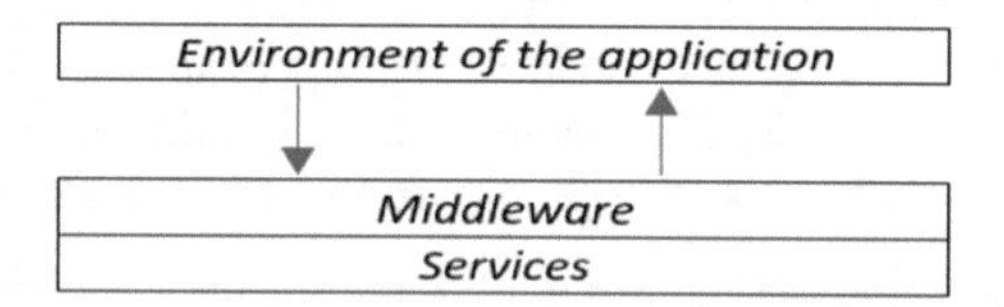

Fig. 2. Architecture of context-aware mobile apps.

This application can be used by every body in the world. App users can be in different countries, different cities, different districts, and so on. In order to ease the application testing, these areas are named with symbols: Zone3 is a city of a country; Zone2 is a district belonging to the city; Zone1 is a ward belonging to the district; and so on.

We consider the following scenario: There is 1 service (e.g., Tourist Spots service or Hotels service or Restaurants service) in Zone1, there are 2 services e.g., (Tourist Spots and Hotels service) or (Hotels and Restaurants service) or (Tourist Spots and Restaurants service) in Zone2 and there are 3 services (e.g., Tourist Spots and Hotels and Restaurants service) in Zone3.

The following test situation will be examined: The user is outside the Zones and moves into Zone1, which has one service (Tourist Spots service); then he/she moves into Zone2, which has two services (Tourist Spots and Hotels service); then he/she moves into Zone3, which has 3 services (Tourist Spots, Hotels and Restaurants service).

4 Test Model for Context-Aware Mobile Apps

Context-aware mobile apps provide adaptive services responding to the dynamically changing contexts in the environment. Context-aware mobile apps often consist of a middleware and a collection of services.

Based on the architecture of context-aware mobile apps (Fig. 2), we proposed a test approach that consists of three phases: modeling the environment of the application; modeling the middleware and the services; and combining the two models above to verify the interactions between the environment and the application [1].

4.1 Modeling the Environment of the Context-Aware Mobile Application

The environment of applications comprises a range of physical facilities, moving entities and sensors or wireless connections to backend systems. For example, in the TripAdvisor application, physical facilities include Zone1, Zone2, Zone3; moving entities may refer to the user who may move through Zones; and wireless connections to backend systems. We proposed to describe the static structure of the environment by bigraphs and to model the dynamics of the environment by reaction rules.

Describing the Static Structure of the Environment. A bigraph consists of two graphs: a place graph that captures notions of locality or containment and a link hypergraph that models connectivity or associations. Therefore, bigraphs are a natural way to model the containments like physical facilities, connections and moving entities of context-aware mobile app.

Figure 3 shows a bigraph, which describes physical facilities (static structure of the environment) of the TripAdvisor application. It consists of three regions (rectangles): Zone3 is a city of a country, Zone2 is a district belonging to the city, Zone1 is a ward belonging to the district. A user may enter or leave a Zone.

Modeling the Dynamics of the Environment. In bigraph models, the mobility of entities in the environment (the dynamics of the environment) is expressed as reactions, with sets of reaction rules that give possible ways in which a system might be reconfigured. The reaction rules describing the user's mobility through Zones of the Tripadvisor application are presented as follows.

Reaction rule r0 (see Fig. 4): The user is outside of the Zone (C0) and moves into Zone1 having one service {Tourist Spots service (C1) or Hotels service (C2) or Restaurants service (C3)}.

Reaction rule r1: User moves from Zone1 having one service {Tourist Spots service (C1)} into Zone2 having two services (including pre-existing service) {Tourist Spots and Hotels service (C4)} or {Tourist Spots and Restaurants service (C5)}.

Reaction rule r2: User moves from Zone1 having one service {Tourist Spots service (C1)} into Zone2 having two services (pre-existing services are not included) {Hotels service and Restaurants service (C6)}.

Reaction rules r3: User moves from Zone2 having two services {Tourist Spots service and Hotels service (C4)} into Zone3 having three services {Tourist Spots service, Hotels service and Restaurants service (C7)}.

We select a bigraph of interest to represent the initial state of the environment and the set of reaction rules as above. Then, by matching reaction rules against the current bigraph, a Bigraph Label Transition System (B-LTS) is created with bigraphs as states and reaction rules as labels.

We model the test situation presented in Sect. 3 by the bigraph and reaction rules as shown in Fig. 5, where reaction rules are labels from r0 to r3 and bigraphs are states from C0 to C7.

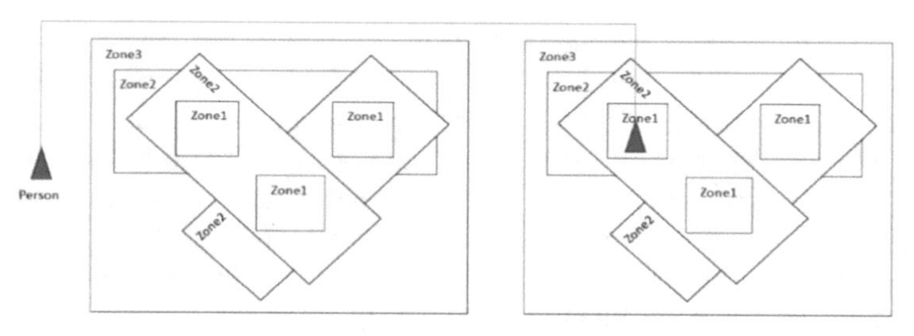

Fig. 3. A bigraph for modeling the TripAdvisor application in a simulated environment

4.2 Modeling Middleware and Services

Middleware is between the environment and a set of services. The middleware often provides three functions: 1) collecting context information from the environment; 2) analyzing and reasoning about the situation; and 3) selecting and invoking proper services to react, which can be atomic or composite services.

Context-aware mobile apps are based the user's context to provide adaptive services. A service in service-oriented architecture may be an atomic service or a composite service [11]. Yu et al. [2] proposed a model-based testing approach based on bigraphical modeling for the context-aware mobile apps with an atomic service. They experimented their model on an airport application and an atomic illumination service. Nevertheless, this model is not implemented for context-aware mobile apps with composite services.

In fact, Yu et al. [2] used an extended finite state machine to model an atomic service. However, when the number of services in context-aware mobile apps grows, then the number of states also increases and using such extended finite state machine model could be impractical. To overcome this limitation, we investigate in our work [1] the use of Dynamic Feature Petri Nets (DFPN) instead of finite state machines, in combination with the bigraph model.

We describe the TripAdvisor application with the following services: Tourist Spots service (T), Hotels service (H), Restaurants service (R) and the location changing of the user among Zones. Each service has 2 states and 2 transitions respectively: Tourist.

Spots service (T) has 2 states of no service (T0) and service (T1) and transitions t0 and t1. For instance, Fig. 6 shows the DFPN for Tourist Spots service.

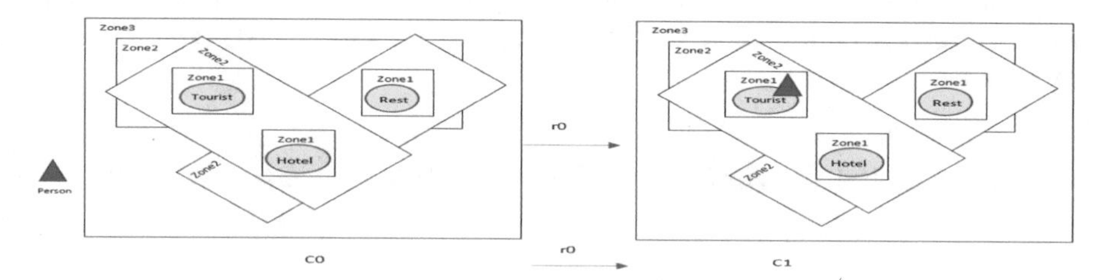

Fig. 4. Reaction rule *r*0.

Basically, a transition is activated if its input states are marked. In the beginning, the token may be consumed by two transitions t0 and t1. If transition t1 fires, then the token moves from state T0 to state T1, and Tourist Spots service is enabled.

Symmetrically, when token moves from state T1 to state T0 (t0 fires), Tourist Spots service is disabled. Similarly, Hotels service (H) has 2 states of no service (H0) and service (H1) and transitions t0 and t2. Restaurants service (R) has 2 states of no service (R0) and service (R1) and transitions t0 and t3. These three services in the TripAdvisor application can be modeled with DFPN (see Fig. 7).

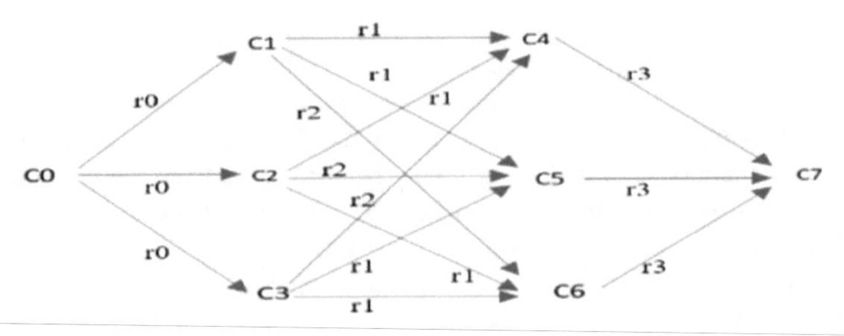

Fig. 5. B-LTS.

We model the test situation with DFPN as shown in Fig. 12 where transition are labels from t0 to t3 and DFPN are states from P0 to P7.

4.3 Combining Two Models to Verify the Interaction Between the Environment and the Application

The combination of BRS and DFPN as a Cartesian product result in a synchronized model. Let B be B-LTS model for the context-aware environment, and D be DFPN model for the context-aware middleware. The interaction between context-aware environment and middleware is modeled as B × D, such that.

The vertex set S is the Cartesian product B × D. It is a finite and non-empty set of states.

Any two vertices (b, d) and (b’, d’) ∈ S are adjacent in S if and only if b = b’ and d is adjacent with d’ in D, or d = d’ and b is adjacent with b’ in B. The result below shows an example of the interactions between the B-LTS in Fig. 5 and the DFPN of the test situation.

$$C0P0 \xrightarrow{t1} C0P1 \xrightarrow{r0} C1P1 \xrightarrow{t2} C1P4 \xrightarrow{r1} C4P4 \xrightarrow{t3} C4P7 \xrightarrow{r3} C7P7$$

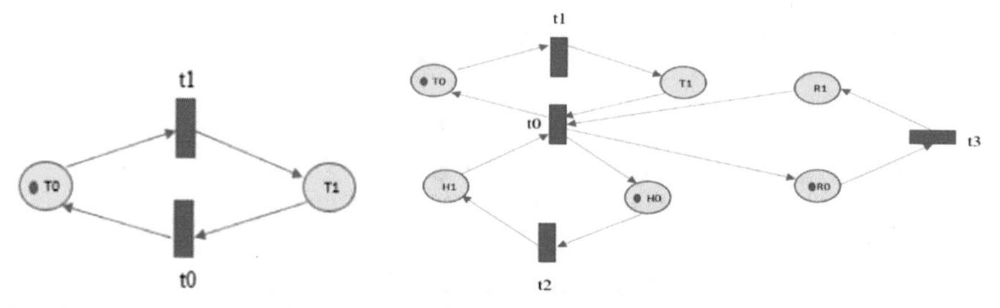

Fig. 6. DFPN for Tourist Spots service.

Fig. 7. DFPN for the TripAdvisor application.

5 Test Criteria for Context-Aware Mobile Apps

5.1 Test Strategy Based on Bigraphical Pattern-Flow Testing

B-LTS may contain a high number of paths. In our previous work, we applied a bigraphical pattern flow testing strategy like in [4] to select a subset of paths. This strategy is similar to data-flow testing strategy as described in [6], but defined on reaction rules in terms of bigraphs instead of variables in data-flow approaches.

In bigraph terms, a reaction rule has the form R - > R′ where R is known as the redex and R′as the reactum. A pattern is a sub-structure in a redex or a reactum of a reaction rule. If a pattern does not appear in redex of reaction rule R, but appears in the reactum, it is called the pattern definition. If a pattern appears in redex of reaction rule R, but does not appear in the reactum, it is called the pattern use. If a pattern appears both in redex and reactum of reaction rule R, it is called the pattern def-use.

A path in a B-LTS is a sequence of $C_0r_0\ldots$Ciri from an initial state to a final state of B-LTS, where Ci stands for bigraph and ri for reaction rule. We select a set of reaction paths such that the set of paths meets all-defs, which ensures that each pattern definition reaches at least a pattern use, or all-uses, which ensures that each pattern definition reaches all pattern uses. Abstract test cases are paths which are generated from the B-LTS in Fig. 5 and the synchronized model with the test situation of TripAdvisor application as follows.

All-Defs. Every def reaches a use.

1. $r0\text{-}r1 \iff C0 \xrightarrow{r0} C1 \xrightarrow{r1} C4$
2. $r1\text{-}r3 \iff C1 \xrightarrow{r1} C4 \xrightarrow{r3} C7$

All-Uses. Every def reaches all possible uses.

1. $r0\text{-}r1 \iff C0 \xrightarrow{r0} C1 \xrightarrow{r1} C4$
2. $r0\text{-}r2 \iff C0 \xrightarrow{r0} C1 \xrightarrow{r2} C6$
3. $r0\text{-}r1\text{-}r3 \iff C0 \xrightarrow{r0} C1 \xrightarrow{r1} C4 \xrightarrow{r3} C7$
4. $r0\text{-}r2\text{-}r4 \iff C0 \xrightarrow{r0} C1 \xrightarrow{r1} C5 \xrightarrow{r3} C7$
5. $r1\text{-}r3 \iff C1 \xrightarrow{r1} C4 \xrightarrow{r3} C7$

All-Du-Paths. All the paths between defs and uses are executed.

1. $r0$-$r1$ <=> $C0 \xrightarrow{r0} C1 \xrightarrow{r1} C4$
2. $r0$-$r1$ <=> $C0 \xrightarrow{r0} C1 \xrightarrow{r1} C5$
3. $r0$-$r2$ <=> $C0 \xrightarrow{r0} C1 \xrightarrow{r2} C6$
4. $r0$-$r1$-$r3$ <=> $C0 \xrightarrow{r0} C1 \xrightarrow{r1} C4 \xrightarrow{r3} C7$
5. $r0$-$r1$-$r3$ <=> $C0 \xrightarrow{r0} C1 \xrightarrow{r1} C5 \xrightarrow{r3} C7$
6. $r0$-$r2$-$r4$ <=> $C0 \xrightarrow{r0} C1 \xrightarrow{r1} C5 \xrightarrow{r3} C7$
7. $r1$-$r3$ <=> $C1 \xrightarrow{r1} C4 \xrightarrow{r3} C7$
8. $r1$-$r3$ <=> $C1 \xrightarrow{r1} C5 \xrightarrow{r3} C7$
9. $C0P0 \xrightarrow{t1} C0P1 \xrightarrow{r0} C1P1 \xrightarrow{t2} C1P4 \xrightarrow{r1} C4P4 \xrightarrow{t3} C4P7 \xrightarrow{r3} C7P7$

5.2 Test Strategy Based on Boundary Values Testing

We explore here the opportunity of using boundary-based coverage criteria [10] as a means to build such partitions and, hence, to help narrowing the input data space.

Boundary values are used to select concrete values from partitions. In test process, faults are often detected around boundaries [13]. Therefore, boundary values should be considered to increase the faults detection. The concrete value selection for partitions are described in [14] as follows: For each partition boundary of a partition, a value should be selected inside the partition close to the boundary, outside the partition close to the boundary or on the boundary. Our approach focuses on values which are selected inside the partition to the boundary.

The location information (including latitude and longitude coordinates) of mobile end user or POIs can be obtained through the Global Navigation Satellite Systems (GNSS), the Geographic Information System (GIS) and the Wireless Communication (WC) technologies [8]. We propose the use of POIs to define partition boundaries for zones in the test model as below. For each POI in a zone, we consider its real-world coordinates $P_i = (long_i; lat_i)$ as provided from GIS, GNSS or WC [8].

Definition 10. Radius R_i (along the real-world coordinates) is maximum distance between the user and a POI. That means approximately Ri(km)/111(km) degree difference along the longitude and the latitude coordinates on the Earth surface [12].

For example, consider the POIs located of a service within 1 km from the user's GPS location. That means approximately 1 km/111 km = 0.009-degree difference along the longitude and the latitude coordinates.

Definition 11. The circle $circle_i$ $(P_i; R_i)$ denotes a circle, the center of the circle is Pi, and the radius of the circle is R_i (along the longitude and the latitude coordinates).

Definition 12. The partition boundaries of the zone having 1 service $service(P_i)$ is a circle $circle_i$.

Definition 13. The partition boundaries of the zone having n services $(service(P_1), service(P_2); \ldots; service(P_n))$ denotes the intersection between n circles $(circle_1; circle_2; \ldots; circle_n)$.

5.3 A Novel Coverage Criterion Based on Pattern-Flow and Boundary Values

We combine a boundary-based criterion and a pattern-flow-based coverage criterion to build a new improved coverage criterion for context-aware mobile apps testing. This criterion makes it possible to define input partitions based on the boundaries of Zones for abstract test cases that are generated from bigraphical pattern-flow-based coverage criteria of test model, as explained in Sect. 4.

Abstract test cases are generated from the test model and bigraphical pattern-flow based testing in Sect. 4, which comprise abstract information about input parameters (e.g., the sequence of positions of the user when he/she moves through Zones). We define partition boundaries for each zone. These partition boundaries are used to select specific information for each abstract test case (e.g. select a proper current coordinates position of user, which is inside partition boundaries of zones when he/she moves through zones). That means, the abstract test cases satisfy the chosen structural, e.g., bigraphical pattern-flow based, coverage criterion. For each abstract test case, proper input values can be selected to satisfy a chosen boundary-based coverage criterion. The result of this approach is a test suite that satisfies a combination of both kinds of coverage criteria.

6 Experiment

We consider context information as location context, and we select test Zones first. TripAdvisor application has been used in many countries and cities, we only test TripAdvisor application in an area of Valence, France as an experiment. We implement the proposed test criterion to the test model presented in Sect. 4.

We test the TripAdvisor application at an area of Valence, France which has location coordinates including longitude from 43.38611 to 43.38736 and latitude from -1.66303 to -1.66113.

There are 38 services in the selected area with 9 services of interest (Tourist Spots, Hotels, Restaurants services). We define the partition boundaries of Zones corresponding to these services, which are proposed in the test model.

We call P_T the location coordinates of service Tourist Spots, P_H the location coordinates of service Hotels and P_R the location coordinates of service Restaurants. Location coordinates of services POIs include (P_T, P_H, P_R). From the area of Valence, we take the location coordinates of services (POIs):

$$P_T.P_{Sas\ Luz\ Voyages}(43.38645;\ -1.66146)\}.$$

$$P_H.\{P_{Hotel\ Relais\ Saint}(43.38629;\quad -1.66158), P_{Brit\ Hotel\ de\ Paris}(43.38624;\quad -1.66118), P_{Saint\ Jean\ de\ Luz\ rentals}(43.38662;\quad -1.66075)\}.$$

Suppose that we use the radius of 50 m. We calculate the partition boundaries of Zone1, Zone2, Zone3 as follows.

Zone1. Zone has 1 service, and is a circle (center is a POI's location, radius of 50 m = 0.00045 degree longitude and latitue).

Tourist Spot Service (P_T). $P_{Sas\ Luz\ Voyages}$(43.38645; –1.66146). Zone1 has Tourist spot service in Zone having longitude from $(longP_T - R)$ to $(longP_T + R)$ and latitude from

$(latP_T - R)$ to $(latP_T + R)$ Equal to Longitude from $(43.38645 - 0.00045) = 43.38600$ to $(43.38645 + 0.00045) = 43.3869$ and atitude from $(-1.66146$ - $0.00045) = -1.66191$ to $(-1.66146 + 0.00045) = -1.66101$.

Hotel Service (P_H). $P_{\text{Hotel Relais Saint}}(43.38629; -1.66158)$. Zone1 has Hotel service in Zone having longitude from 43.38584 to 43.38674 and latitude from −1.66203 to −1.66113.

Restaurant Service (P_R). $P_{\text{Boulangerie Magri Christophe Ogi-labea}}(43.38635; \ -1.66169)$. Zone1 has Restaurant service in Zone having longitude from 43.3859 to 43.3868 and latitude from −1.66214 to −1.66124.

Zone2. Zone has 2 services, and is the intersection of 2 circles (2 Zone1 as in Fig. 8).

$C1$ is Zone1 has Tourist spot service $P_{Sas\ Luz\ Voyages}$. $C1$ is a circle (center is $P_{Sas\ Luz\ Voyages}$, radius is 50 m) which has equation:

$$x^2 + y^2 - 86.7729x + 3.32292y + 1885.14449 = 0 \tag{1}$$

$C2$ is Zone1 has Hotel service $P_{Hotel\ Relais\ Saint}$. $C2$ is circle (center is $P_{Hotel\ Relais\ Saint}$, radius is 50 m) which has equation:

$$x^2 + y^2 - 86.77258x + 3.32316y + 1885.13101 = 0 \tag{2}$$

Intersection of $C1$ and $C2$ is the results of two Eqs. (1) and (2):

$$\begin{cases} x^2 + y^2 - 86.7729x + 3.32292y + 1885.14449 = 0 \\ x^2 + y^2 - 86.77258x + 3.32316y + 1885.13101 = 0 \end{cases}$$

$$\begin{cases} x_1 = 43.38662, y_1 = -1.66185 \\ x_2 = 43.38612, y_2 = -1.66119 \end{cases}$$

Intersection of $C1$ and $C2$ is area (x_1, x_2) (Fig. 8).

Zone3. Zone has 3 services, and is the intersection between 3 circles (3 Zone1 as in Fig. 9).

$C1$: is in Zone1 and has $P_{Sas\ Luz\ Voyages}$ service

$$x^2 + y^2 - 86.7729x + 3.32292y + 1885.14449 = 0 \tag{3}$$

$C2$: is in Zone1 and has $P_{Hotel\ Relais\ Saint}$ service

$$x^2 + y^2 - 86.77258x + 3.32316y + 1885.131001 = 0 \tag{4}$$

$C3$: is in Zone1 and has $P_{Boulangerie\ Magri\ Christophe\ Ogi\text{-}labea}$ service

$$x^2 + y^2 - 86.77270x + 3.32338y + 1885.13658 = 0 \tag{5}$$

Intersection of $C1$ and $C2$ is the results of two Eqs. (3) and (4):

$$\begin{cases} x_{x1} = 43.38662, y_{x1} = -1.66185 \\ x_{x2} = 43.38612, y_{x2} = -1.66119 \end{cases}$$

Intersection of $C2$ and $C3$ is the results of two Eqs. (4) and (5):

$$\begin{cases} x_{z1} = 43.41555, y_{z1} = -1.64589 \\ x_{z2} = 43.35713, y_{z2} = -1.67775 \end{cases}$$

Intersection of $C1$ and $C3$ is the results of two Eqs. (3) and (5):

$$\begin{cases} x_{y1} = 43.62799, y_{y1} = -1.76661 \\ x_{y2} = 43.14745, y_{y2} = -1.55768 \end{cases}$$

Intersection of $C1$, $C2$ and $C3$ is area $(z1, x2, y2)$ (Fig. 9).

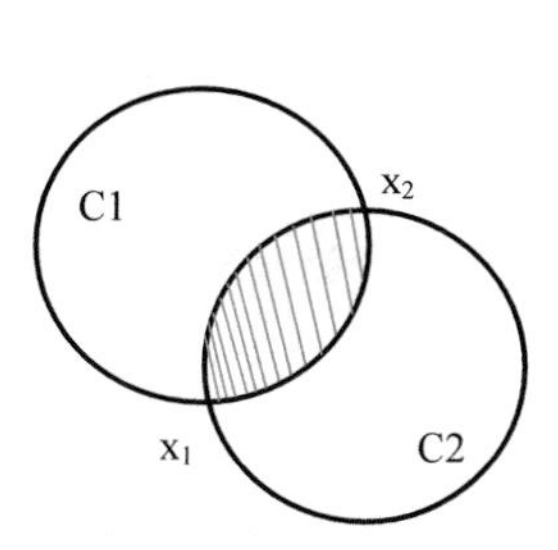

Fig. 8. Intersection between 2 circles

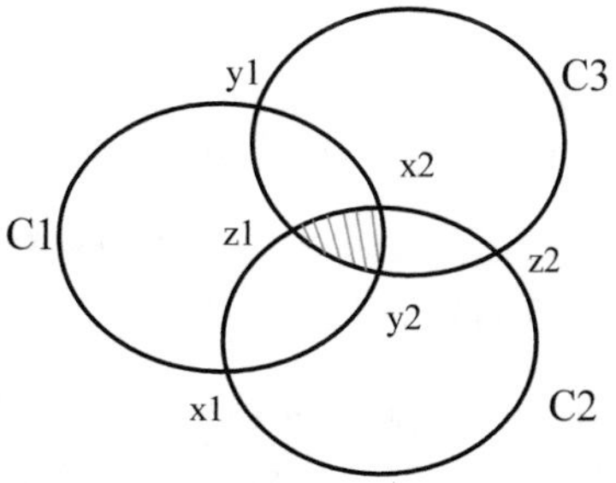

Fig. 9. Intersection between 3 circles

The abstract test cases are generated from t he test model in Sect. 4. They can be expressed in another way as follows.

1. The user is outside the Zones and moves into Zone1.
2. The user is outside the Zones and moves into Zone1 and then moves into Zone2.
3. The user is outside the Zones and moves into Zone1 and then moves into Zone2 and after that moves into Zone3.
4. The user is in Zone1 and then moves into Zone2.
5. The user is in Zone1 and then moves into Zone2 and then moves into Zone3.
6. The user is in Zone2 and then moves into Zone3

Based on Table 1, we can select a proper current coordinates position of the user who is inside the partition boundaries of Zones when he/she moves through Zones corresponding to the abstract test cases.

1. The user is outside the Zones and moves into Zone1.

$$L_0(43.38611;\ -1.66303) -> L_1(43.38605;\ -1.66197)$$
$$L_0(43.38611;\ -1.66303) -> L_1(43.38590;\ -1.66214)$$

2. The user is outside the Zones and moves into Zone1 and then moves into Zone2.

$$L_0(43.38611;\ -1.66303) -> L_1(43.38605;\ -1.66197) -> L_2(43.38622,\ -1.66130)$$
$$L_0(43.38611;\ -1.66303) -> L_1(43.3859;\ -1.66214) -> L_2(43.38652;\ -1.66170)$$

Table 1. Partition boundaries of Zones.

Zones	Partition boundaries of Zones	
	Longitude	Latitude
Zone1 has Tourist spot service	From 43.38600 to 43.3869	From −1.66191 to −1.66101
Zone1 has Hotel service	From 43.38584 to 43.38674	From −1.66203 to −1.66113
Zone1 has Restaurant service	From 43.38590 to 43.38680	From −1.66214 to −1.66124
Zone2 has Tourist spot service and Hotel service	From 43.38612 to 43.38662	From −1.66185 to −1.66119
Zone3 has all three services	From 43.35713 to 43.62799	From −1.76661 to −1.66119

3. The user is outside the Zones and moves into Zone1 and then moves into Zone2 and after that moves into Zone3.

$$L_0(43.38611;\ -1.66303) -> L_1(43.38605;\ -1.66197) -> L_2(43.38622;\ -1.66130) -> L_3(43.35805;\ -1.67890)$$

4. The user is in Zone1 and then moves into Zone2.

$$L_1(43.38605;\ -1.66197) -> L_2(43.38622;\ -1.66130)$$
$$L_1(43.38600;\ -1.66205) -> L_2(43.38632;\ -1.66165)$$

5. The user is in Zone1 and then moves into Zone2 and then moves into Zone3.

$$L_1(43.38605;\ -1.66197) -> L_2(43.38622;\ -1.66130) -> L_3(43.35805;\ -1.67890)$$
$$L_1(43.38600;\ -1.66205) -> L_2(43.38632;\ -1.66165) -> L_3(43.35895;\ -1.67990)$$

6. The user is in Zone2 and then moves into Zone3.

$$L_2(43.38622;\ -1.66130) -> L_3(43.35805;\ -1.67890)$$
$$L_2(43.38632;\ -1.66165) -> L_3(43.35895;\ -1.67990)$$

Instead of randomly selecting all location coordinates in the test execution process for context-aware mobile apps, we select the user's location coordinates inside partition boundaries. This saves time and cost in test execution process.

7 Conclusion and Future Work

Testing context-aware mobile apps is challenging due to the complexity of context variability. Current testing approaches cannot efficiently handle dynamic variability of context-aware mobile apps. To solve this problem, we present a model-based testing approach of mobile apps that uses a combination of a Bigraph Reaction System and a Dynamic Feature Petri Net for automatic generation of test cases. This model addresses the mobile app testing challenges related to the context location of context-aware mobile apps. In this paper, we propose a test criterion combining pattern flow-based coverage criteria and boundary-based coverage criteria and illustrate its use on the TripAdvisor application. This criterion helps narrowing the input data space and cover all test situations in test process when selecting the user's location coordinates inside partition boundaries. As future work, we plan to study the automation of the coverage assessment for any location dependent mobile application.

References

1. Nguyen, T.B., Le, T.T.B., Aktouf, O., Parissis, I.: Mobile applications testing based on Bigraphs and Dynamic feature Petri nets. In: Nguyen, N.T., Dao, N.N., Pham, Q.D., Le, H.A. (eds.) Intelligence of Things: Technologies and Applications 2022, vol. 148, pp. 215–225. Springer, Heidelberg (2022)
2. Yu, L., Tsai, W.-T., Perrone, G.: Testing context-aware applications based on bigraphical modeling. IEEE Trans. Reliab. **65**, 1584–1611 (2016)
3. Siqueira, B.R., Ferrari, F.C., Souza, K.E., Camargo, V.V., Lemos, R.J.S.T.: Testing of adaptive and context-aware systems: approaches and challenges. Verification and Reliability **1772**, 1–46 (2021)
4. Ammann, P., Offutt, J., Huang, H.: Coverage criteria for logical expressions. In: 14th International Symposium on Software Reliability Engineering, pp. 99–107 (2003)
5. Muschevici, R., Clarke, D., Proenca, J.: Feature Petri nets. In: 14th International Conference, pp. 13–17. Jeju, Korea (2010)
6. Heng, L., Chan, W.K., Tse, T.H.: Testing context-aware middleware-centric programs: a data flow approach and an RFID-based experimentation. In: 14th International symposium on Foundations of software engineering, pp. 242–252 (2006)
7. Zhai, K., Jiang, B., Chan, W.K.: Prioritizing test cases for regression testing of location-based services: metrics, techniques, and case study. IEEE Trans. Serv. Comput. **7**, 54–67 (2014)
8. Qun, R., Dunham, M.H.: Using semantic caching to manage location dependent data in mobile computing. In: 6th Annual International Conference on Mobile Computing and Networking, pp.210–222 (2000)
9. Zhang, T., Gao, J., Aktouf, O., Uehara, T.: Test Model and Coverage Analysis for Location-based Mobile Services, SEKE, pp. 80–86 (2015)
10. Kosmatov, N., Legeard, B., Peureux, F., Utting, M.: Boundary coverage criteria for test generation from formal models. In: 15th International Symposium on Software Reliability Engineering, pp. 139–150 (2004)
11. http://docs.oasisopen.org/wsbpel/2.0/CS01/wsbpel-v2.0-CS01.html. Accessed 2021/10/21
12. https://gkchronicle.com/world-geography/Locating-points-on-earths-surface.php. Accessed 2023/1/21
13. White, L.J., Cohen, E.I.: A domain strategy for computer program testing. IEEE Trans. Software Eng. **6**, 247–257 (1980)
14. Beizer, B.: Software Testing Techniques. John Wiley & Sons, USA (1990)

Author Index

N. T. Nguyen et al. (Eds.): CITA 2023, LNNS 734, pp. 405–406, 2023.
https://doi.org/10.1007/978-3-031-36886-8

Printed in the United States
by Baker & Taylor Publisher Services